D1822930

**Library of Congress Cataloging in Publication Data**

Main entry under title:

Nucleic acids in plants.

Bibliography: p.
Includes indexes.
1. Nucleic acids. 2. Botanical chemistry.
3. Plant viruses. I. Hall, Timothy C. II. Davies,
Jeffrey W.
QK898.N8N83 581.8'732 78-25602
ISBN 0-8493-5291-6 (Volume I)
ISBN 0-8493-5292-4 (Volume II)



International Standard Book Number 0-8493-5291-6 (Volume I)
International Standard Book Number 0-8493-5292-4 (Volume II)

Library of Congress Card Number 78-25602
Printed in the United States

# Nucleic Acids in Plants

## Volume II

Editors

**Timothy C. Hall, Ph.D.**

Professor
Department of Horticulture
University of Wisconsin
Madison, Wisconsin

**Jeffrey W. Davies, Ph.D.**

Head
Virology Department
John Innes Institute
Norwich, England

CRC PRESS, INC.
Boca Raton, Florida 33431

# PREFACE

These are exciting times in the biological sciences. Each day brings new discoveries that help to explain how nucleic acid sequence and structure molds the form and function of every living organism. The intelligence that distinguishes man from other creatures has always made him curious about himself and the universe around him. In retrospect, one can see the gathering concepts of biology stimulated by Linnaeus who initiated the classification of living things in a logical fashion, and accelerated by Darwin's recognition of evolution as the process of slow change and diversification. Mendel, whose work was nearly overlooked, pioneered the science of genetics. The studies of Avery and of Hershey elegantly demonstrated that DNA is the carrier of genetic information. The transition to the era of molecular biology was made by the brilliant insight of Watson and Crick. They correctly deduced the structure of DNA, and succinctly showed how it could provide a simple chemical code for genetic information while also serving as its own template for replication from generation to generation.

It is upon these foundations that the rapid succession of todays discoveries are built. Almost all of the innovative work of the 1960s flowed from studies on microorganisms because of their short life cycles and their relatively simple genetic content. Fortunately, results from these studies can now be used towards advancing animal and plant biology.

Our ambition in the organization of this book was to explore the current status of knowledge about nucleic acids in plants. We wanted the reader to be able to learn how this research is being undertaken. Therefore, we asked the contributing authors to include details of approaches and methods. Where feasible, they have provided protocols that can be followed by those who wish to repeat results, extend data, make improvements, or use them in new applications. We have done our best to accentuate work in plants. Results from studies on animals and microorganisms have been used only as they underlie techniques in plant research or as they point to new directions applicable to plant studies.

The title *Nucleic Acids in Plants*, rather than *Nucleic Acids of Plants,* was chosen deliberately. We felt that it was essential to include information on the nucleic acids of plant viruses and viroids. Strictly, these are nucleic acids that are *in*, not *of* plant cells. Viral nucleic acids are not rarities in plant cells; their presence is widespread in natural plant populations. Viruses and viroids have evolved (and continue to evolve) in adaptation with the biosynthetic systems of plant cells. Therefore, they represent appropriate model systems for the scientist whose ultimate interest is the more complicated metabolism of proteins and nucleic acids of the plant cell. Consequently, much topical research on nucleic acids in plants follows from techniques developed in connection with viral studies.

We greatly appreciate the efforts of all our contributors. We were fortunate in attracting young and active researchers to describe their study areas. They have endeavored to make their chapters readable and to reflect their enthusiasm for the subject. We learned that for the active researcher it is truly hard to find the time to write a review. Inevitably, some of the data that were novel when submitted will be almost out of date by the time this book is published; despite the dedicated cooperation of CRC Press in expediting its appearance. The impression that this is regrettable is tempered by the realization that this is merely a symptom of the rapid progress being made in nucleic acid research. Indeed, we hope that this book will stimulate its readers to participate in the rapid advancement of knowledge concerning nucleic acids in plants.

**T. C. Hall**
**J. W. Davies**

# THE EDITORS

**Timothy C. Hall, Ph.D.**, is a Professor of Horticulture at the University of Wisconsin, Madison. He is a member of the American Society for Microbiology, the Biochemical Society (U.K.), and is chairman of the Recombinant DNA Committee of the American Society of Plant Physiologists. Dr. Hall received his B.Sc. in 1962 and his Ph.D. in 1965 from the University of Nottingham, England. Following a year at the University of Minnesota as Louis W. and Maud Hill Foundation Research Fellow in the Horticultural Science Department, Dr. Hall was appointed to the University of Wisconsin faculty in 1966. During 1977 he undertook sabbatical research-study at the Université Louis Pasteur, Strasbourg, France.

Throughout his career Dr. Hall has been interested in fostering the use of molecular biological approaches towards the enhancement of crop plant productivity. He has published many scientific articles in the fields or virology, cold stress physiology, protein synthesis, and nucleic acids in plants.

**Jeffery W. Davies, Ph.D.**, is the Head of the Virology Department of the John Innes Institute, Norwich, England.

Dr. Davies received his B.Sc. in 1963, and Ph.D. in 1966, at the University of Nottingham, England, where he began his research into plant protein synthesis, using protoplasts and cell-free extracts. From 1966 to 1970 he was Lecturer in Molecular Biology at the University of Edinburgh. During this time he developed an interest in the structure and translation function of RNA, especially of viruses. After an E.M.B.O. Fellowship in Leiden, The Netherlands, he was appointed as Assistant Scientist in the Biophysics Laboratory at the University of Wisconsin, Madison, where he continued this research from 1972 to 1976 with plant viruses.

Dr. Davies is a member of the Institute of Biology, a member of the Biochemical Society, the Society for General Microbiology, and the American Society for Microbiology. In 1975, he was elected as a Fellow of the Linnean Society of London.

## CONTRIBUTORS

**Wayne M. Becker, Ph.D.**
Professor of Botany
University of Wisconsin
Madison, Wisconsin

**Edwin T. Bingham, Ph.D.**
Professor of Agronomy
University of Wisconsin
Madison, Wisconsin

**Jeffery W. Davies, Ph.D.**
Head, Virology Department
John Innes Institute
Norwich, England

**Elizabeth Dickson, Ph.D.**
Assistant Professor
Laboratory of Genetics
The Rockefeller University
New York, New York

**Robert B. Goldberg, Ph.D.**
Associate Professor of Biology
University of California
Los Angeles, California

**Timothy C. Hall, Ph.D.**
Professor of Horticulture
University of Wisconsin
Madison, Wisconsin

**Roger Hull, Ph.D.**
Research Scientist
Department of Virology
John Innes Institute
Norwich, England

**Leslie C. Lane, Ph.D.**
Assistant Professor of Plant Pathology
University of Nebraska
Lincoln, Nebraska

**Christopher J. Leaver, Ph.D.**
Lecturer
Department of Botany
University of Edinburgh
Edinburgh, Scotland

**Jeff Schell, Ph.D.**
Professor
Department of Genetics
State University
Gent, Belgium

**Krishna K. Tewari, Ph.D.**
Professor of Biochemistry
Department of Molecular Biology and Biochemistry
University of California
Irvine, California

**Virginia Walbot, Ph.D.**
Assistant Professor of Biology
Washington University
St. Louis, Missouri

**Jacques-Henry Weil, Sc.D.**
Professor of Biochemistry
Universite Louis Pasteur
Strasbourg, France

**Milton Zaitlin, Ph.D.**
Professor of Plant Pathology
Cornell University
Ithaca, New York

# TABLE OF CONTENTS

## Volume I

I. PLANT DNA



II. PLANT RNA

**Volume II**

## III. PLANT VIRUS NUCLEIC ACIDS



## IV. VIROIDS, PLASMIDS, AND GENETIC ENGINEERING

# Section III
# *Plant Virus Nucleic Acids*

# THE DNA OF PLANT DNA VIRUSES*

**R. Hull**

## TABLE OF CONTENTS



* The literature reviewed for this chapter was up to November 1977. The guidelines for containment of genetic engineering experiments have since changed. For current guidelines in the U. S. A. the reader should refer to the relevant *Federal Register.*

# I. INTRODUCTION

Most plant viruses studied contain single-stranded RNA. It was only in the late 1960s that a plant virus, cauliflower mosaic virus (CaMV), was shown to contain DNA.[1] Since then, four (or possibly seven) other viruses have been reported to resemble CaMV in many characteristics; these are all grouped as caulimoviruses.[2] The three that have been most studied and that will be referred to in this chapter are CaMV, dahlia mosaic virus (DaMV), and carnation etched ring virus (CERV).

Four further viruses of higher plants, which do not resemble members of the caulimovirus group, have recently been reported to contain DNA.These are potato leaf roll virus (PLRV),[3] bean golden yellow mosaic virus (BGYMV),[4] and maize streak (MSV) and cassava latent viruses (CLV);[4a] it is being proposed that BGYMV, MSV, and CLV be grouped together under the name "Gemini viruses".

The subject of DNA containing viruses of higher plants has recently been reviewed by Shepherd.[5] He presents evidence showing that the nucleic acids of the caulimoviruses are DNA, and, therefore, this point will not be discussd here. The evidence for the DNA content of PLRV and the gemini viruses will be mentioned in this chapter.

Interest in DNA containing viruses is rapidly awakening, primarily because of their possible use in so-called 'genetic engineering' experiments involving recombinant DNAs. Therefore, this chapter will be mainly concerned with the properties of these viruses and their potential for use in such experiments.

# II. PREPARATION TECHNIQUES

## A. Purification of Viruses

All the known plant DNA containing viruses are obtained in yields lower than many of the RNA containing ones. The main problem to be overcome in obtaining purified preparations of members of the caulimovirus group is that the virus particles occur in stable inclusion bodies (Figure 1A and B). The original purification technique for CaMV involved n-butanol[6] and gave relatively low and variable yields. Hull et al.[7] and Hull and Shepherd[7a] studied the effects of various treatments on the stability of CaMV inclusion bodies. They found that treatment of turnip leaf homogenate (*Brassica rapa* cv. Just Right, infected for 24 to 35 days), blended in 0.5 *M* potassium phosphate buffer pH 7.2 containing 0.75% sodium sulphite, with 2.5% Triton® X-100 and 1 *M* urea at 4°C for 18 hr gave optimum release of virus; such treatment can be effected in large containers rotated in the manner of roller cultures. After low-speed centrifugation (10 min at 7000 r/min in a Sorval® SS 34 rotor) the virus can be concentrated by high-speed centrifugation (1.5 hr at 27,000 r/min in a Spinco® R30 rotor or 2.5 hr at 20,000 r/min in a Spinco R21 rotor) and the pellets dispersed in water overnight. The virus can be further purified by using 10 to 40% sucrose gradients (0.01 *M* potassium phosphate pH 7.2; 2 hr at 23,000 r/min in Spinco SW 25.1 rotor) and by using isopycnic CsCl centrifugation (36% w/v CsCl in 0.01 *M* potassium phosphate pH 7.2; 16 hr at 40,000 r/min in Spinco R50 rotor). All operations should be at 4°C.

The yield is affected by the isolate of CaMV and by the host.[7] Using the Triton/urea technique, the Cabbage B isolate (ATCC PV147) grown in turnip cv. Just Right yields 6 to 10 mg/kg leaf tissue. This technique has been used to purify CERV[8], but apparently not quite as successfully as CaMV; it is not known if it is suitabe for other members of the caulimovirus group. The main difficulty with extraction of PLRV is that it occurs mainly in vascular tissues and is in very low concentration. Purification procedures have been published by Peters[9] and by Sarkar.[10] A technique for purifying BGYMV is published by Goodman et al.,[11] for MSV by Bock et al.,[47] and CSV by Harrison et al.[4A]

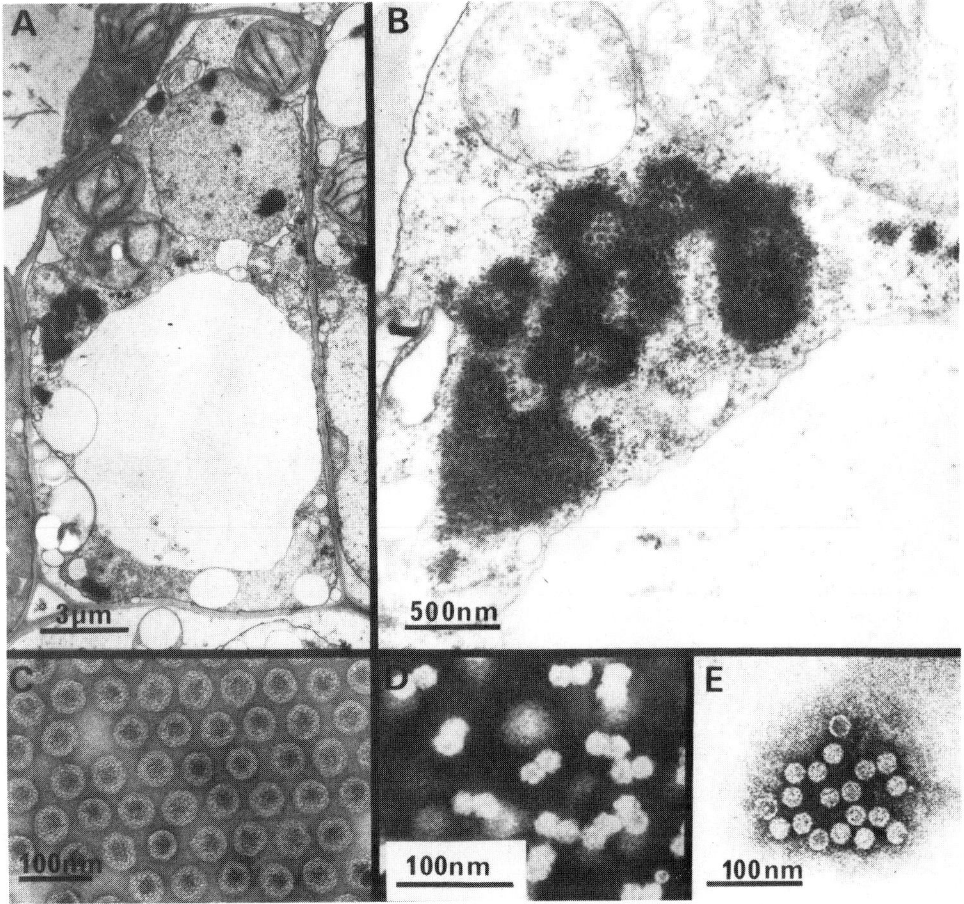


FIGURE 1. A and B. Electron micrographs of thin section of *Brassica rapa* c.v. Just Right infected with CaMV. A. Whole cell showing inclusion bodies within the cytoplasm. B. Inclusion body showing that most of the virus particles are within the matrix. C. Particles of CaMV from purified preparation negatively stained with uranyl acetate. D. Particles of BGYMV negatively stained with potassium phosphotungstate. E. Particles of PLRV negatively stained with potassium phosphotungstate. (A and B courtesy of A. Plaskitt; C courtesy of R. W. Horne; D courtesy of R. M. Goodman; E courtesy of D. Peters.)

Some of the physical properties of the nucleoproteins are listed in Table 1. Figure 1C, D, and E show electron micrographs of the particles of CaMV, BGYMV, and PLRV, respectively.

### B. Extraction of Nucleic Acids

The particles of members of the caulimovirus group (CaMV, DaMV, and CERV) are remarkably resistant to disruption by the normal procedures.[6,8,12] The only reliable method for extracting nucleic acid fom these viruses is the one devised by Shepherd et al.[6] which involves the digestion of the viral coat proteins with pronase (50 $\mu$g nuclease free pronase/mg virus) in 0.01 *M* sodium citrate, 0.15 *M* NaCl (1 × SSC) and sodium dodecyl sulphate (SDS) (0.25 to 0.5%). Pronase and residual undigested proteins can be removed with either phenol, sodium perchlorate,[12] or by isopycnic centrifugation in CsCl.[13] Isopycnic centrifugation in CsCl (55.5% w/w CsCl in 1 × SSC; 35,000 r/min for 60 to 72 hr in Spinco R40 or R50 rotor at 20°C) is necessary at some stage if clean nucleic acid is required.

TABLE 1

Some Physical Properties of the Nucleoproteins of DNA Containing Plant Viruses

| Virus | $s_{20w}$ | Nucleic acid content (%) | Molecular weight (× $10^{-6}$) | 0.1% $E_{260nm}$ | Size (nm) and shape | Ref. |
|---|---|---|---|---|---|---|
| Caulimoviruses | | | | | | |
| Cauliflower mosaic | 208 | 17 | 22.8 | 4.4[a]<br>7.0 | 50 sphere | 7, 50 |
| Dahlia mosaic | 254 | 16 | — | — | 50 sphere | 51—54 |
| Carnation etched ring | — | — | — | — | 47 sphere | 55, 56 |
| Mirabilis mosaic | 254 | 16—17 | — | — | 45—50 sphere | 57 |
| Other viruses | | | | | | |
| Potato leafroll | — | 40 | 1.4[b] | 8.0 | 23 sphere | 9, 10 |
| Bean golden yellow mosaic | 69.4 | — | — | — | 18 sphere, frequently germinate pairs | 5, 11, 58 |
| Maize streak | 76, 54 | — | — | — | c20 sphere, frequently germinate pairs | 47 |
| Cassava latent | 76, 54 | — | — | — | c20 sphere, frequently germinate pairs | 47a |

[a] The value of 4.4 allows for light scattering, that of 7.0 does not.[7]
[b] Estimate from nucleic acid content and weight published by Sarkar.[10]

Nucleic acid has been extracted from PLRV by treatment of the virus with SDS and phenol.[10] Goodman[4] prepared nucleic acid from BGYMV by disrupting the virions using 2% SDS at 20°C for 38 min, precipitating the dodecyl sulphate with 0.04 *M* KCl at 4°C and dialysing the nucleic acid in the supernatant against 1 × SSC. PLRV and BGYMV nucleic acids give positive reactions with diphenylamine.[5,10]

## III. PROPERTIES OF THE VIRAL NUCLEIC ACIDS

The nucleic acid of CaMV has been studied more extensively than that of any of the other viruses; therefore, much of the information given below refers to this virus.

### A. Native Nucleic Acids

#### *1. Base Composition*

Base composition of CaMV nucleic acid determined by two methods (enzymatic and acid hydrolysis), is A = 29.1 (moles/100 mol nucleotides); G = 20.5; C = 21.4; T = 29.[6] Thus, there is molar equivalence between A and T and between G and C indicating that the DNA is double-stranded. Using nearest neighbour frequency analysis, Russell et al.[14] confirmed this and suggested that there are similarities between the base composition of CaMV nucleic acid and that of host cauliflower DNA.

The base composition the RNA found associated with CaMV nucleic acid is discussed in Section III. B. 2.

#### *2. Melting Behavior*

CaMV nucleic acid was shown to have a cooperative melting curve with a Tm of 87.2°C (range 83 to 89°C) in 1 × SSC (Figure 2A) and of 69.5 to 70°C in 0.1 × SSC by Shepherd et al.;[6] in 1 × SSC the hyperchromicity was 33 to 36%. Shepherd et al.[6] observed that this melting behavior was consistent with the nucleic acid being double-stranded DNA with a G+C content of 43.5%. A close analysis of the melting curve of CaMV nucleic acid in 1 × SSC indicates some base composition heterogeneity[15] (Figure 2B); there appear to be three or perhaps four regions of the nucleic acid which

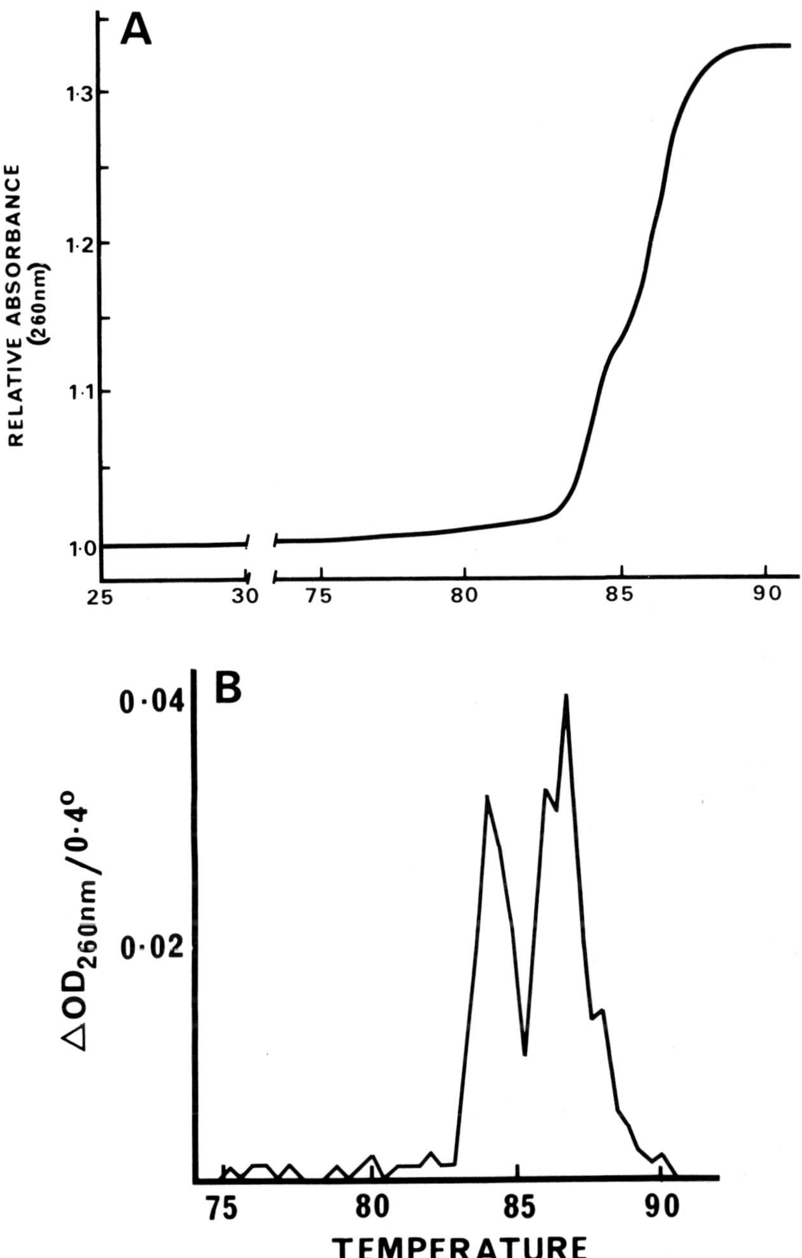


FIGURE 2. Melting behavior of CaMV nucleic acid in 1 × SSC. A. Melting curve of the nucleic acid. B. Differential melting curve derived from A at 0.4°C intervals. (Taken from Hull, R. and Howell, S. H., unpublished data.)

melt differentially and which presumably differ in G + C content. Similar differential melting of CaMV nucleic acid can be seen in 0.1 × SSC and in 0.01 *M* Tris/1m*M* EDTA pH 8.5 (1 × TE) and 10 × TE (Table 2).

CaMV nucleic acid renatures rapidly after melting, and a snap-back mechanism has been suggested.[6] However, no evidence of type I renaturation[15a] (commonly called snap-back) was found on studying the reassociation of CaMV nucleic acid under var-

TABLE 2

**Melting of CaMV Nucleic Acid[a]**

| Buffer | Melting temperatures 1[b] | 2 | 3 | 4 | Total hyperchromicity (%) |
|---|---|---|---|---|---|
| 1 × SSC[c] | 86.0 | 88.0 | 88.8 | 90.0 | 33.5 |
| 0.1 × SSC | 68.8 | 70.6 | 72.0 | 73.4 | 33.1 |
| 10 × TE | 80.0 | — | 83.2 | 84.4 | 32.0 |
| 1 × TE | 61.4 | 63.0 | 64.2 | 65.8 | 28.9 |

[a] Hull and Howell[15]
[b] Peak in differential melting curve (Figure 2B).
[c] SSC = 0.15 *M* NaCl, 0.015 *M* sodium citrate. TE = 0.01 *M* Tris, 0.001 *M* EDTA pH 8.5.

ious conditions;[15] in all cases it reassociated with second order kinetics resembling type II renaturation.[15a] A preparation of CaMV nucleic acid which had a hyperchromicity of 35.8% on melting in 1 × SSC had a relative hypochromicity of 33.8% when subjected to a temperature drop of Tm−20° (68°C) for 30 min.[15] The reassociation kinetics had a $Cot_{1/2} = 8.7 \times 10^{-3} M\ sec^{-1}$ which when compared wth the values for *Escherichia coli* chromosomal DNA indicate a kinetic complexity of 4 to 5 × $10^6$.[15]

The nucleic acid of DaMV has a melting behavior characteristic of double-stranded DNA.[16]

PLRV nucleic acid in 1 × SSC melts over the range 80 to 92°C with a Tm of about 87.4°C (equivalent to a G+C content of 44.2%) and a hyperchromicity of approximately 32%.[10] BGYMV nucleic acid melts over a wide temperature range (20 to 70°C) with a hyperchromicity of about 15%;[16a] the nucleic acids of MSV and CLV have similar melting behavior.[4a]

### *3. Isopycnic Banding*

CaMV nucleic acid has a buoyant density in CsCl of 1.702 g/cc[6] consistent with a GC content of 43%; Russell et al.[12] report a buoyant density of 1.6996 g/cc (equivalent to a GC content of 40.6%) and Hull[12a] found the buoyant density to be 1.704 ± 0.004 g/cc (10 observations). Civerolo and Lawson[17] note that as well as a major band at 1.700 to 1.701 g/cc, their CaMV nucleic acid preparations had a minor band at 1.637 to 1.687 g/cc; they suggest that the major band is the circular molecules and the minor band the linear molecules. CaMV nucleic acid forms a single band in ethidium bromide per CsCl[6,17] at a density of 1.557 g/cc[12] which indicates that there are no supercoiled molecules with high helix densities.

The nucleic acid of PLRV bands in CsCl gradients at 1.698 g/cc which corresponds to a GC content of approximately 38%.[10] The nucleic acid of BGYMV bands at approximately 1.717 g/cc in CsCl gradients.[16a]

### *4. Sedimentation Behavior*

Native CaMV nucleic acid sediments in neutral sucrose gradients as two major components with some faster sedimenting material[12,14,17] (Figure 3A and B). The faster of the major components, with an $S_{20}w$ of 19.0*S*, appears to be relaxed, (nicked) circular

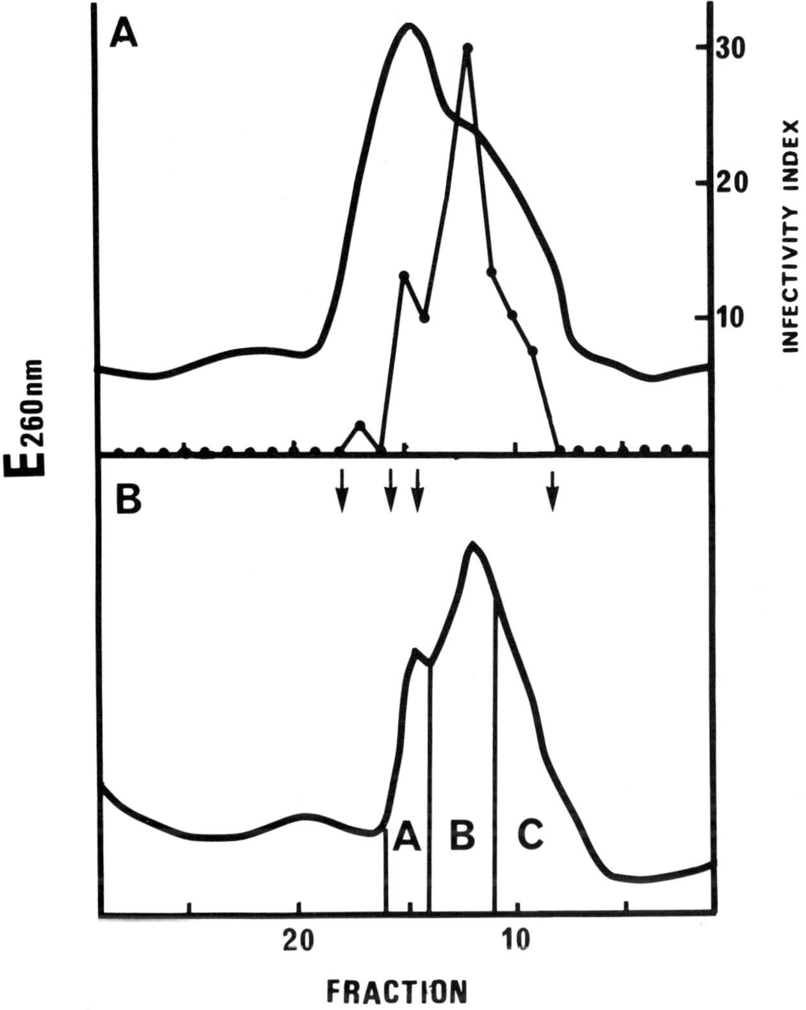


FIGURE 3. Sedimentation profiles of two preparations of native CaMV nucleic acid in 5 to 20% neutral sucrose gradients. The gradients were centrifuged for 2¼ hr at 45,000 rpm in a Spinco® SW 50.1 rotor and fractionated by dripping from the bottom of the tubes. Sedimentation is from left to right. A. —— = OD; •—• = infectivity. B. CaMV-nucleic acid was co-sedimented with $^{32}P$-Polyoma virus DNA forms II and III and with $^{3}H$ ColE1-DNA. The arrows indicate the depth to which, from left to right, Polyoma form III, Polyoma form II, ColE1, 17S and ColE1, 23S DNA sedimented. The divisions of the CaMV nucleic acid peak show bulk fractions referred to in the text. (From Hull, R. and Shepherd, R. J., *Virology*, 79, 218, 1977. With permission.)

molecules (for molecular forms see Section III. A. 8) and the slower sedimenting component, 17.1*S*, is composed of linear molecules[12,14,17] (Table 3). The ratio of the two major components varies, however, depending on the previous history of the virus and nucleic acid preparation. Nucleic acid from freshly prepared virus from plants infected for the optimum period (25 to 35 days) always had more of the 19.0*S* component than the 17.1*S* component (Figure 3B); nucleic acid from virus either from old infections or from preparations stored for some time sometimes had more of the 17.1*S* than the 19.0*S* component (Figure 3A). As was suggested by Russell et al.[14] and Hull and Howell,[21] the linear form is probably derived from the nicked circular form. This is also

TABLE 3

**Molecular Forms of CaMV Nucleic Acid as identified by Electron Microscopy**

| | Number of Molecules | Linear (%) | Circular (%) | Average aai of circular molecules[a] |
|---|---|---|---|---|
| Control | 102 | 27.5 | 72.5 | 11.3 |
| Fraction A[b] | 64 | 81.2 | 18.8 | 16.3 |
| Fraction B | 52 | 25.0 | 75.0 | 13.2 |
| Fraction C | 61 | 9.8 | 90.2 | 10.5 |
| Ethidium bromide/CsCl | 62 | 26.9 | 73.1 | 11.1 |

[a] AAI = apparent area index described in Section III. A. 5.

[b] Fractions from sucrose gradient similar to that shown in Figure 3B.

From Hull, R. and Shepherd, R. J., *Virology*, 79, 216, 1977. With permission.

supported by the observation that infectivity is associated with the nicked circular form and not with the linear molecules (Figure 3A).[12]

The sedimentation of CERV nucleic acid appears to be similar to that of CaMV nucleic acid.[8] The sedimentation coefficient of PLRV nucleic acid is reported to be approximately 8.4*S*;[10] the author does not give any indication of the homogeneity of the preparation. The nucleic acid of BGYMV sediments as a single component at approximately 16*S* in 1 × SSC;[4] the sedimentation coefficient of this nucleic acid however is dependent on ionic strength and pH.[16a]

### 5. *Electron Microscopy*

The molecules of CaMV and CERV nucleic acids have been visualized in the electron microscope[8,12,14,17,18] after being prepared using modifications of the basic film protein technique of Kleinschmidt and Zahn.[19] The modification described by Davis et al.[20] has worked well in the author's hands. It is important to remove ethanol (by dialysis)[18] and to dilute the nucleic acid sample to approximately 1 μg/mℓ in spreading solution (0.5 *M* ammonium acetate, 1 m*M* EDTA) before adding cytochrome C (to give 0.1 mg/mℓ).[12]

Examination of CaMV nucleic acid in the electron microscope shows two forms of molecules — linear and circular[12,14,17,18] (Figure 4). The lengths reported for the molecules ranges from 2.28 to 2.47 μm (Table 4); no evidence of any oligomeric forms has been found.

The circular molecules of CaMV nucleic acid show a peculiar property which has been described as twistedness.[12,14,17] Hull and Shepherd[12] suggested a measure of this twistedness which they called the apparent area index (AAI). The outline of each molecule was cut out from photocopies of the micrographs at a magnification of 70,000, weighed and the weight expressed as a percentage of that expected from the area of a circle of the same contour length (2.45 μm × 70,000). Thus the smaller the AAI, the more twisted the molecule; values of aai for the circular molecules in Figure 4 are listed in the legend. Statistical tests showed that the AAI was a reproducible measure of twistedness. From Figure 5 it can be seen that preparations of CaMV nucleic acid contain molecules with a wide range of AAI.

As has been noted above (Table 3), the slower sedimenting major component of CaMV nucleic acid (Fraction A of Figure 3A) consists mainly of linear molecules, and the faster sedimenting major component (Fraction B of Figure 3A), is mainly open

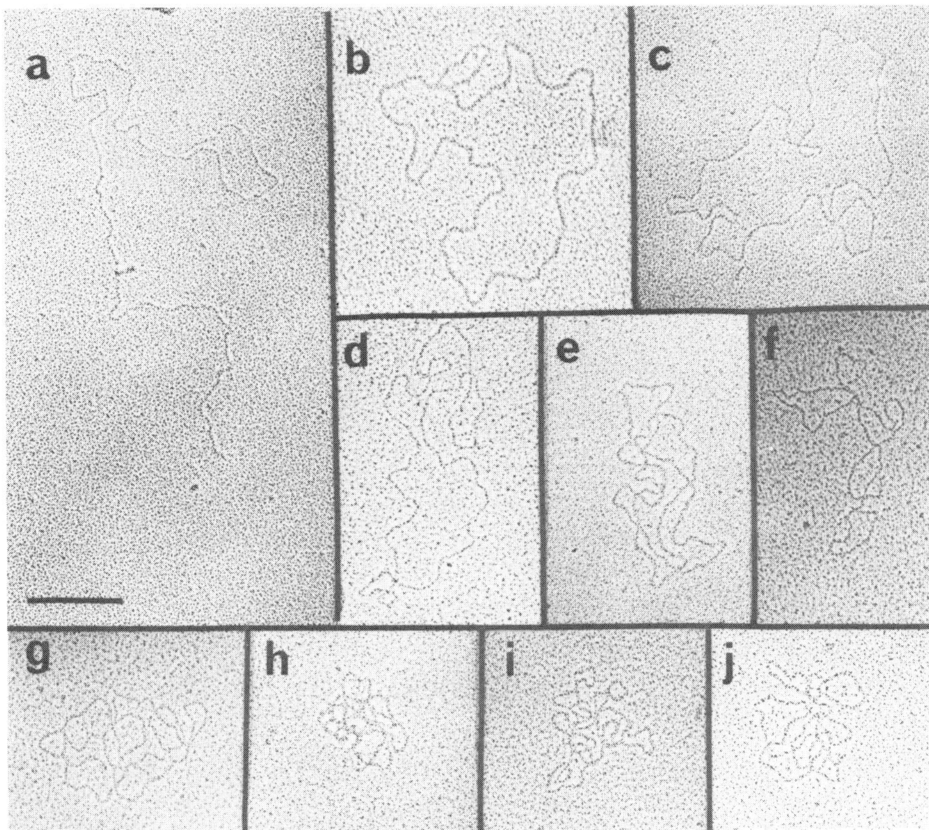


FIGURE 4. Electron micrographs of CaMV nucleic acid. A. Linear form from sucrose gradient fraction A (See Figure 1B). B. to J. Circular forms with apparent area indices (see text) in parentheses, from B ethidium bromide/CsCl band (27.9). C. Sucrose gradient fraction A (26.5). D. Control (14.8). E. Ethidium bromide/CsCl band (10.1). F and G. Control (6.7 and 11.5). H. Ethidium bromide/CsCl band (4.4). I. Sucrose gradient fraction B (6.3). J. Sucrose gradient fraction C (7.8). Bar = 200 nm (From Hull, R. and Shepherd, R. J., *Virology*, 79, 220, 1977. With permission.)

TABLE 4

**Reported Lengths of Plant Viral DNAs**

| Virus | Length (μm) | Ref. |
|---|---|---|
| CaMV | 2.47 ± 0.14 | 14 |
| | 2.31 ± 0.24 | 18 |
| | 2.28—2.46 | 17 |
| | 2.42±0.08 | 12 |
| CERV | 2.1—2.3 | 8 |

circular molecules.[12,14,17] The nucleic acid which sediments faster than the major peaks (Fraction C of Figure 3A) contains circular molecules with an average AAI significantly smaller than those of Fraction B (Figure 5);[12] Civerolo and Lawson[17] also observed that faster sedimenting CaMV nucleic acid molecules were more twisted than the slower sedimenting ones.

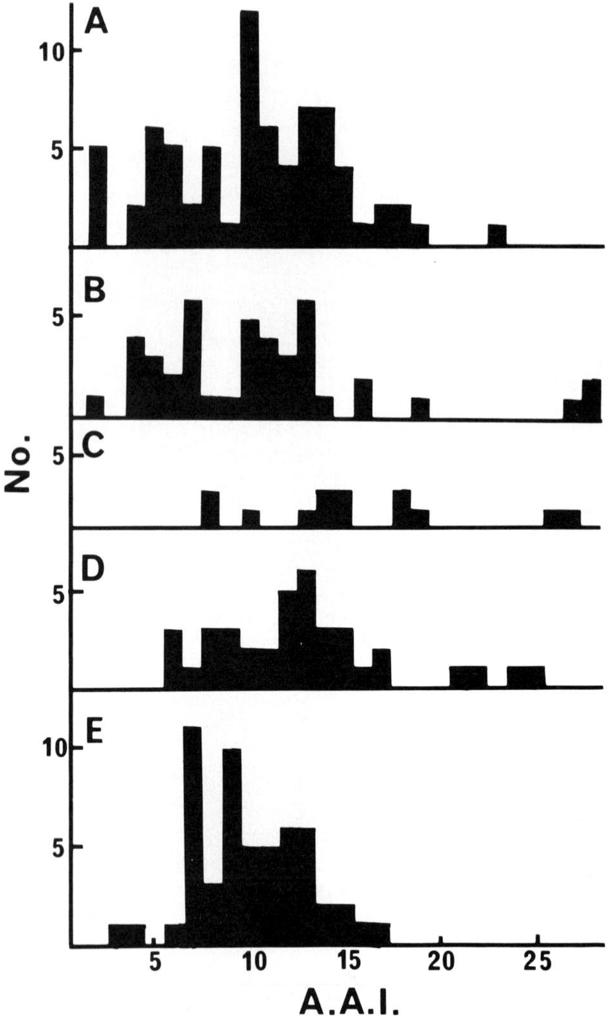


FIGURE 5. Histograms of distribution of apparent area indices (AAI) (see text for definition) of circular CaMV nucleic acid molecules in (A) control, (B) ethidium bromide/CsCl gradient band, (C) sucrose gradient fraction A (see Figure 3B), (D) sucrose gradient fraction B, and (E) sucrose gradient fraction C. (From Hull, R. and Shepherd, R. J., *Virology*, 79, 221, 1977. With permission.)

Electron microscopy of CERV nucleic acid showed both linear and circular forms, with the linear forms being the most common. Most of the circular molecules were very twisted; those that could be measured were approximately 2.1 to 2.3 $\mu$m in length[8] (Table 4).

The nucleic acids of MSV and CLV appear as a mixture of circular and linear molecules in the electron microscope[16a] with the circular molecules predominating in most preparations. The molecular weight of the circular molecules of MSV DNA was $0.71 \pm 0.01 \times 10^6$ and of CLV DNA was $0.80 \pm 0.1 \times 10^6$ (assuming they were single-stranded — see Section IV. C).

### 6. Molecular Weights

The molecular weight estimates for the nucleic acids of CaMV, CERV, and PLRV

TABLE 5

**Molecular Weight Estimates of Plant Viral DNAs**

| Virus | Method | Molecular weight ($\times 10^{-6}$) | Ref. |
|---|---|---|---|
| CaMV | Sedimentation circular form | 4.12—4.64 | 12 |
| | Sedimentation linear form | 4.10—5.60 | 12 |
| | Electron microscopy | 4.74 | 14 |
| | | 4.4 | 18 |
| | | 4.48—4.70 | 17 |
| | | 4.74 | 12 |
| | Particle molecular weight | 3.9 | 7 |
| | Restriction endonucleases | 4.90 | 23 |
| | | 4.86—4.89 | 22 |
| | Kinetic complexity | 4.20 | 15 |
| CERV | Electron microscopy | 4.1—4.5 | 8 |
| PLRV | Sedimentation | 0.56 | 10 |

are given in Table 5. It can be seen that the caulimoviruses have nucleic acids of molecular weight 4.5 to 5.0 × $10^6$. It is interesting to note that the value for the kinetic complexity of CaMV nucleic acid is close to the hydrodynamic molecular weight, indicating that there are few, if any, repeating segments. The molecular weight of PLRV nucleic acid will be discussed in Section VII.

The molecular weight of BGYMV DNA has been estimated to be in the range 0.66 to 0.95 × $10^6$ from sedimentation velocity and CsCl equilibrium data.[16a] It has been noted above (Section III. A. 5) that electron microscopy of MSV and CLV DNAs give molecular weight estimates of 0.71 and 0.80 × $10^6$ respectively.[4a]

*7. Gel Electrophoresis*

When CaMV nucleic acid is electrophoresed in 2.5% polyacrylamide gels, it forms three bands[12] (Figure 6). The slowest migrating band is the circular form (with infectivity associated with it). The next band is the linear form, and the rapidly migrating band is probably small degraded pieces of nucleic acid.

CaMV nucleic acid migrates as three distinct components (A, B, and C in order of migration) (Figure 7A) on electrophoresis in composite 1.7 to 2.2% polyacrylamide — 0.5% agarose gels;[8,17] preparations of CERV nucleic acid form up to three bands (which correspond with those of CaMV) in polyacrylamide/agarose gels.[8] Civerolo and Lawson[17] showed that CaMV bands A and B are circular forms with those in band A being very twisted and those in band B being predominantly open circular; the nucleic acid in band C was linear molecules. Although the gel patterns of CERV nucleic acid resembled those of CaMV, linear forms were found in all three bands and circular forms only in band B;[8] there was no explanation for these differences.

In 0.7% agarose gels and in 1.8% acrylamide — 0.5% agarose gels CaMV nucleic acid forms multiple bands, with generally two major sharp bands and a series of faster moving ones (Figure 7B).[17,22,23] The faster moving bands are thought to be twisted circular molecules.[17] In 1% and 1.4% agarose gels, the order of the bands is reversed (Figure 7C and Figure 8).[21,24]

Hull and Howell[21] found that the fastest moving pair of bands in 1% and 1.4% agarose gels were linear molecules and that the slower moving group of multiple bands were circular molecules (Figure 8). From Figure 5 it can be seen that there are apparently discrete classes of twistedness of molecules (as measured by AAI); if this is so, it is a possible explanation for the multiple banding of CaMV in agarose gels. However,

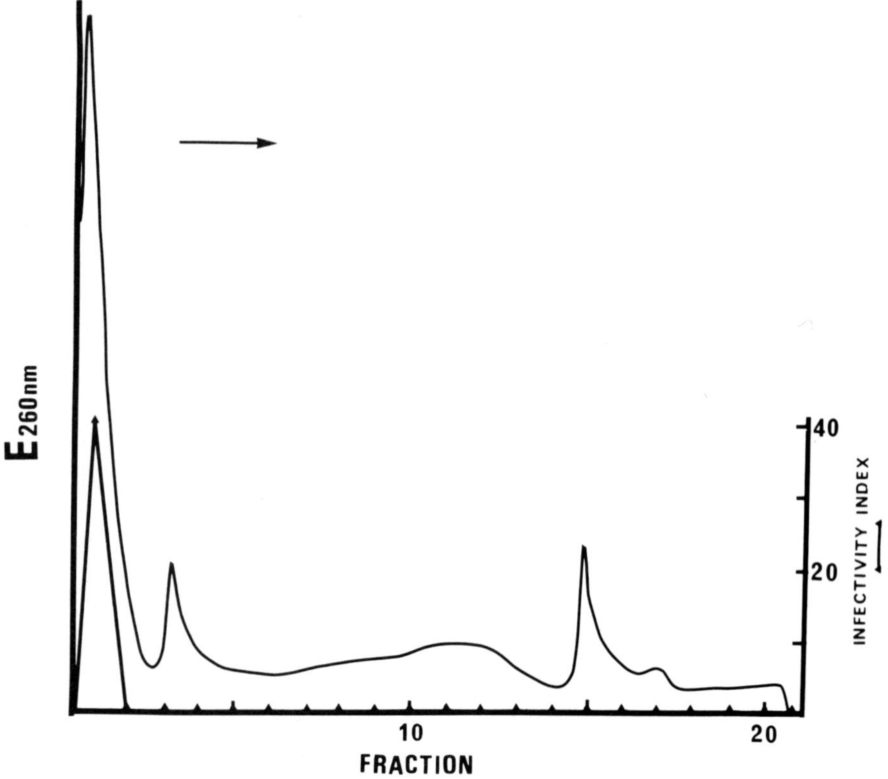


FIGURE 6. Densitometer trace of CaMV nucleic acid after electrophoresis in 2.5% polyacrylamide gels of 1½ hr at 12 V/cm. The gel was cut into slices each of which was disrupted in 0.25 mℓ 1 × SSC and the infectivity tested. (From Hull, R. and Shepherd, R. J., *Virology*, 79, 219, 1977. With permission.)

Yot[24] found that the populations of molecular forms in the two major bands and in the multiple bands were the same (75% circular, 25% linear), but that there were slight size differences. However, he considered that the size differences were too small to account for the differences in migration.

Harrison et al.[4a] report that MSV and CLV nucleic acids are resolved into two components on electrophoresis in formamide containing polyacrylamide gels. They found that the nucleic acid in the faster migrating band was linear molecules only whereas, that in the slower migrating band was mainly circular molecules.

### *8. Molecular Forms of CaMV Nucleic Acid*

There is no doubt that CaMV nucleic acid is composed primarily of double-stranded DNA (see Section III. B. 2 for comments about RNA content). From electron microscopy and from the behavior in rate-zonal sucrose gradients and in ethidium bromide/CsCl gradients, both linear and relaxed circular forms can be recognized. CaMV nucleic acid preparations also contain twisted circular molecules which migrate in agarose gels (Section III. A. 7) in a manner similar to that of populations of supercoils with different helix densities.[25] If the molecules were supercoiled, one would expect them to contain approximately 7 to 10 superhelical turns which should give a separation in CsCl/ethidium bromide gradients of 1.10 to 1.15 relative to that of relaxed molecules. (SV 40 DNA I with 24 superhelical turns[25] has a relative buoyant density separation

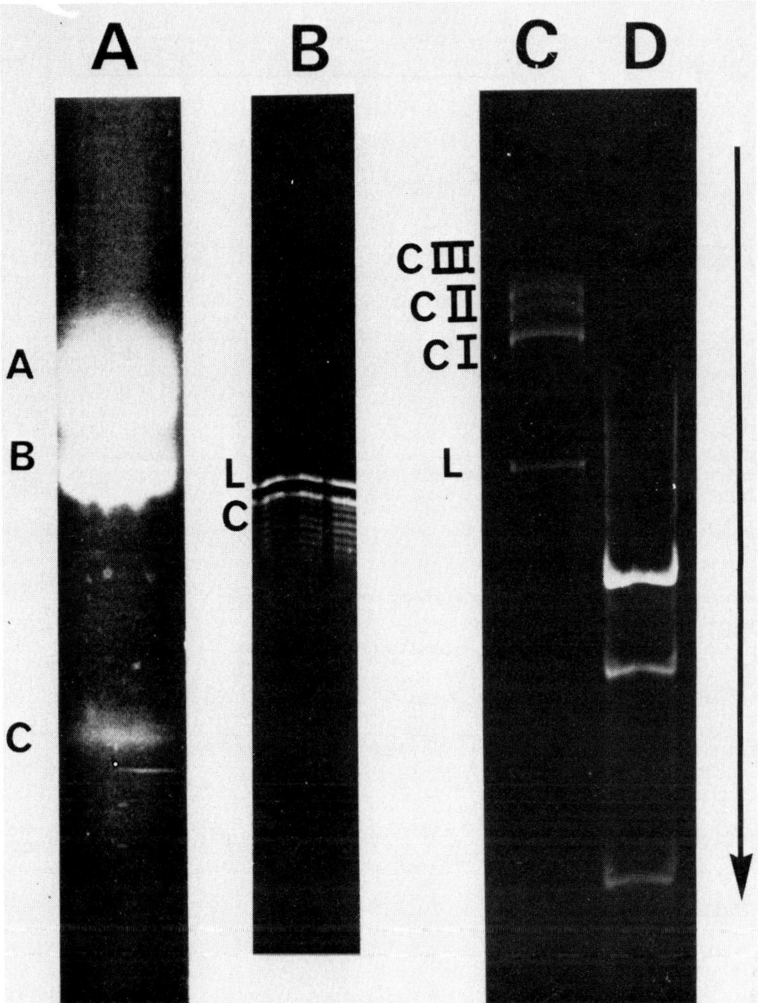


FIGURE 7. Gel electrophoresis of CaMV nucleic acid. A. Unfractionated nucleic acid electrophoresed in composite 2.2% polyacrylamide — 0.5% agarose gel at 2.5 V/cm for 10 hr using 0.5 *M* Tris, 0.02 *M* sodium acetate, 2m *M* $Na_2$ EDTA. 18 m *M* NaCl buffer. (Adapted from Civerolo, E. L. and Lawson, R. H., *Phytopathology*, 68, 101, 1978.) B. Nucleic acid electrophoresed in 1.8% acrylamide-0.5% agarose gel at 3 V/cm for 16 hr using 40 m*M* Tris, 20 m*M* sodium acetate, 2 m*M* EDTA pH 7.8. (Courtesy of P. Yot.) C. Nucleic acid electrophoresed in 1% agarose gel, conditions similar to B. CI, II, and III refer to three apparently different populations of circular molecules, L denotes linear molecules. The reverse order of the bands in B relative to that of C was determined by eluting the bands from B and re-electrophoresing of 1% agarose gels. (Courtesy of P. Yot.) D. Nucleic acid denatured using 0.2 *N* NaOH at 0°C for 1 min and then electrophoresed on 1% agarose gel. (Courtesy of P. Yot.)

of 1.26.)[25a] However, CaMV nucleic acid does not show this separation (Section III. A. 3) in CsCl/ethidium bromide gradients which together with the lack of a fast sedimenting component in alkaline rate-zonal gradients (Section III. B. 1) the lack of effect of yeast relaxing enzyme,[24] and the presence of nicks or gaps in the molecules[21,23] suggest that it is unlikely that these molecules are covalently closed supercoils of high helix density. This does not rule out the possibility that the nucleic acid in the virus particles

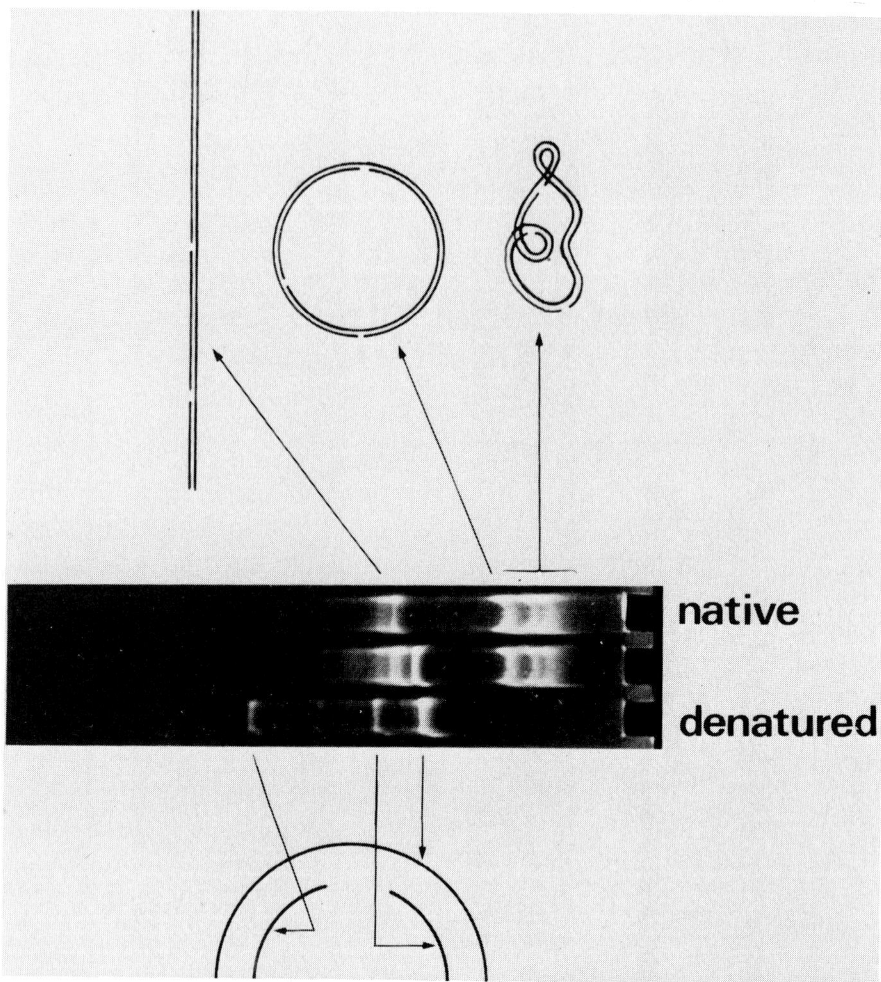


FIGURE 8. Diagram showing the various forms of native CaMV nucleic acid and the result of denaturing the nucleic acid. On the left hand side are the twisted circles, the open circles and the linear molecules which are separated on a 1% agarose gel (left hand channel). In the right hand channel of the gel are three bands derived from the denaturation of the DNA containing three single strand breaks, one in one strand and two in the other. The center channel shows partially denatured DNA. (Data taken from Hull, R. and Howell, S. H., *Virology*, 86, 482, 1978.)

might be in a supercoil form stabilized by a protein in the manner of the relaxation complex of ColE1 plasmid;[26] the use of pronase and SDS necessary for the extraction of CaMV nucleic acid would remove such a protein if it existed.

Vollenweider et al.[27] showed that, in ionic strengths greater than that of 100 m$M$ sodium acetate, linear $T_7$ DNA and relaxed circular PM2 DNA showed some of the properties of covalently closed DNA (e.g., electron microscope appearance). However, because the ends of the strands were not fixed, such molecules did not have the properties of superhelices in the presence of intercalating dyes; no reason was given for this change in tertiary structure. The twisted CaMV nucleic acid molecules resemble in some respects the $T_7$ and PM2 molecules at high ionic strength. They differ, however, in that the twisted molecules of CaMV nucleic acid are found at low ionic strength (e.g., 0.1 × SSC). It is possible that the single-stranded gaps in the molecules[21,23] (Section III. B) might give internal sticky areas. However, further work is needed on CaMV nucleic acid to elucidate the structure of these twisted molecules.

### B. Denatured and Renatured Nucleic Acids

#### *1. Products of Denaturation and Renaturation*

The melting properties of plant viral DNAs have been discussed in Section III. A. 2. Civerolo and Lawson[17] showed that after treatment of CaMV nucleic acid in either 0.01 *M* phosphate or in 1 × TE buffer (see Table 2) at temperatures of 75°C or higher, followed by either slow or rapid cooling; the nucleic acid formed several bands on gradients which sedimented more rapidly than did untreated nucleic acid. Analysis in sucrose gradients of CaMV nucleic acid which had been melted and then renatured at Tm-20°C showed that if this was done in 1 × SSC, all the nucleic acid sedimented rapidly; but if done in 0.1 × SSC, a certain proportion sedimented at the same rate as the untreated control. Examination of these preparations in the electron microscope revealed that melting and renaturation in 1 × SSC gave rise to large aggregates of nucleic acid. In 0.1 × SSC, the aggregates were smaller, and there were some unaggregated molecules.[17a]

CaMV nucleic acid can be fully and irreversibly denatured by heat and formaldehyde. In 1 × SSC and 1/10 volume 40% HCHO, the Tm is 61.6°C with a hyperchromicity of 40.5%; there is no decrease in optical density on cooling.[6,12] Formaldehyde denatured CaMV nucleic acid sediments as two major components (Figure 9A) with $S_{20}w$'s of 16.1 and 13.5*s* considered to be single-stranded circular and linear molecules, respectively.[12] Figure 9A shows that there is no rapidly sedimenting CaMV nucleic acid (hence, no evidence for covalently closed supercoiled molecules) and that there is only very little slow sedimenting material. However, using electron microscopy and gel electrophoresis (Figure 7D and Figure 8) on CaMV nucleic acid which had been heat- or alkali denatured, Volovitch et al.[23] and Hull and Howell[21] suggested that the native nucleic acid contained two or three single-stranded interruptions (depending upon the isolate) which were not randomly located; these were also detected by electrophoresis of the digestion products of $S_1$ nuclease (see Section IV. C). Thus, the sedimenting components of denatured CaMV nucleic acid noted above are probably single-stranded molecules of different lengths.

Volovitch et al.[23] and Hull and Howell[21] suggest that one of the strands of CaMV DNA contain one interruption and the other strand contains two interruptions. This would give the three bands (for most isolates) on denaturation and is illustrated in Figure 8. Isolate CM4-184 DNA which has two interruptions, one in each strand, gives only one major band upon denaturation.[21]

Addition of alkali (0.3 *N* KOH) to CaMV nucleic acid gives a hyperchromicity of 34.5% at 37°C and melting analysis indicates that this nucleic acid is fully denatured.[12]

The sedimentation behavior of alkali-treated CaMV nucleic acid depends on the length of time that the nucleic acid is exposed to alkali. If the nucleic acid is alkali-treated for a short period (37°C for 10 min) and then centrifuged rapidly into alkaline gradients, the sedimentation pattern resembles that after formaldehyde denaturation.[12] After slightly longer alkali treatment (30 min and longer time for gradients), three peaks are found;[17] two probably correspond to the two peaks mentioned earlier and the third is slower sedimenting. After prolonged exposure to alkali (18 hr), the nucleic acid sediments as a broad peak covering the range 10 to 13*S* (Figure 9B). If formaldehyde treatment follows the alkali treatment, the majority of the material sediments in the range 6.8 to 9.2*s* but some UV absorbance sediments as fast as 16*S*.[12] Thus, with prolonged exposure to alkali, breaks are induced in some, if not many, of the nucleic acid strands. Using the molecular weight of fragmented nucleic acid estimated from

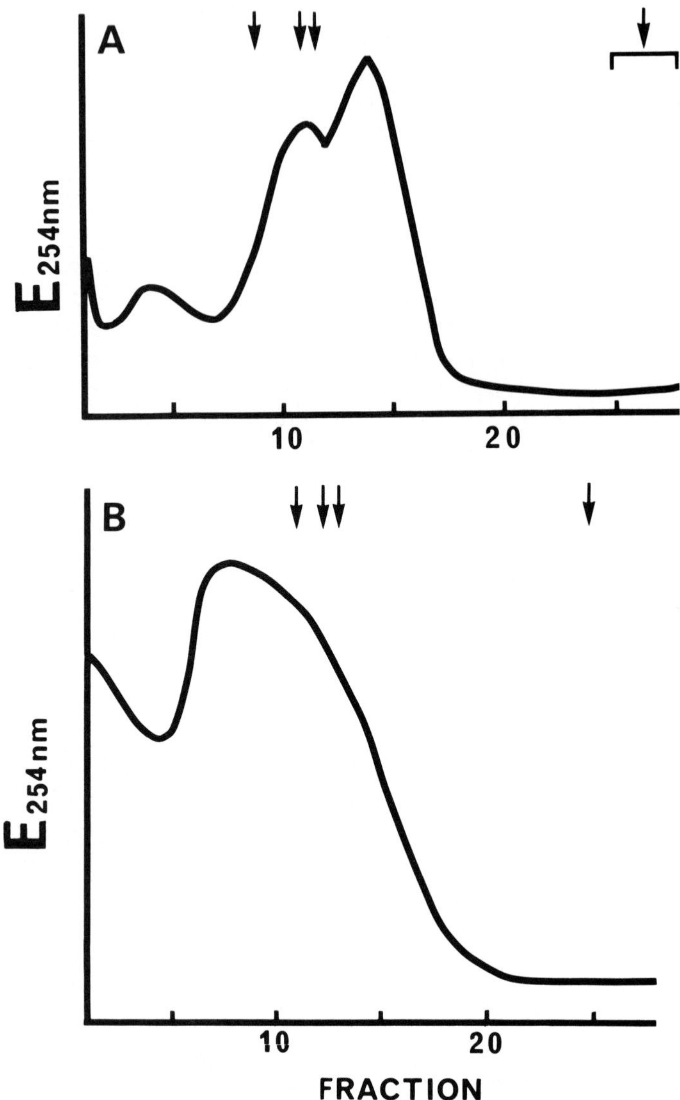


FIGURE 9. A. Sedimentation profile of formaldehyde denatured CaMV nucleic acid in a neutral sucrose gradient. The gradient conditions are as in Figure 3. The arrows indicate the depth to which, from left to right, formaldehyde denatured $^{32}$P-Polyoma, $^{3}$H ColE1, $^{32}$P-Polyoma and $^{3}$H ColE1 DNAs sedimented. (From Hull, R. and Shepherd, R. J., *Virology*, 79, 222-223, 1977. With permission.) B. Profile of CaMV nucleic acid treated with 0.3 *N* KOH for 18 hr at 37°C and sedimented into an alkaline sucrose gradient for 2¼ hr, 45,000 rpm in a SW 50.1 rotor. The arrows indicate the depth to which, from left to right, $^{32}$P-Polyoma, $^{32}$P-polyoma, $^{3}$H ColE1 and $^{3}$H ColE1 DNAs sedimented. (From Hull, R. and Shepherd, R. J., *Virology*, 79, 216, 1977. With permission.)

the median sedimentation coefficient, Hull and Shepherd[12] considered that there were on the average four or five breaks per single strand. The faster sedimenting fraction of native nucleic acid (Fraction C, Figure 3A) contains a higher proportion of molecules with alkali-labile regions than does the slower sedimenting material (Fractions A and B of Figure 3A) (Figure 10).[12]

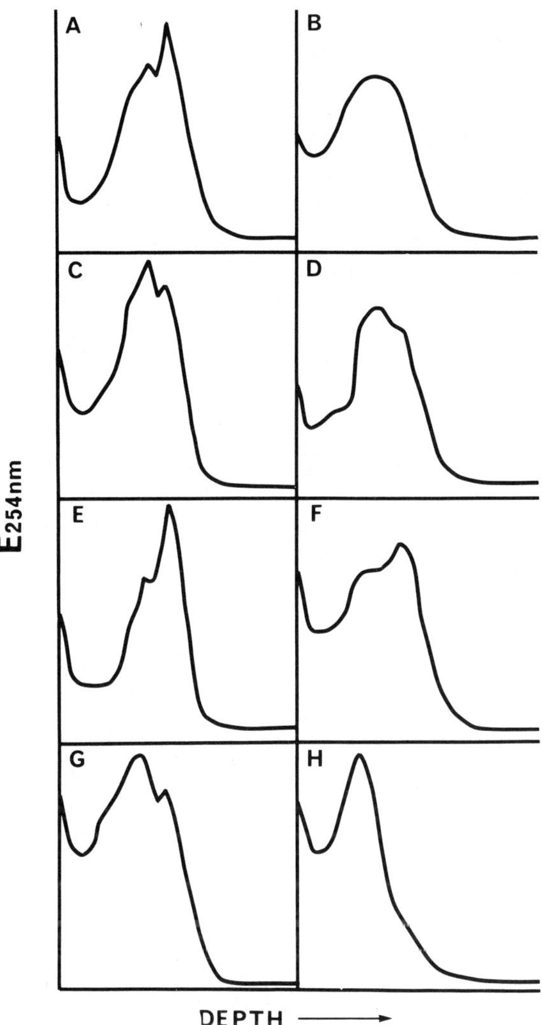


FIGURE 10. Sedimentation profiles in neutral sucrose gradients of CaMV nucleic acid; A, C, E, and G were formaldehyde denatured and B, D, F and H were alkali treated (18 hr 0.3 *N* KOH) before formaldehyde denaturation. A and B are from an unfractionated nucleic acid preparation; C and D are from the bulk fraction designed A in Figure 3B. E and F are from bulk fraction B of Figure 3B. G and H are from bulk fraction C of Figure 3B. The gradient conditions are as described in Figure 3. (From Hull, R. and Shepherd, R. J., *Virology*, 79, 222, 1977. With permission.)

Attempts to renature CaMV nucleic acid after alkali denaturation produced material showing a considerable amount of mismatching and which sedimented very rapidly in sucrose gradients.[12]

### 2. Nature of Alkali Labile Regions of CaMV Nucleic Acid

Hull and Shepherd[12] reported a series of observations on the nature of the alkali labile regions of CaMV nucleic acid. The alkali-labile regions were not due to the presence of apurinic acid, but they were susceptible to a certain but varying extent to pancreatic ribonuclease. Using $^{35}S$ labeling of alkali extracts of CaMV nucleic acid, they demonstrated the presence of ribonucleotides which comprised about 0.6% of the total nucleic acid and had molar ratios of 33.1% G, 18.3% A, 27.5% C, and 21.1% U. Some recent studies carried out on $^{125}I$-labeled CaMV nucleic acid[28,28a] indicate that while no oligoribonucleotides appear in two-dimensional fingerprints of undigested nucleic acid, pretreatment with RNase T1 or pancreatic RNase (Figure 11) gives fingerprints with complexities comparable to those of HeLa cell 5*s* RNA[29].

If this RNA is covalently linked into the DNA strands, it could explain the alkali labile regions. Repeated melting and quenching of the nucleic acid followed by CsCl gradient centrifugation failed to remove the RNA indicating that it is, at least, closely associated with the DNA.[12]

### *3. Discussion*

Further unusual properties of CaMV nucleic acid are revealed on denaturation. The sedimentation properties of the nucleic acid after formaldehyde or short alkali treatment[12,17] or after heat denaturation at low ionic strength (0.1 × SSC)[17] did not indicate a significant amount of very small nucleic acid pieces. However Volovitch et al.[23] and Hull and Howell[21] have shown that CaMV nucleic acid does contain two or three (depending on the isolate) nicks or gaps at specific sites (see Section IV. C). From studies[15] on the incorporation of deoxyribonucleotides into CaMV nucleic acid using *Micrococcus luteus* polymerase it appears that these sites are gaps of some tens of nucleotides. Labeling using polynucleotide kinase and γ labeled $^{32}P$ ATP shows that the gaps contain 5′hydroxyl groups. The origin of these gaps is not yet known.

The simplest explanation of the formation of shorter pieces of CaMV nucleic acid on prolonged alkali treatment[12,17] is that a proportion of the nucleic acid strands contain ribonucleotide regions which give rise to the alkali lability. Hull and Shepherd[12] estimated that approximately 40% of the molecules in a nucleic preparation contained alkali labile regions and that these regions would be composed of approximately 20 to 25 ribonucleotides.

It is interesting to note the apparent correlation between the prevalence of twisted circular CaMV nucleic acid molecules and those with alkali-labile regions. It is possible that the twisting might be in some way associated with the presence of ribonucleotides.

RNA is involved as primers in the initiation of DNA synthesis.[31-33] Usually the RNA primer is removed, but there are numerous reports of at least some of the RNA remaining within DNA.[34-36] The RNA associated with CaMV-DNA is of a similar size and base ratio to that found in supercoiled ColE1 plasmid DNA grown in the presence of chloramphenicol.[37]

Another possible reason for the presence of segments of RNA would be that they are the result of misincorporation. This would explain them apparently being joined to both the 5′- and 3′-ends of DNA segments; residual primers would need to be joined to the 3′-end of the DNA enzymically.

## IV. EFFECT OF NUCLEASES

### A. Pancreatic RNase and DNase

One of the first observations indicating that CaMV genome was DNA was the abolition of infectivity upon treatment with pancreatic DNase but not with pancreatic RNase.[1] Similarly, the genomes of CERV, DaMV, PLRV, BGYMV, MSV, and CLV have been shown to be susceptible to DNase but not to RNase.[4,4a,8,12,38,39] However, as noted above (Section III. B. 2) pancreatic RNase can have some effect on a certain proportion of CaMV nucleic acid molecules.

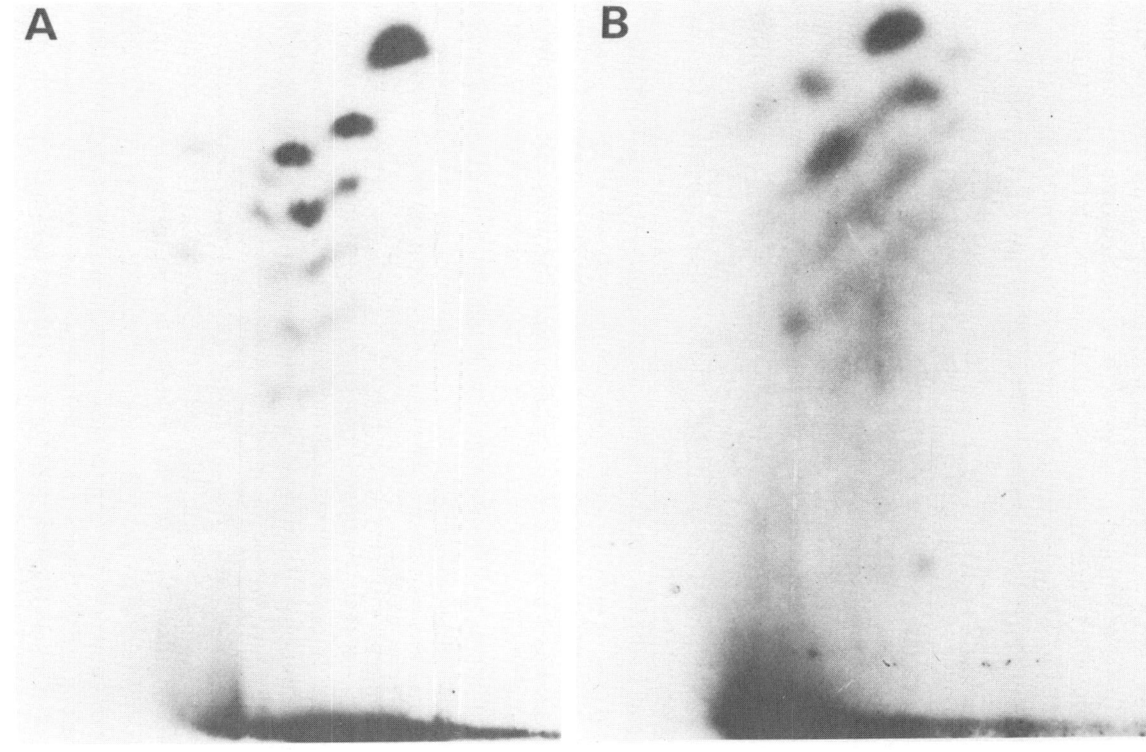


FIGURE 11. Ribonuclease T1 and Pancreatic RNase fingerprints of $^{125}$I-labeled CaMV nucleic acid. Aliquots of $5 \times 10^6$ cpm of CaMV nucleic acid iodinated in vitro to a specific activity of $10 \times 10^6$ cpm/$\mu$g were digested either with RNase T1 (1 mg/ml) or with Pancreatic RNase (1 mg/ml) in 0.01 *M* Tris, pH 7.5, 0.001 *M* EDTA for 45 min at 37°C and subjected to two-dimensional fingerprinting analysis. The first dimension (right to left in the figure) involved separation of oligonucleotides on the basis of charge and was accomplished by high voltage electrophoresis at pH 3.5 in the presence of 7 *M* urea. The second dimension (bottom to top in the figure) involved ascending homochromatography, a process in which the smallest oligonucleotides move the furthest. A. Pancreatic RNase fingerprint. B. Ribonuclease T1 fingrprint.[28] For further details concerning applications of RNA fingerprinting analysis and for comparison with other RNA fingerprints, see Chapter 12.

### B. Restriction Endonucleases

Recently, there have been several studies started on the cleavage of CaMV nucleic acid using restriction endonucleases.[21-23,40,40a] The products of the endonuclease digest of the nucleic acid of type strain (Cabbage B) are listed in Table 6.

Using double and partial digestion techniques, Volovitch et al.,[23] Meagher et al.,[22] and Hull and Howell[21] have produced maps of the endonuclease digestion sites of CaMV nucleic acid. Hull and Howell[21] showed that the nucleic acids from different isolates of CaMV differed in their restriction endonuclease maps, and they recognized four different groups of isolates. Figure 12 shows the combined data from the maps of the nucleic acid of the type strain of CaMV (called CaMV Cabb B-D by Hull and Howell[21]). The nomenclature of the EcoR1 fragments is that used by Hull and Howell[21] which differs from that used by Meagher et al.[22] The origin is taken as being the EcoR1 cleavage site between the largest fragment (b + c) (fragment a on the Meagher et al. nomenclature) and the smallest fragment (d) (fragment c on the Meagher et al. nomenclature); for ease of comparison of the cleavage sites of the different endonucleases, the nucleic acid molecule is represented as being linear.

The differences between the maps of the four groups of CaMV isolates[21] are shown in Figure 13. It can be seen that most of the differences lie in EcoR1 fragment c which in CM4-184 DNA is smaller than in that of Cabb B, and in Australian isolate DNA is slightly larger (Table 6). The portion missing from CM4-184 DNA EcoR1 fragment c contains the two Bam HI sites and an $S_1$ nuclease site.

Meagher et al.[22] reported that although digestion with the endonucleases they used was complete within 5 min, some of the fragments were produced in nonstoichometric amounts. They list the following sites on native nucleic acid (Cabb B-D isolate) as showing variable cleavage: — Bam HI site A/C, Hind III sites A/B, B/H, and E/F and the Sal I site; the majority of the nucleic acid molecules are not cleaved at these sites. This could explain why Volovitch et al.,[23] Langridge et al.,[40] and Hull and Howell[21] could not detect the Bam HI site A/C (Table 6, Figure 12). Most of these sites are fully cleaved in molecularly cloned nucleic acid (see Section V) and such nucleic acid was used in the map produced by Meagher et al.[22] However, Hull and Howell[21] showed that the nucleic acids from isolates of CaMV differed in their susceptibility to endonucleases such as Sal I. Thus, it is possible that Meagher et al.[22] selected molecules with susceptible sites during their cloning. However, they also suggest other possible causes of heterogeneity such as single base substitutions, small deletions or inversions; they found no evidence for large deletions or inversions in the isolate they used. The heterogeneity might be due to the high multiplicity of infection inherent with the production of virus in plants. Hull and Howell[21] noted three forms of heterogeneity as expressed in the restriction endonuclease digest patterns. As well as the variable cleavage sites discussed above, they found possible evidence for mutation or change of an isolate (from Cabb B-D to Cabb B-JI). This, however, could be due to contamination and needs to be investigated further. Preliminary attempts to find host-induced changes have been unsuccessful.[15] They[21] also found that some of the minor bands (in less than stoichometric amounts) of EcoR1, Sal I, and Bam HI digests were derived from linear molecules which appeared to arise from double-stranded breaks at specific points.

### C. Single-strand Specific Nuclease

Volovitch et al.[23] and Hull and Howell[21] have shown that CaMV-DNA is susceptible to nuclease $S_1$ (the single-strand specific nuclease from *Aspergillus oryzae*) at the sites of the single-strand breaks (see Section III. B. 1). The map positions of these sites are shown in Figures 12 and 13. It has been noted above (Section V. B) that one of the

TABLE 6

**Molecular Weights of Products after Treatment of CaMV DNA with Restriction Endonucleases**

| CaMV[a] isolate | Enzyme | Molecular weight (× $10^{-6}$) of fragments | | | | | | | | | | Ref. |
|---|---|---|---|---|---|---|---|---|---|---|---|---|
| | | A[a] | B | C | D | E | F | G | H | I | J | |
| Cabb B-D | EcoR1 | 2.0 | | 2.7(B+C)[b] | 0.30 | | | | | | | 21 |
| | EcoR1 | 1.90 | | 2.7(B+C) | 0.30 | | | | | | | 23 |
| | EcoR1 | 1.95 | | 2.7(B+C) | 0.32 | | | | | | | 22 |
| | Bam HI | 3.1 | 1.65 | 0.135 | | | | | | | | 22 |
| | Bam HI[c] | 4.69 | | 0.185 | | | | | | | | 23 |
| | Hpa1[c] | 1.39 | 1.10 | 0.83 | 0.34 | 0.32 | 0.25 | 0.24 | 0.15 | 0.13 | 0.13 | 23 |
| | Hpa2[c] | 3.49 | 1.39 | | | | | | | | | 23 |
| | Hind III | 1.34 | 0.80 | 0.65 | 0.49 | 0.35 | 0.33 | 0.31 | 0.29 | 0.14 | 0.135 | 22 |
| | Sal GI | 5.0 | | | | | | | | | | 21 |
| | Sal GI | 4.88 | | | | | | | | | | 22 |
| Cabb B-JI | EcoR1 | 2.0 | 1.5 | 1.2 | 0.3 | | | | | | | 21 |
| | Sal GI | 5.0 | | | | | | | | | | 21 |
| CM4-184 | EcoR1 | 2.0 | 1.5 | 0.95 | 0.3 | | | | | | | 21 |
| | Sal GI | 4.8 | | | | | | | | | | 21 |
| Australian | EcoR1 | 2.0 | 1.5 | 1.3 | 0.3 | | | | | | | 21 |
| | Sal GI | 5.0 | | | | | | | | | | 21 |

[a] For details of the isolates and the convention for naming fragments see Hull and Howell.[21] For map positions of fragments see Figure 13.

[b] On the convention of Hull and Howell[21] the largest EcoR1 fragment of this isolate is designated B + C.

[c] Based on Figure 5 of Volovitch et al.[23] and nucleic acid molecular weight of $4.88 \times 10^6$.

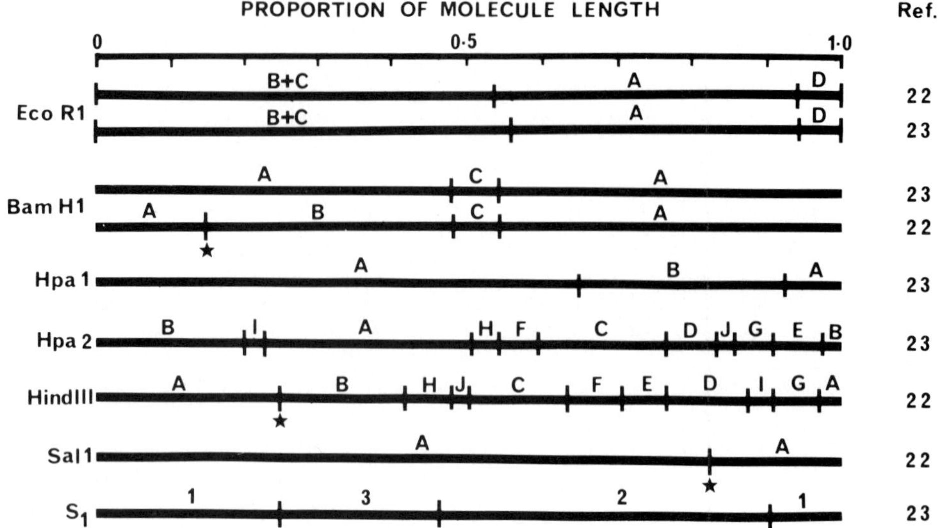


FIGURE 12. Map of restriction endonuclease and nuclease $S_1$ sites in CaMV nucleic acid. For convenience the molecule is represented as being linear. * sites which are difficult to digest or are only partially digested in nucleic acid from the virus. Nomenclature of EcoR1 fragments is as in Hull and Howell.[21]

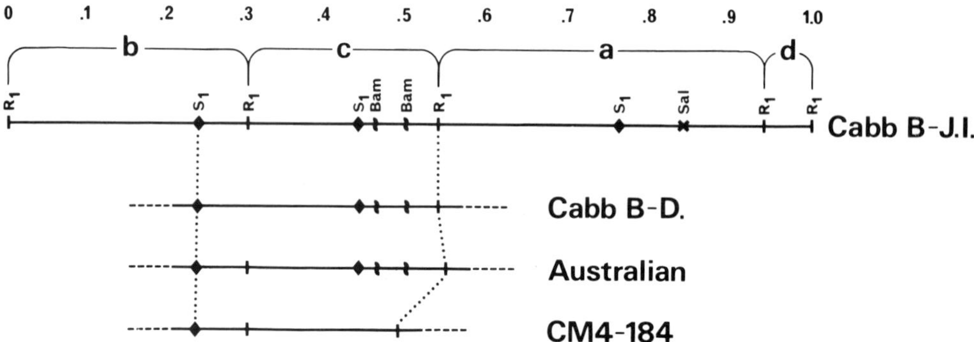


FIGURE 13. Diagram showing the differences in the restriction endonuclease maps of the DNAs of four isolates of CaMV. The numbers along the top are map positions and the letters a through d are the four EcoR1 fragments of Cabb B-JI isolate. R1 denotes EcoR1 sites; $S_1$ denotes single stranded nuclease $S_1$ sites; Bam denotes Bam HI sites; and Sal denotes Sal GI sites. (Data taken from Hull, R. and Howell, S. H., *Virology*, 86, 482, 1978.)

nuclease $S_1$ sites is missing in the DNA from CM4-184 isolate; the other two sites are diametrically opposite ne another and thus $S_1$ digestion results in half length molecules.

Nuclease $S_1$ destroys the nucleic acids of BGYMV,[16a] MSV, and CLV[4a] confirming that they are predominantly single-stranded.

## V. MOLECULAR CLONING

Meagher et al.[41] used the plasmids ColE1, pMB9, and pBR 313 as vehicles for cloning EcoRI and Bam HI restriction endonuclease fragments of CaMV nucleic acid in *Escherichia coli* strain HB 101. They cloned recombinant plasmids containing all three EcoRI fragments and others containing one or both of the larger two EcoRI fragments

TABLE 7

**Polypeptides Synthesized in *Escherichia coli* Minicells Containing Molecularly Cloned Fragments of CaMV Nucleic Acid**

| DNA fragment | DNA size ($\times 10^{-6}$) | Proteins detected ($\times 10^{-3}$) |
|---|---|---|
| CaMV | 2.6 | 52 and 37 |
| CaMV | 1.9 | 54, 43 and 40 |
| CaMV | 2.6, 1.9, 0.3 | 42, 43, 40, and 37 |
| pRm 23 CaMV | 2.6 | 37 |
| pRM 23 CaMV (pRM 41) | 1.9 | 43 and 40 |
| pRM 23 CaMV (pRM 44) | 1.9 | 43, 40, and 16 |

Adapted from Meagher, R. B., Tait, R. C. and Boyer, H. W., *Cell,* 10, 521, 1977.

and one or both of the larger Bam HI fragments. As has been noted above (Section IV. B), these fragments were used in constructing the restriction map; they contained restriction endonuclease sites which were found to be variable in the viral nucleic acid. The CaMV nucleic acid fragments were cloned with great difficulty when compared with fragments of bacterial DNA; approximately 100 times as much CaMV nucleic acid was needed to produce the same frequency of recombinant plasmids as did *E. coli* lambda phage DNA.

Szeto et al.[40a] used the plasmid pGM 706 to clone Sal I digested CaMV DNA in *E. coli.* The cloned viral DNA gave identical or near identical fragments after digestion with EcoRI, Bam HI, Hind II and III, and Hae III endonucleases as did CaMV DNA which had not been cloned. They were unable to clone the Bam HI fragments of CaMV.

CaMV nucleic acid has the coding capacity for 8 to 10 proteins. As Meagher et al.[22] point out, the only gene function known with certainty for CaMV genome is the aphid acquisition factor.[47] It is fair to assume that, at least, some of the viral coat proteins are coded for by the virus. Kelley et al.,[43] Brunt et al.,[44] and Hull and Shepherd[45] have shown that the isolated virus possesses about four polypeptide species, two of which (64,000 and 37,000) account for over 90% of the total protein.

Meagher et al.[41] have examined the polypeptides synthesized in *E. coli* minicells containing cloned fragments of CaMV nucleic acid. Three unique polypeptide species were found to be synthesized in relatively large amounts (Table 7). From Table 7 it can be seen that the $2.6 \times 10^6$ CaMV nucleic acid EcoRI fragment (fragment B + C Figure 13) codes for the 37,000 polypeptide species and the $1.9 \times 10^6$ fragment (fragment A, Figure 13) codes for the 43,000 and 40,000 polypeptide species. In one recombinant (pRM44) the $1.9 \times 10^6$ fragment was also found to code for a 16,000 polypeptide species; this fragment was found to be inserted in the plasmid in the opposite orientation to that in prRM 41. Using isoelectric focusing, cyanogen bromide digestion, and serological tests, Meagher et al.[41] showed that none of the three main polypeptide species synthesized in *E. coli* minicells resembled the major coat proteins of CaMV.

## VI. GROWTH IN PROTOPLASTS

Howell and Hull[46] have recently reported on the growth of CaMV in protoplasts isolated from turnips. Using sucrose density gradient centrifugation, they were able to detect virus produced in protoplasts in the presence of $^{32}P$ label. They found that virus

TABLE 8

**Comparison of Properties of Caulimoviruses with Those of Papovavirus Groups A and B**

| | | Papovavirus[b] | |
|---|---|---|---|
| Nucleoprotein | Caulimovirus[a] | A | B |
| Shape | Sphere | Sphere | Sphere |
| Size (nm) | 45—50 | 55 | 45 |
| $S_{20}w$ | 208—254 | 296—300 | 240 |
| Molecular weight ($\times 10^{-6}$) | 22.8 | 28 | 20—25 |
| Nucleic Acid (%) | 16—17 | 12 | 12 |
| DNA Forms ($S_{20}w$) | Circular (19)<br>Linear (17) | Supercoil (28)<br>Circular (21)<br>Linear (18) | Supercoil (21)<br>Circular (16)<br>Linear (14) |
| Molecular weight ($\times 10^{-6}$) | 4.5—5.0 | 5.0 | 3.0 |
| % GC | 43 | 41—47 | 41—48 |
| Protein types and molecular weight | One major, molecular weight 37,000<br>3 minor | One major, molecular weight about 60,000<br>4—5 minor | One major, molecular weight about 43,000<br>5 minor |
| Number per capsid | 420 of the major protein | 420 | 420 |

[a] Data from Hull and colleagues,[7] Hull and Shepherd,[12] and this chapter.
[b] Data from Melnick and colleagues.[59]

production was very slow with little virus being detectable before 72 hr post infection. The main virus production was between 72 and 96 hr post infection. They detected an RNA transcript of apparent molecular weight 1.5 to 1.8 $\times 10^6$ which is polyadenylated. Using hybridization techniques, they showed that this transcript was from only one DNA strand, that with the single break in it (see Section III. B. 1). The RNA transcript was produced 48 to 72 hr post infection.

## VII. DISCUSSION

The caulimoviruses have generally been regarded as a discrete group of viruses differing from other groups of double-stranded DNA containing viruses. However, their properties do resemble, to a certain extent, those of the papovavirus A group (Table 8). Although it is likely that this resemblence is fortuitous and is based on some structural and functional limitations, more needs to be known about caulimoviruses before one can be certain that these two groups are completely different.

The nucleic acid of at least some of the caulimoviruses has some unusual properties. The topology of the circular molecules discussed in Section III. A. 8 differs from that of the papovaviruses and other DNA viruses. The occurrence of alkali-labile regions in a significant proportion of the nucleic acid molecules from native CaMV and the single strand gaps at specific locations are other properties not found in most other DNA viruses.

The data presently available on the nucleic acid of PLRV are difficult to interpret. The evidence that the nucleic acid is DNA needs further substantiation. Furthermore, it is difficult to envisage how a virus with a double-stranded nucleic acid of molecular weight 0.56 $\times 10^6$ can function by itself.

The geminiviruses are a very recently recognized group of plant viruses, each of the three present members having single-stranded DNA which in at least two of the members is circular. The nucleic acid is small (0.6 to 0.9 × $10^6$), and it will be interesting to determine if all the genome is carried on a single molecule. The capsid proteins of these viruses have molecular weights of approximately 30,000[16a,47a] which, as noted by Goodman,[16a] would account for more than 30% of the coding capacity of a single DNA molecule.

The techniques involved in genetic engineering are best performed using double-stranded DNA. This means that as far as plant viral DNAs are concerned, the nucleic acids of the caulimoviruses are the only ones that can be used. As has been described in Section V, experiments to investigate the expression of eucaryotic DNA in a procaryotic system have been attempted. Of more fundamental interest is the use of CaMV nucleic acid as a vector to carry procaryotic and eucaryotic genes into plant systems. However these types of experiments reveal gaps in our knowledge of CaMV nucleic acid. Among the questions which will have to be answered are what proportion of the total nucleic acid is needed, and in what form should it be for infection and replication; furthermore, what are the site(s) and mechanisms of replication? These areas of research are likely to attract most of the attention on plant viral DNAs in the future. One further consideration concerning the genetic engineering type of experiment is the safety and containment problem. The present guidelines suggest Category II containment in the United Kingdom[48] and P3 + EKI or P2 + EK2 containment in the USA.[49] However, researchers using recombinant DNA techniques should be aware that as well as the possibilities of producing new pathogens of plants, there are also similarities between caulimoviruses and members of the papovavirus group A (Table 8).

## ACKNOWLEDGMENTS

I wish to thank Drs. E. L. Civerolo, R. M. Goodman, B. D. Harrison, J. Langridge, R. H. Lawson, R. B. Meagher, and P. Yot for sending me unpublished manuscripts and data.

## REFERENCES

1. **Shepherd, R. J., Wakeman, R. J., and Romanko, R. R.,** DNA in cauliflower mosaic virus, *Virology,* 36, 150, 1968.
2. **Harrison, B. D., Finch, J. T., Gibbs, A. J., Hollings, M., Shepherd, R. J., Valenta, V., and Wetter, C.,** Sixteen groups of plant viruses, *Virology,* 45, 356, 1971.
3. **Sarkar, S. and Blessing, J.,** DNA-like properties of the nucleic acid of potato leafroll virus, *Naturwissenschaften,* 60, 480, 1973.
4. **Goodman, R. M.,** Infectious DNA from a whitefly-transmitted virus of *Phaseolus vulgaris, Nature (London),* 266, 54, 1977.

4a. **Harrison, B. D., Barker, H., Bock, K. R., Guthrie, E. J., Meredith, G., and Atkinson, M.,** Plant viruses with circular single-stranded DNA, *Nature (London),* 270, 760, 1977.

5. **Shepherd, R. J.,** DNA viruses of higher plants, *Adv. Virus Res.,* 20, 305, 1976.
6. **Shephered, R. J., Bruening, G. E., and Wakeman, R. J.,** Double-stranded DNA from cauliflower mosaic virus, *Virology,* 41, 339, 1970.
7. **Hull, R., Shepherd, R. J., and Harvey, J. D.,** Cauliflower mosaic virus: an improved purification procedure and some properties of the virus particles, *J. Gen. Virol.,* 31, 93, 1976.

7a. **Hull, R. and Shepherd, R. J.,** unpublished observation.

8. **Lawson, R. H. and Civerolo, E. L.,** Purification of carnation etched ring virus and comparative properties of CERV and cauliflower mosaic virus nucleic acids, *Acta Horti.,* 59, 49, 1976.

9. **Peters, D.**, The purification of potato leafroll virus from its vector *Myzus persicae*, *Virology*, 31, 46, 1967.
10. **Sarkar, S.**, Potato leafroll virus contains a double-stranded DNA, *Virology*, 70, 265, 1976.
11. **Goodman, R. M., Bird, J., and Thongmeearkom, P.**, An unusual virus-like particle associated with golden yellow mosaic of beans, *Phaseolus vulgaris*, *Phytopathology*, 67, 37, 1977.
12. **Hull, R. and Shepherd, R. J.**, The structure of cauliflower mosaic virus genome, *Virology*, 79, 216, 1977.
13. **Langridge, J.**, personal communication.
14. **Russell, G. J., Follett, E. A. C., Subak-Sharpe, J. H., and Harrison, B. D.**, The double-stranded DNA of cauliflower mosaic virus, *J. Gen. Virol.*, 11, 129, 1971.
15. **Hull, R. and Howell, S. H.**, unpublished observation.
15a. **Geiduschek, E. P.**, On the factors controlling the reversibility of DNA denaturation, *J. Mol. Biol.*, 4, 467, 1962.
16. **Gömec, B.**, unpublished observation.
16a. **Goodman, R. M.**, Single-stranded DNA genome in a whitefly-transmitted plant virus, *Virology*, 83, 171, 1977.
17. **Civerolo, E. L. and Lawson, R. H.**, Topological forms of cauliflower mosaic virus nucleic acid, *Phytopathology*, 68, 101, 1978.
18. **Shepherd, R. J. and Wakeman, R. J.**, Observation on the size and morphology of cauliflower mosaic virus deoxyribonucleic acid, *Phytopathology*, 61, 188, 1971.
19. **Kleinschmidt, A. K. and Zahn, R. K.**, Über Desoxy-ribonucleinsaüre- Molekeln in Protein - Mischfilmen, *Z. Naturforsch.*, 14b, 770, 1959.
20. **Davis, R. W., Simon, M., and Davidson, N.**, Electron microscope heteroduplex methods for mapping regions of base sequence homology in nucleic acids, *Methods Enzymol.*, 21, 413, 1971.
21. **Hull, R. and Howell, S. H.**, Structure of the cauliflower mosaic virus genome. II. Variations in DNA structure and sequences between isolates, *Virology*, 86, 482, 1978.
22. **Meagher, R. B., Shepherd, R. J., and Boyer, H. W.**, The structure of cauliflower mosaic virus: I. A restriction endonuclease map of cauliflower mosaic virus DNA, *Virology*, 80, 362, 1977.
23. **Volovitch, M., Dumas, J. P., Drugeon, G., and Yot, P.**, Single-stranded interruptions in cauliflower mosaic virus DNA, in *Acides Nucléiques et Synthèse des Proteines chez les Végêteux*, Bogorad, L. and Weil, J. H., Eds., Centre National de la Reserche Scientifique, Paris, 1977, 635, 639.
24. **Yot, P.**, personal communication.
25. **Keller, W. and Wendel, I.**, Stepwise relaxation of supercoiled SV40 DNA, *Cold Spring Harbor Symp. Quant. Biol.*, 39, 199, 1974.
25a. **Gray, H. B., Upholt, W. B., and Vinograd, J.**, A buoyant density method for the determination of superhelix density of closed circular DNA, *J. Mol. Biol.*, 62, 1, 1971.
26. **Clewell, D. B. and Helinski, D. R.**, Supercoiled circular DNA-protein complex in *Escherichia coli*: purification and induced conversion to an open circular DNA form, *Proc. Natl. Acad. Sci. USA.*, 62, 1159, 1970.
27. **Vollenweider, H. J., Koller, Th., Parello, J., and Sogo, J. M.**, Superstructure of linear duplex DNA, *Proc. Natl. Acad. Sci. USA.*, 73, 4125, 1976.
28. **Dickson, E. and Hull, R.**, unpublished observation.
28a. **Dickson, E.**, Viroids: Infectious RNA in Plants, in Nucleic Acids in Plants, Hall, T. C. and Davies, J., Eds., CRC Press, Cleveland.
29. **Robertson, H. D., Dickson, E., Model, P., and Prensky, W.**, Application of fingerprinting techniques to iodinated nucleic acids, *Proc. Natl. Acad. Sci. USA.*, 70, 3260, 1973.
30. **Koch, J. and Bruhn, A.**, Nuclease $S_1$ cleavage and the primary structure of mitochondrial DNA, *Eur. J. Biochem.*, 63, 147, 1976.
31. **Chargaff, E.**, Initiation of enzymic synthesis of deoxyribonucleic acid by ribonucleic acid primers, *Prog. Nucleic Acid Res. Mol. Biol.*, 16, 1, 1976.
32. **Dressler, D.**, The recent excitement in the DNA growing point problem, *Annu. Rev. Microbiol.*, 29, 525, 1975.
33. **Gefter, M. L.**, DNA replication, *Annu. Rev. Biochem.*, 44, 45, 1975.
34. **Rosenkranz, H. S.**, RNA in coliphage T5, *Nature (London)*, 242, 327, 1973.
35. **Hirsch, I. and Vonka, V.**, Ribonucleotides linked to DNA of herpes simplex virus, *J. Virol.*, 13, 1162, 1974.
36. **Babiuk, L. A. and Rouse, B. T.**, Ribonucleotides in infectious bovine rhinotracheitis virus DNA, *J. Gen. Virol.*, 31, 221, 1976.
37. **Williams, P. H., Boyer, H. W., and Helinski, D. R.**, Size and base composition of RNA in supercoiled plasmid DNA, *Proc. Natl. Acad. Sci. USA.*, 70, 3744, 1973.
38. **Fujisawa, I., Rubio-Huertos, M., and Matsui, C.**, Deoxyribonuclease digestion of the nucleic acid from carnation etched ring virus, *Phytopathology*, 62, 810, 1972.

39. **Fujisawa, I., Rubio-Huertos, M., and Matsui, C.,** Deoxyribonucleic acid in dahlia mosaic virus, *Phytopathology,* 64, 287, 1974.
40. **Langridge, J., Brock, R. D., Scowcroft, W. R., and Merriam, V.,** Preparation of a molecular vector for plant cells, *CSIRO Div. Plant Industry Genetic Section Annu. Rep.,* p. 25, 1976.
40a. **Szeto, W. W., Homer, D. H., Carlson, P. S., and Thomas, C. A.,** Cloning of cauliflower mosaic virus (CLMV) DNA in *Escherichia coli, Science,* 196, 210, 1977.
41. **Meagher, R. B., Tait, R. C., and Boyer, H. W.,** Protein expression in *E. coli* minicells by recombinant plasmids containing eucaryotic DNA fragments, *Cell,* 10, 521, 1977.
42. **Lung, M. C. Y. and Pirone, R. P.,** Studies on the reason for differential transmissibility of cauliflower mosaic virus by aphids, *Phytopathology,* 63, 910, 1973.
43. **Kelley, D. C., Cooper, V., and Walkey, D. G. A.,** Cauliflower mosaic virus structural proteins, *Microbios,* 10, 239, 1974.
44. **Brunt, A. A., Barton, R. J., Tremaine, J. H., and Stace-Smith, R.,** The composition of cauliflower mosaic virus protein, *J. Gen. Virol.,* 27, 101, 1075.
45. **Hull, R. and Shepherd, R. J.,** The coat proteins of cauliflower mosaic virus, *Virology,* 70, 217, 1976.
46. **Howell, S. H. and Hull, R.,** Multiplication of cauliflower mosaic virus and transcription of its genome in turnip leaf protoplasts, *Virology,* 86, 468, 1978.
47. **Bock, K. R., Guthrie, E. J., and Woods, R. D.,** Purification of maize streak virus and its relationship to viruses associated with streak disease of sugar cane and *Panicum maximum, Annu. Appl. Biol.,* 77, 289, 1973.
47a. **Bock, K. R., Guthrie, E. J., Meredith, G., and Barker, H.,** RNA and protein components of maize streak and cassava latent viruses, *Annu. Appl. Biol.,* 85, 305, 1977.
48. Genetic manipulation: guidelines out, *Nature (London),* 263, 4, 1976.
49. Genetic manipulation: guidelines issued, *Nature (London),* 262, 2, 1976.
50. **Pirone, T. P., Pound, G. S., and Shepherd, R. J.,** Properties and serology of purified cauliflower mosaic virus, *Phytopathology,* 51, 541, 1961.
51. **Brunt, A. A.,** Partial purification, morphology, and serology of dahlia mosaic virus, *Virology,* 28, 778, 1968.
52. **Brunt, A. A.,** Some hosts and properties of dahlia mosaic virus, *Annu. Appl. Biol.,* 67, 357, 1971.
53. **Brunt, A. A.,** Dahlia mosaic virus, Commonwealth Mycological Institute/Association of Applied Biologists, descriptions of plant viruses No. 51, 1971.
54. **Gömec, B.,** Ph.D. Thesis, University of California, Davis, 1973.
55. **Hollings, M. and Stone, O. M.,** Carnation etched ring, *Glasshouse Crops Res. Inst. Annu. Rep.,* 1968, 102, 1969.
56. **Fujisawa, I., Rubio-Huertos, M., and Matsui, C.,** Incorporation of thymidine $^3$H into carnation etched ring virus, *Phytopathology,* 61, 681, 1971.
57. **Brunt, A. A. and Kitajima, E. W.,** Intracellular location and some properties of *Mirabilis* mosaic virus, a new member of the cauliflower mosaic group of viruses, *Phytopathol. Z.,* 76, 265, 1973.
58. **Galvez, G. E. and Castano, M.,** Purification of the whitefly-transmitted bean golden mosaic virus, *Turrialba,* 26, 205, 1976.
59. **Melnick, J. L., Allison, A. C., Butel, J. S., Eckhart, W., Eddy, B. E., Kit, S., Levine, A. J., Miles, J. A. R., Pagano, J. S., Sachs, L., and Vonka, V.,** Papovaviridae, *Intervirology,* 3, 106, 1974.

# THE RNAs OF MONOPARTITE PLANT VIRUSES

**M. Zaitlin**

## TABLE OF CONTENTS

# I. INTRODUCTION

Monopartite plant viruses contain all of their genomic information in a single strand of nucleic acid, either RNA or DNA. Only RNA-containing viruses are considered here; DNA viruses are discussed in Volume II, Section III by Hull. Redundant viral genetic information may also be present in other species of viral nucleic acid contained within infectious virus particles (termed virions), but these RNAs are not necessary to generate a complete, productive infection. Further, these redundant pieces of nucleic acid might be encapsidated in separate particles or in the particle containing the genomic nucleic acid. This concept is important because of the recent findings that infections involving tobacco mosaic virus (TMV) or turnip yellow mosaic virus (TYMV) generate, in addition to the genomic RNA, small RNA molecules which are subsets of the genomic RNA and which are messenger RNAs (mRNAs) for specific viral translational products — as will be discussed in Section VI and in more detail in Volume II, Section III by Davies.

A list of viruses that are considered to be monopartite is given in Table 1; here they are classified either by their accepted grouping[1] or, in several cases, as unclassified monotypic viruses. The virus considered to be "type" is given first in each specific group. Only a few examples are given for each, as some contain large numbers of viruses and strains. For example, in 1974 the potyvirus group consisted of 87 viruses and 15 strains,[2] and there are at least 45 strains of TMV and an even greater number of laboratory produced and spontaneous mutants.

Strictly speaking, to qualify as a monopartite virus, virus preparations should contain at least one population of homogeneous RNA molecules which contain the complete viral genome. As examples for this table,[3-33] viruses have been selected in which the RNAs have been examined either by density gradient centrifugation or preferably by polyacrylamide gel electrophoresis where the spectrum of RNA sizes may be examined and where the above conditions could be satisfied. Surprisingly, in only a few instances has the infectivity been assigned unequivocally to the large principal RNA; often, and without documentation, minor bands are considered degradation products. We now know however, in the case of TMV and TYMV that this is not true because some of the minor RNAs are generated during replication and probably have functional roles, even though these RNAs are not required for infection per se. Further, some viruses such as southern bean mosaic virus (SBMV) and the tombusviruses contain significantly less RNA (1.3 to 1.4 × $10^6$ daltons) than the majority of plant viruses in which the total genomic RNA is typically at least 2 × $10^6$ daltons. Thus, it is not inconceivable that the populaton of such virions may contain more than one species of similarly sized RNAs. In these groups, the structure of the virions is well understood; only one RNA molecule may be accommodated per virion. Thus, if more than one type of genomic RNA did exist, each would have to be encapsidated separately. An answer to this supposition awaits either methodology to separate such putative RNAs or sequencing studies on the RNAs which would give an unambiguous answer.

TABLE 1

**Properties of the RNAs from Representative Plant Viruses Considered to be Monopartite**

| Virus or group | MW ($\times 10^{-6}$) | % RNA in virion | Base composition G | A | C | U | Remarks | Ref. |
|---|---|---|---|---|---|---|---|---|
| A. Virions of tubular morphology | | | | | | | | |
| Tobamovirus | | | | | | | | |
| Tobacco mosaic (Type strain) | 2.0—2.1 | 5 | 25 | 30 | 18 | 26 | Infectivity associated with major band, but other virion-related RNAs can be encapsidated. | 3,4 |
| Potexvirus | | | | | | | | |
| Potato virus X | 2.1—2.2 | 5—6 | 22 | 32 | 24 | 22 | | 5,6 |
| Papaya mosaic | 2.2 | 5—7 | 21 | 34 | 23 | 22 | | 7 |
| Cymbidium mosaic | 2.5 | 5.5 | 21 | 29 | 24 | 26 | | 8 |
| Potyvirus | | | | | | | | |
| Potato virus Y | 3.1 | ∼5 | Varies marked with strain | | | | Infectivity associated with major RNA band. | 9, 10 |
| Tobacco etch | 3.1—3.3 | 5 | 23 | 30 | 20 | 27 | | 11, 12 |
| Turnip mosaic | 3.1—3.3 | 5 | 22 | 35 | 22 | 21 | | 11, 13 |
| Closterovirus | | | | | | | | |
| Sugar beet yellows | 4.3, 4.6 | 6 | 28 | 27 | 22 | 23 | Infectivity associated with major RNA band; minor components present. | 14,15 |
| Apple chlorotic leafspot | 2.3 | 5 | — | — | — | — | | 16 |
| Carlavirus | | | | | | | | |
| Carnation latent | 6 | — | — | — | — | — | Very little is known of the RNAs of this group of viruses. | 17 |

TABLE 1 (continued)

**Properties of the RNAs from Representative Plant Viruses Considered to be Monopartite**

| Virus or group | MW ($\times 10^{-6}$) | % RNA in virion | Base composition | | | | Remarks | Ref. |
|---|---|---|---|---|---|---|---|---|
| | | | G | A | C | U | | |
| B. Virions of isometric morphology | | | | | | | | |
| Tymovirus | | | | | | | | |
| Turnip yellow mosaic | 1.9—2.0 | 33—34 | 17 | 22 | 38 | 22 | Purified preparations contain several classes of particles; $2\times10^{6}$ MW RNA is infectious; virions also contain smaller mRNAs. | 18, 19, 20, 21, 22 |
| Eggplant mosaic | 2.0 | 36 | — | — | — | — | Nucleoprotein particles also contain various tRNAs. | 23,24 |
| Tombusvirus | | | | | | | | |
| Tomato bushy stunt | 1.4 | 17 | 29 | 26 | 21 | 26 | | 25 |
| Turnip crinkle | 1.4 | 17 | 28 | 26 | 24 | 22 | | 26 |
| Luteovirus | | | | | | | | |
| Barley yellow dwarf | 2.0 | — | — | — | — | — | Studies difficult because of very low virus yields and obligate requirement for aphids for transmission. Some "degraded" RNA seen on gels. | 27 |
| Tobacco necrosis virus group | | | | | | | | |
| Tobacco necrosis | 1.2—1.6 | 19 | 23 | 29 | 24 | 24 | | 22, 28 |

| | | | | | | | | |
|---|---|---|---|---|---|---|---|---|
| Ungrouped | | | | | | | | |
| Carnation mottle | 1.4 | | 27 | 30 | 19 | 24 | | 26 |
| Cymbidium ringspot | 1.7 | 15 | 28 | 25 | 21 | 26 | Two lower MW RNAs considered to be "degradation products". | 28a |
| Hibiscus chlorotic ringspot | 1.55 | 14.1 | 24 | 26 | 26 | 24 | Some virion heterogenecity, but only 1 species of RNA. | 29 |
| Saguaro cactus | 1.4 | 17.2 | 29 | 24 | 21 | 27 | | 26 |
| Southern bean mosaic | 1.3—1.4 | 21 | 27 | 22—24 | 22—24 | 27 | | 22, 26 |
| C. Virions of bacilliform morphology | | | | | | | | |
| Plant rhabdovirus | | | | | | | | |
| Lettuce necrotic yellows | 3.5—4.5 | — | — | — | — | — | | 30, 33 |
| Potato yellow dwarf | 4.6 | — | 21 | 29 | 21 | 29 | | 31 |
| Sonchus yellow net | 4.4 | — | — | — | — | — | Minor RNAs seen in sucrose gradient analyses. | 32 |

## II. SINGLE-STRANDED VIRAL RNAs

By convention, the RNA which contains the genetic information and from which that information is translated is termed the "plus" strand. Its complement is the "minus" or "negative" strand. The RNAs in the virions of most monopartite plant viruses are "plus," with the exception of the plant rhabdoviruses, in which the RNA is "minus".

### A. Virus Isolation

Most viral RNAs are isolated from the virions themselves although free viral RNA may often be found in tissues; low concentration and the problems associated with the separation of viral RNA from other RNAs in the plant make its isolation impractical. It should be pointed out however, that there is a known case where a specific fragment of viral RNA, the mRNA for TMV coat protein does not occur in nucleoprotein particles and must of necessity be isolated from plant tissue extracts.[34]

Thus, one can start with highly purified virions in anticipation of isolating viral RNA, free of host RNA. Obviously, thorough virus purification is vital to realize this expectation. In most cases, the isolated RNAs are essentially free of host contaminants; some circumspection is in order however, as there are at least two examples where host RNAs are encapsidated with viral protein, and such particles are isolated as a part of the virion population. Siegel[35] observed that a small proportion of the TMV particles (2% of the $U_2$ strain) are not viruses but "pseudovirions", i.e., they contain host RNA encapsidated in viral protein. It has also been shown recently that eggplant mosaic virus (EMV) preparations contain encapsidated RNA molecules with various transfer RNA activities; they are probably tRNAs of host origin.[24]

It is beyond the scope of this chapter to consider the purification of the virions used as starting material to prepare RNA. Fortunately, very thorough reviews of this subject have been published.[36,37] It must be emphasized, however, that the purification procedure may influence the size heterogeneity of the isolated RNAs. For example, long filamentous virions, such as those in the closterovirus group which can reach 1500 nm, are subject to breakage during purification thereby yielding broken RNAs. In order to get unbroken virions, cells must be ruptured gently and high *g* forces must be avoided.[38] Freezing and thawing can also break elongate virions.[39] On the other hand, freezing and thawing has been used to help isolate the RNA of TYMV.[40] Extremes of pH should also be avoided.[41]

### B. RNA Extraction

In general, extraction of viral RNAs does not require techniques different or more esoteric than those used for RNA isolation in general — i.e., a conventional two-phase phenol extraction,[42] usually with sodium dodecyl sulfate (SDS) as an aid to virus disruption and as a nuclease inhibitor. Sometimes other nuclease inhibitors, such as macaloid,[43] bentonite,[44] or diethylpyrocarbonate[45] are added; although with purified viruses, they generally are not necessary if care is taken to minimize conditions favoring nuclease activity by working quickly and by keeping the solutions cold after virus disruption. It is important to note however, that diethylpyrocarbonate can bind viral RNA to protein, preventing nucleic acid extraction from TMV.[46]

As indicated by Ralph and Bergquist,[41] no RNA extraction procedure is universal. The RNAs of some viruses are not extractable with phenol, whether it be the two-phase or a single-phase method.[47] In the case of wheat streak mosaic virus[48] and several potyviruses,[11] RNA extraction is accomplished with a high pH-SDS method.[48] In general, viruses are stabilized by weak interactions between protein and protein or protein and RNA. The nature and strengths of these interactions for different viruses have been deduced from the ease with which they are dissociated with SDS. Those very sensitive to SDS disruption are considered to be stabilized by electrostatic protein-RNA interactions; whereas, those resistant to such disruption are apparently stabilized by hydrophobic protein-protein interactions.[49]

*1. A Procedure for Phenol Extraction of Viral RNA*

Nuclease contamination must be avoided when working with nucleic acids. All glassware should be heated or acid washed, and gloves should be worn to avoid fingerborne RNases. All reagents (except phenol) should be boiled or autoclaved. As a safety measure, gloves and goggles should be worn during the phenol treatment.

The following is a generalized procedure which permits much flexibility in buffers, reagents, etc., while still yielding an acceptable product. For a general discussion of the procedures, refer to the reviews by Ralph and Bergquist[41] and by Bruening.[50] The essential ingredient is phenol which is used as a protein denaturing agent, with SDS as an aid in virus disruption. Some recipes also call for the addition of about 15% m-cresol to preclude the crystallization of phenol at low temperatures and to aid in deproteinization, and 0.1% 8-hydroxyquinoline to increase RNA yield, acting as a chelating and reducing agent.[41,42] To extract RNA, the virus — phenol — buffer mixture is agitated vigorously, the phases are separated by centrifugation, and the RNA is precipitated from the aqueous phase with ethanol or salt.

Reagents:

| | |
|---|---|
| Buffers: | Tris 0.1 *M* containing $Na_2$ EDTA, 0.01 *M*, pH 7.5. For viruses with RNAs containing poly (*A*) sequences, the pH should be about 9.0;[51] 3 *M* Na acetate, pH 4.0. |
| Phenol: | Either crystals or liquid phenol may be used; a suitable product should give a colorless solution. If colored, distillation is recommended. Saturate in a separatory funnel three times, each time with ¼ volume of buffer, or until the pH of the equilibrated aqueous layer is 6.5 or higher.[43] Discard the aqueous phases. |
| SDS: | A 20% solution should be clear and colorless. |

To virus in buffer contained in a glass centrifuge tube, add sufficient 20% SDS solution to make the final concentration 1%. Heat tube in a water bath at 37°C to disrupt the virions; the time required is judged by the clearing of the solution. The temperature for disruption can also vary, depending on the virus; for example, the cowpea strain of TMV requires 80°C.[52] Cool to room temperature. Add an equal volume of buffer-saturated phenol, stopper the tube tightly, and shake vigorously for approximately 5 min. Cool the solution on ice; at this stage, temperatures above 20°C may damage the RNAs.[39] Centrifuge at ∿ 10,000 *g* for 5 to 10 min to separate the phases, preferably in a swinging bucket rotor. With a pipet or syringe, remove the upper aqueous phase and place it in a clean centrifuge tube. (When removing the aqueous phase take care to avoid the interface.) Add ½ volume of buffer-saturated phenol to this aqueous phase and ½ volume of buffer to the original tube. Shake both tubes again, centrifuge, and remove both aqueous phases and combine them in a flask.

Discard the phenol phases. Add a few drops of 3 *M* pH 4.0 acetate buffer and two volumes of ice-cold 95% ethanol. Put the flask in a freezer for at least 4 hr, preferably overnight. Collect the RNA precipitate by centrifuging at 10,000 *g* for 10 min. Resuspend in < ¼ volume of water or appropriate buffer, depending on the ultimate use of the RNA. Residual phenol may be removed either by dialysis in the cold, ether extraction or by washing the alcohol precipitate with ethanolic sodium acetate.[50] The final preparation is clarified by centrifugation. Concentration may be estimated by absorbance $E^{0.1\%,\ 1\ cm}_{260\ nm} = 25$ in solutions without $Mg^{++}$ and 20 for those containing $Mg^{++}$. Yields vary considerably, ranging from 10 to 70% of the theoretical expectation with different viruses.[52a] RNAs may be stored as frozen solutions, but are more stable when stored at −20°C as precipitates under the ethanolic solution.[43]

The integrity and size distribution of the RNAs are best determined by gel electrophoresis.[53] For high molecular weight RNAs, low concentration gels stabilized by agarose are recommended.[54,55] The details of the electrophoretic procedures have been carefully summarized by Adesnik.[56]

### 2. Other Procedures for RNA Extraction

In addition to the single-phase phenol extraction[47] and the high pH-SDS method[48] mentioned above, RNA may also be extracted from virions with a simple sodium perchlorate procedure.[52a] In this procedure, the proteins of the virus are complexed with SDS and then are made insoluble by the addition of sodium perchlorate. The RNA remains in solution and may be precipitated with ethanol. Aside from its simplicity, this method gave a higher yield than the two-phase phenol system and was effective with potato virus X (PVX) and SBMV where the two-phase system was unsatisfactory. Here again though, this method was not effective with all viruses.

Viral RNAs have also been obtained by heating viruses with detergents, salts, or various protein denaturing agents such as guanidine hydrochloride or urea. These methods are detailed by Ralph and Bergquist.[41]

## III. DOUBLE-STRANDED VIRAL-RELATED RNAs

It is axiomatic that all nucleic acids replicate by engendering a complementary copy of themselves. These complementary strands can be isolated while hydrogen-bonded to the template RNA on which they are patterned. There is ample evidence, however, to suggest that base-pairing is not complete in vivo, and protein (replicase enzyme?) is associated with the RNAs; the structures as isolated are thus considered to result from the extraction procedure[57] and have a different form in vivo. When the isolated dsRNA is composed of a completely base-paired structure, it is termed the replicative form (RF); when it has a backbone of double-stranded RNA with single-stranded free ends, it is the replicative intermediate (RI). With the viruses under consideration here, the virion RNA (+ strand) is the template for the complementary RNA (− strand), forming the RF, and the free ends on the RI are considered to be plus (+) strands.

In the monopartite group of viruses, double-stranded RNAs (dsRNAs) have been investigated only in TMV, TYMV, and tobacco necrosis virus (TNV) infections, although it can be safely assumed that all viral infections engender dsRNAs. An RF for SBMV has been observed.[58]

### A. Tobacco Mosaic Virus

Double-stranded forms of TMV RNA were described in 1964 [59,60,61] shortly after the first discovery of viral-engendered dsRNA in encephalomyocarditis virus-infected mouse ascites tumor cells by Montagnier and Sanders.[62] In each of these TMV studies, only RF-like molecules were detected, as phenol-extracted nucleic acid was treated with DNase and RNase under conditions designed to preserve dsRNAs. Such treatment would preclude the detection of RI which contains RNase-sensitive, single-stranded RNA.

The most definitive investigations, using procedures without RNase[55,63,64] have utilized CF-11 cellulose chromatography[65] to resolve the RF and RI molecules. Several of these studies demonstrated that the RF is largely RNase resistant in high salt and contains complete strands of the size of TMV RNA. RF is not infectious until it is denatured[55] confirming the integrity of the plus (+) strand in the duplex. The RF has a buoyant density in $Cs_2SO_4$ of 1.615 g/cm$^3$.[55]

The heat denaturation reaction for dsRNAs of TMV has been studied in detail.[64] TMV dsRNA exhibited a very sharp thermal transition, with a $T_m$ of 97°C in SSC buffer (NaCl, 150 m*M*; sodium citrate 15m*M*, pH 7.0 at room temperature), a value consistent with that found with other dsRNAs and with dsTMV RNA in an earlier study.[59] The molecular weight as determined by polyacrylamide gel electrophoresis using dsRNAs as marker was $3.8 \pm 0.2 \times 10^6$, close to the expected theoretical value of $\sim 4 \times 10^6$.[66]

TMV RI has also been isolated by CF-11 cellulose chromatography, and its properties are consistent with the postulated structure of a double-stranded backbone with single-stranded tails. Its higher buoyant density (1.630 g/cm$^3$ in cesium sulfate) than the RF is also predictable from its partial single-stranded nature. Double-stranded TMV RNA has been found in the membrane rich fraction of the leaf homogenates,[55,67,68] and some of the RI has been shown to be associated with membrane-bound but not free polyribosomes in leaf homogenates.[69]

The molecular weight of TMV RI is not easy to assess by polyacrylamide gel electrophoresis because the behavior of single- and double-stranded RNAs differ on gels with each class of RNA requiring the appropriate single- or double-stranded RNAs as molecular weight markers. The molecular weight of TMV RI has been estimated as $5 \times 10^6$ using single-stranded RNA markers[55] and $6.7 \pm 0.6 \times 10^6$ using double-stranded RNA markers.[66] The true size is probably somewhere between these two extremes, although electrophoretic migration of dsRNAs of this size depart from linearity with respect to the log of the molecular weight,[70] making the determination tenuous.

The following evidence indicates that RF and RI or the corresponding in vivo forms are progenitors for the synthesis of TMV RNA:

1. The nascent chains of single-stranded RNA in RI contain more label than the double-stranded backbone in pulse label experiments.[63]
2. Pulse-chase experiments with separated cells from infected leaves indicate that radioactivity may be readily chased from RI, and to a lesser extent from RF into TMV RNA.[71]
3. In protoplasts infected in vitro, RF and RI incorporate $^{32}P$ early in the infection, and their synthesis and viral RNA synthesis stop at approximately the same time. Some of the radioactivity could be chased from these molecules in a pulse-chase experiment, although much more label accumulated in TMV RNA than was lost from RF together with RI.[66]
4. Free minus (−) strands could not be detected in homogenates of diseased leaves.[72] However, they are synthesized early in the infection as part of a dsRNA, and their level remained constant during the period of rapid TMV synthesis.[73] It is important to note, however, that free minus (−) strands might be difficult to detect because of their expected tendency to anneal to the large excess of free plus (+) strands in the extract.

### B. Turnip Yellow Mosaic Virus

Investigations on the dsRNAs associated with TYMV infection have not been as detailed as those of TMV. The principal studies involved isolation using RNase,[74,75] which would preclude seeing RI, although Pinck et al.[76] reported as part of another study that they had isolated an RI-like molecule by hydroxyapatite chromatography. More recently, columns of CF-11 cellulose have been used to isolate the dsRNA products of the TYMV replicase. These RNAs had properties of RF and RI.[77]

TYMV dsRNA has been found both in the nuclear and chloroplast fractions of leaf homogenates, with most in the latter. It is interesting that the nuclear dsRNA incorporates label early in the replication process and that this label is most probably in the minus (−) strand. The bulk of the dsRNA ($\sim$95%) is in the chloroplast fraction, however, and it becomes labeled during the time of maximal virus synthesis.[78] The buoyant density of RNase-resistant TYMV dsRNA is 1.617 g/cm$^3$ in cesium sulfate.[75]

### C. Tobacco Necrosis Virus

Both TNV and Satellite associated with tobacco necrosis virus (STNV) which is dependent on TNV for its replication, engender dsRNAs in mixedly-infected plants.[79] STNV RNA contains coding information for its coat protein[80] but requires the presence of replicating TNV for its replication.[81] Thus, STNV RNA contains unique genetic information not found in TNV RNA. Hence, it is not unexpected that dsRNA specific to STNV RNA would be present. By molecular hybridization, TNV RNA and STNV RNA have only a 10% sequence homology,[79] although in a more recent study, a value of less than 2% was observed.[82]

### D. A Procedure for the Isolation of dsRNAs from Virus-infected Plants Using Chromatography on CF-11 Cellulose

This method was develoed by Franklin[65] and was modified somewhat by Bishop and Koch.[83] It is based on the preferential binding of dsRNAs to cellulose in ethanol containing buffers. Mixtures of single-stranded RNA (ss) and dsRNAs, usually containing some DNA, are applied to a column in buffer mixed with 35% ethanol or, in some cases, 15% ethanol. The column is washed with buffers containing progressively less ethanol and finally with a buffer with no ethanol, which elutes the dsRNAs. The amount of ssRNA which contaminates the dsRNA in the 100% buffer fraction is influenced by the temperature at which the columns are run, as low temperatures encourage more structure in ssRNAs and thus they behave more like dsRNAs on the CF-11 column.[84] It has been estimated that an RNA must have at least 10 to 15% double-strandedness to chromatograph with the dsRNA; RNAs with less double-strandedness will elute in the 15% ethanol:buffer.[85]

Reagents are as follows:

- STE Buffer — 0.1 *M* NaCl, 1 m*M* EDTA, 50 m*M* Tris, pH 7.0 at 25°. This buffer is mixed with ethanol to give buffer: ethanol (v/v) mixtures of 65:35 and 85:15.
- Cellulose — Whatman CF-11 cellulose powder. (Whatman, Inc., Clifton, N.J.) Some lots do not work well for reasons not understood.[55]

#### *1. Preparation of RNA*

A total nucleic acid extract is prepared from diseased leaf tissue with a conventional phenol extraction similar to that described earlier for viral RNA. To disrupt the tissue, grinding in a mortar with liquid $N_2$ is preferred as blendor grinding may shear dsRNA.[58] See Jackson et al.[55] for a typical protocol. The final preparation should be suspended in STE buffer and clarified by centrifugation.

**Enrichment of RF and RI** — It is possible to fractionate the RNA by taking advantage of the differential solubility of RF and RI in either 1 *M* NaCl or 2 *M* LiCl. Sufficient single-strandedness in the RI will cause it to precipitate in those salts in the cold; whereas, the RF will remain soluble. Fractionation is not always complete, but a considerable degree of separation of RF from RI may be accomplished.

RF may be further enriched by adjusting the 2 *M* LiCl supernatant to 4*M*. After incubation in the cold for 8 to 18 hr, the precipated RF may be recoveed by centrifugation. DNA and low molecular weight single-stranded RNAs remain in the supernatant.[181]

**CF-11 and column preparation**[83] — CF-11 is suspended in water, fine particles are removed by successive decantations, and the slurry soaked for 12 hr in STE buffer, but supplemented with 10 m*M* EDTA and 1% 2-mercaptoethanol. It is then soaked for another 12 hr in STE buffer containing 35%-v/v ethanol. Columns are poured under gravity and washed with the STE buffer containing the appropriate amount of ethanol (15% in the example given below). Columns may be reused repeatedly and between uses are stored at room temperature in buffer with 35% ethanol. CF-11 has a capacity of 0.2 mg nucleic acid per ml of packed column volume.[83] If samples are applied to the column in STE buffer with 15% ethanol rather than 35%, much of the unwanted nucleic acid is not adsorbed, and the effective capacity for dsRNA is increased.

**CF-11 chromatography of dsRNAs** — Nucleic acid in STE buffer containing 15% ethanol (v/v) is applied to a column (approximately 2.5 × 20 cm) equilibrated with that solution. Buffer is applied to the top of the column through a constant head device or by hand applications to the top of the column, both with gravity flow.

The effluent is best monitored at 260 nm through a flow cell. When the absorbance has returned to its original value, the eluting buffer is changed to STE alone, and dsRNA, contaminated with some ssRNA, is collected. The dsRNA containing solution is made 15% with respect to ethanol, and the process is repeated, usually on a smaller column (approximately 1 × 10 cm). This process is repeated a second time yielding dsRNA with very little ssRNA contaminant. The final product is precipitated from the STE with 2 volumes of ethanol and is resuspended in water or desired buffer. It may be stored as a frozen solution.

### E. Other Methods Used in the Isolation or Detection of ds Viral RNAs

Although chromatography on CF-11 is apparently the method of choice for dsRNA isolation, alternative methods have utilized chromatography on hydroxyapatite,[76,86] (reviewed by Kothari and Shanker[87]), hybridization with labeled viral RNA[88], gel filtration on columns of Sephadex® G-200,[89] or selective LiCl fractionation.[181] A method using agarose gel chromatography at high ionic strength[90] has been used to prepare dsRNA from plants infected with the cowpea strain of TMV,[52] although the precise nature of the products was not analyzed.

## IV. REPLICATION OF VIRAL RNA: ROLE OF RNA-SYNTHESIZING ENZYMES

### A. General Considerations

As expected from experiences with RNA phages, and consistent with the finding of double-stranded viral RNA in plants, RNA synthesizing enzymes which generate RNA

| | Step I | | Step II | |
|---|---|---|---|---|
| Viral RNA (+) | → | Complementary viral RNA (−) | → | Viral RNA (+) |
| | Enzyme | | Enzyme | |

FIGURE 1. Enzymes possibly involved in RNA virus replication.

complementary to viral RNA must be involved in plant viral replication. By several criteria,[91] neither DNA nor DNA synthesis seems to be involved. The relevant enzymes thus are RNA-dependent RNA polymerases, termed either polymerases, synthetases, or replicases — the latter term is preferred by most workers in the field. Such enzymes have indeed been observed associated with plant virus infections, and although the discussion here emphasizes monopartite plant viruses, the same principles would pertain to multipartite RNA plant viruses as well. According to dogma, these enzymes use the viral (+) RNA strand as a template, first catalyzing the synthesis of a complementary (−) strand (Step I in Figure 1) which then serves as a template for the synthesis of nascent viral (+) strands (Step II).

Complementary viral (−) RNA, when isolated, is hydrogen-bonded to the viral (+) RNA to a greater or lesser extent, resulting in the RF and RI discussed in the previous section. It is not known whether the enzymes postulated at Steps I and II are the same molecule, whether one is a modified form of the other, or if they are entirely different molecules.

There is, furthermore, a reasonable doubt that some of the enzymes isolated from plant virus-infected tissues are replicases in the true sense of the term, in that they have not been shown to engender the viral (+) RNA or even a portion of it. When virus-infected tissues are homogenized, particulate fractions (presumably containing much membrane material) yield a bound enzyme, which under appropriate conditions, promotes the synthesis of (+) RNA, probably by completing RNA molecules initiated prior to tissue disruption. Bona fide replicative structures have been shown to result from such reactions, and thus one could justify calling these enzymes replicases, even though, in some cases, no free ss viral RNA molecules are synthesized. On the other hand, the enzymes found unbound in the cell homogenate (usually taken from the supernatant after freeing the homogenate of the bound enzyme, or sometimes by solubilizing the bound enzyme) have different properties from the bound enzyme, and there is a suspicion that they could be host enzymes (discussed below), or perhaps they represent only the enzyme involved in Step I. It is premature to call these enzymes viral replicases until their function(s) in viral RNA synthesis is established. I have been guilty of this exaggeration myself.[92]

Regardless of the virus, the forms of the RNA synthesizing enzymes seem to fall into two distinctive categories:

*1. Bound Enzyme*

Reaction rates here are rapid and generally plateau within an hour (Figure 2), often sooner.[77] The enzyme shows little or no response to exogenous RNA; in all probability, the template RNA is bound to the enzyme when isolated from virus-infected tissue. Thus, it catalyzes the completion of synthesis of viral (+) strands and sometimes (−) strands. The kinetics of this reaction suggest that reinitiation of RNA synthesis does not take place.

*2. Unbound Enzyme*

Unbound enzymes have more or less linear kinetics and can synthesize RNA for long periods (Figure 2). They require RNA templates, but synthesize only RNA that

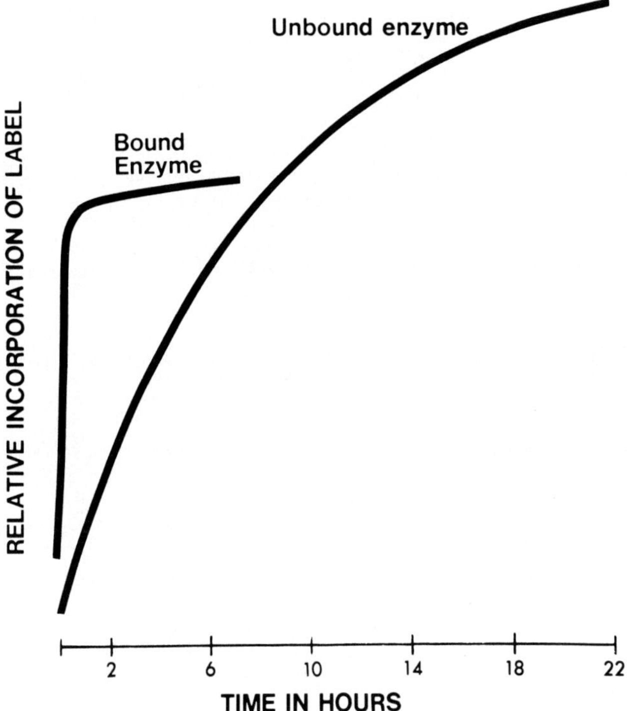


FIGURE 2. Stylized presentation of the kinetics of RNA synthesis shown by "bound" and "unbound" RNA-dependent RNA polymerases. The characteristics of these two forms of the enzyme(s) is given in the text. The relative levels of incorporation shown by the two enzymes could vary according to the conditions of the assay and are not intended to imply that the unbound enzyme necessarily can synthesize more RNA than the bound.

is complementary to the RNA used as template. As given below, two procedures yield unbound enzymes, although it is not certain that in both cases the same enzyme is obtained; evidence with TMV infections suggests that they are identical.[93]

**a. *Soluble Enzyme***

This enzyme is obtained from leaf homogenates in an unbound form; i.e., it is obtained from the supernatant fraction of the leaf homogenate clarified at relatively high *g* forces. It is not known, however, whether the enzyme is truly unbound in vivo or if it derives from the bound enzyme (or vice versa, for that matter!)

**b. *Solubilized Enzyme***

Starting with a preparation containing bound enzyme (usually a 20,000 to 30,000 *g* pellet), it is possible to solubilize an enzyme either by the use of various detergents (Lubrol W[77] Nonidet P-40,[92] Triton®X-100[94]) with magnesium deficient buffers[95] or by high speed centrifugation over a sucrose cushion, leaving the enzyme in the supernatant.[96]In all but the latter case, such treatments do not yield a true unbound enzyme with the characteristics given above, but require further purification, presumably to remove the solubilized membranes and the RNA from the enzyme preparations; the resultant preparations then have the characteristics of the unbound enzyme. Such procedures have employed liquid polymer phase separation,[77] DEAE Bio-Gel® column chromatography[95] and glycerol gradient sedimentation.[92]

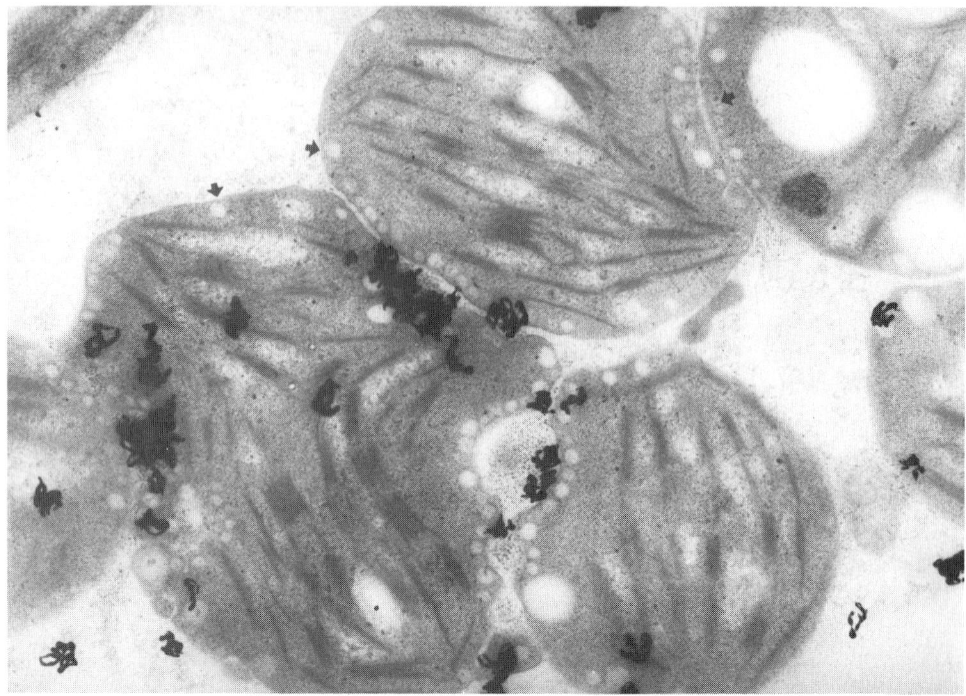

FIGURE 3. Electron micrograph and overlying autoradiograph of a "polyplast" from a Chinese cabbage cell infected with TYMV. As a consequence of TYMV infection, the individual chloroplasts associate with one another to form the polyplasts.[101,151] The tissues were allowed to incorporate $^3$H-uridine after a pretreatment with actinomycin D. At the periphery of the chloroplasts, small vesicles may be seen (arrows); the $^3$H accumulates there and at the chloroplast membranes, as judged by the accumulation of silver grains in the emulsion. These areas are considered to be the principal site of viral RNA synthesis, much of which is dsRNA.[152] (Photograph courtesy of J. M. Bové.)

## B. Turnip Yellow Mosaic Virus

The RNA-dependent RNA-polymerases associated with TYMV infections were the first of this class of enzymes successfully demonstrated with plant viruses.[97,98,99] The most comprehensive studies have been those of J. M. Bové and his colleagues[77,100] who have shown that the enzymatic synthesis of TYMV RNA takes place in virus-specific vesicles associated with the outer membrane of the chloroplasts; these chloroplasts frequently associate with one another as polyplasts[101] (see Figure 3).

As in the case of TMV and TNV, one can distinguish between a membrane-bound replicase which does not require a template because it is template associated, and a template-free, soluble, unbound enzyme extracted from those membranes, which requires RNA for activity. The bound enzyme activity is found in the fraction of the leaf homogenate sedimenting between 100 and 10,000*g*.

The unbound enzyme, prepared from the bound enzyme by treatment with Lubrol W, followed by a polyethylene glycol (PEG)-dextran two-phase system and DEAE cellulose chromatography of the PEG phase yields an unbound enzyme which shows a degree of preference for TYMV RNA and the eggplant mosaic virus (EMV) RNA as template.[100] Furthermore, the product of the reaction is full-length (−) RNA, hydrogen-bonded to TYMV (+) RNA. These workers also showed that the enzyme could bind to [$^{32}$P]-TYMV RNA; further demonstration of the specificity of the enzyme was shown in experiments where unlabeled TYMV and EMV-RNAs competed with [$^{32}$P]-

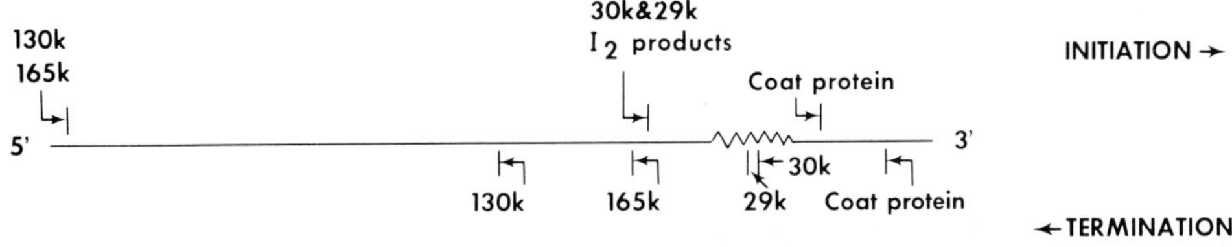


FIGURE 4. A genetic map of TMV RNA. The composite information used to construct this map is derived from in vitro translation studies with the common strain coupled with partial removal of protein from the virus and exposing the 5′-end of the RNA,[34,142] and translation, molecular hybridization, and aminoacylation studies with the cowpea strain.[136] It has recently been established that the 130,000 and the 165,000 mol wt proteins have a common initiation site,[182] although uncertainty applies to the common origin given for the $I_2$ products. The jagged area on the map represents the region containing the nucleation site for assembly of the common ($U_1$) strain of TMV from protein and RNA.[155] The map is drawn to approximate scale, assuming an average value of 110 for an amino acid to determine the number of amino acids/protein molecule. The position of each protein initiation site is not known with precision, with the exception of the site nearest the 5′-end of the RNA.[144,183]

TYMV RNA more effectively for binding to the enzyme than did TMV RNA or brome mosaic virus (BMV) RNA. The molecular weight of the enzyme was estimated to be 400,000 on sucrose gradients.

No enzyme activity was observed in extracts prepared in a similar way from uninfected leaf tissue, which seems to set this preparation apart from most of the other soluble RNA-dependent polymerases of monopartite and multipartite viruses (see Section C and D and Volume II, Section III by Lane).

## C. Tobacco Mosaic Virus

Cell-free synthesis of TMV RNA was attempted first in 1962 by Bandurski and Maheshwari[102] and in 1964 by Semal et al.[103] who sought, unsuccessfully, to demonstrate a preferential incorporation of nucleotides into RNA in subcellular extracts prepared from TMV-infected plants. The first demonstration of an RNA-dependent RNA polymerase activity related to TMV infection was by Ralph and Wojcik in 1969[67] who showed the synthesis of dsRNA in a membrane-rich fraction of homogenates of TMV-infected leaves. Using essentially their methodology, Bradley and Zaitlin[104] characterized the products of this bound enzyme on polyacrylamide gels and demonstrated products which co-migrated with authentic RF and RI, but observed no synthesis of single-stranded TMV RNA. Bound enzyme showed a peak in activity ∿ 3 days after inoculation with plants held at 25°C[92].

In an extension of this study, Zaitlin et al.[92] prepared an unbound form of the enzyme by extracting the homogenate fraction containing the bound enzyme with the detergent Nonidet P-40 + 0.5 *M* KC1, followed by sedimentation on glycerol gradients. This enzyme required an RNA template for activity, but exhibited no specificity towards any of the RNAs tested. Its product was a (−) strand, which when isolated was hydrogen-bonded to the RNA used as template. On glycerol gradients, the molecular weight of the enzyme was estimated to be approximately 160,000. This puts it in the size range of the major viral translation product, variously reported in different laboratories (using SDS-polyacrylamide gel electrophoresis) as having a molecular weight of 130,000 to 140,000,[105] and which has been suggested as the TMV replicase.[92] In recent studies, however,[93] this protein was not present in partially-purified, enzymatically-active preparations of the soluble enzyme, so that it cannot be involved in catalysis mediated by that enzyme. In infected protoplasts, it is synthesized relatively early in the replication process;[117,118] the gene for it is near the 5′ end of TMV RNA,[34] (Figure 4) and thus it is most probably the earliest viral gene to be translated. These observations suggest the 130,000 to 140,000 molecular weight protein should be involved in an early viral function such as RNA replication. From the above, however — if we can extrapolate from in vitro studies — it does not appear to catalyze the formation of (−) strand (Step I enzyme, Figure 1), but it is not ruled out as a constituent of the Step II enzyme where viral RNA (+) strands are synthesized using the (−) strands generated by the Step I enzyme as template.

Unbound enzymes have also been isolated from leaf homogenates without the use of detergent solubilization. Brishammar and Juntii,[106] using gel filtration of leaf homogenates followed by affinity chromatography, found that the enzyme required an RNA template and showed some preference for TMV RNA among the nine RNAs tested. There are two other studies with "replicase" from TMV;[107,108] in the latter, enzyme was extracted from virus-infected callus tissue.

### D. Tobacco Necrosis Virus

Enzyme studies with this virus are complicated by the extensive necrosis induced in the tissue by infection and the occasional presence of a small satellite virus ($2 \times 10^6$ mol wt) which depends on the TNV for its replication.[81] RNA replication has been investigated by Stussi-Garaud et al.[109] Bound enzyme activity peaked at 4 days post infection (in plants grown at 19°C) and then fell abruptly. The findings with this virus are similar in many ways to those observed with TMV, in that both bound and soluble enzymes are found. Low levels of enzyme activity were observed in uninfected plants. On glycerol gradients, the solubilized enzyme was estimated to have a molecular weight of ∿ 200,000.[94]

### E. Plant Rhabdoviruses

Plant rhabdoviruses such as lettuce necrotic yellows and potato yellow dwarf are negative-strand viruses. The virions contain a constitutive RNA-dependent RNA polymerase, a **transcriptase** which can catalyze the synthesis of a transcript of the virion RNA. This transcript, from analogy with animal rhabdoviruses,[110] and by convention, is the (+) or mRNA; the RNA in the virion is the (−) strand.

To date, transcriptase activity has been reported from only one plant rhabdovirus, lettuce necrotic yellows virus (LNYV).[111] The products of the reaction of the enzyme are several small ssRNAs which are, for the most part, complementary to the LNYV RNA.[30] The requirement for this virion-associated enzyme explains why isolated viral LNYV RNA is not infectious; it is the (−) strand, and must be transcribed before viral-directed proteins can be synthesized. Detergent-treated virions lose some protein but retain an infectious nucleocapsid with a specific infectivity of only 1/20 that of the intact virion.[112]

The technology used to detect transcriptase activity[111] is very similar to that described below for the soluble plant RNA-dependent RNA polymerases, but with two exceptions: (1) the virions must be treated with a nonionic detergent (Nonidet P-40) which can be included in the enzyme reaction mixture. The detergent serves to remove the viral envelope and to expose the enzyme; and (2) no template is required, as the viral RNA is retained in the disrupted virions and serves that function. The single-stranded products of the enzyme of (+) polarity are released from the nucleocapsid during the reaction. As would be expected, the enzyme requires all four nucleoside triphosphates and $Mg^{++}$ and is insensitive to DNase and actinomycin D. The time course of the reaction resembles that of the soluble RNA-dependent RNA polymerases, suggesting some reinitiation may be occurring.

### F. RNA-dependent RNA Polymerases from Uninfected Plants

Low levels of RNA-dependent RNA polymerizing activity in uninfected plants was noted by Gilliland and Symons in 1968[113] and Ralph and Wojcik in 1969.[67] The phenomenon was essentially rediscovered by Astier-Manifacier and Cornuet[114] and by Duda et al.[115] and studied in more detail. More recently, enzymes of this sort have been found in uninfected plant controls used in RNA-dependent RNA polymerase studies with several plant virus infections[93,96,109,116] and a possible role for them in viral replication has been suggested.[93,94] In addition to its detection in Chinese cabbage and tobacco, an enzyme with similar properties has been demonstrated in pea, beet, and lettuce leaves.[93]

The host enzymes of this type are characterized as template-dependent (except Duda et al.,[115] where endogenous template is apparently associated with the enzyme) RNA polymerases that synthesize a strand of RNA complementary to the template; the isolated product is hydrogen-bonded to the template forming an RNase-resistant ds molecule. The enzyme is usually unbound in the leaf homogenate, but considerable activity can be released from the particulate fractions of leaf homogenates by treatment with Lubrol (Bol et al.[116]). These workers maintain that a virus-specific (alfalfa mosaic virus (AMV) in this case) enzyme may be separated from this extract when it has been prepared from infected tissues. On the other hand, other workers can find no differences in the soluble or solubilized enzymes isolated from infected or uninfected leaf tissue. The levels of these enzyme activities increase markedly in diseased tissue, and it is reasonable to ask if the soluble or solubilized enzyme isolated from diseased and healthy leaves are the same. With TMV, there is reasonable suspicion that they are: the soluble and bound enzyme activities rise and fall in parallel as a function of time after inoculation.[115] Further, when the unbound enzymes were purified ~ 50-fold from both TMV-infected and uninfected tissues, they behaved identically upon $(NH_4)_2SO_4$ fractionation, gel filtration, and phosphocellulose and DEAE-Bio Gel chromatography. They also shared common kinetics and cofactor requirements; and with a given RNA as template, the products were the same.[93]

It is not at all certain that these host enzymes play a role in plant viral RNA replication, but judging from their in vitro properties, it is not hard to visualize that they could be involved in the formation of the complementary strand to the infecting viral RNA (Step I enzyme in Figure 1). It is interesting that there is an early actinomycin D-sensitive step in TMV[119] and potato virus X (PVX)[120] replication. Could this involve the enhanced production of this enzyme?

There is no known role for an RNA-dependent RNA polymerase in healthy plants, and only a hint that the alleged products of such an enzyme (i.e., dsRNA) exist in plants.[121] Remarkably similar enzymes exist in animal tissues,[122] but suggestions that they might amplify mRNAs have been discounted.[123]

### G. A Procedure for Detection of RNA-dependent RNA Polymerase from Plant Tissues[92]

**Reagents and stock solutions (all stored frozen).**

- **Grinding buffer (GB) — Sucrose 0.4** *M*, $KCl$, 10 m*M*, $MgCl_2$ 5 m*M*, Tris buffer, 50 m*M*, adjusted for pH 8.1 at 4°C (pH ~ 7.5 at 25°C), Glycerol 20% (v/v). Immediately before use, 2-mercaptoethanol is added to 10 m*M*.
- Resuspending buffer (RB) — $KCl$, 10 m*M*, $NH_4Cl$, 25 m*M*, Tris buffer, 50 m*M* adjusted for pH 8.1 at 4°C (pH ~ 7.5 at 25°C), Glycerol 20% (v/v). Immediately before use, 2-mercaptoethanol is added to 10 m*M*.
- Nucleoside triphosphates (NTPs) — 20 μmol/mℓ each of ATP, GTP, and CTP (sodium salts) in water.
- Actinomycin D, 400 μg/mℓ

- $MgCl_2$, 0.1 *M*
- $(NH_4)_2SO_4$, 1.0 *M*
- Template RNA, ∿ 1.20 mg/mℓ
- DL-Dithiothreitol (DTT, Sigma) 0.1 *M*
- [$^3$H] uridine 5′ triphosphate (10 to 25 Ci per milimole)
- Tris buffer, 0.1 *M* pH 8 at 33°C

*1. Preparation of Enzyme Extracts (after [92])*

**a. Bound Enzyme**

In the following description, all reagents are cold and all procedures done in ice. Glassware is flamed or acid washed to eliminate contaminating nucleases. To help minimize contamination from nuclei which would contribute DNA-dependent RNA polymerase, tissue is usually disrupted in a relatively gentle manner in solutions designed to preserve nuclei and allow their elimination by centrifugation. (Of course, if the virus under consideration replicates in the nucleus, the nuclei are saved, and the other fractions are discarded.) Tissue disruption is achieved by chopping with a razor blade or by grinding in a mortar rather than by blending. These methods are less efficient than blending, but cause less nuclear disruption.

In the following example, tissue is disrupted by chopping leaves with a modified 2-bladed electric carving knife in which the distal half of each knife blade is removed and replaced by a holder for a single-edge razor blade. Washed leaf tissue is minced with this device in a plastic ice box dish placed on ice, using 1.25 mℓ of GB per gram of tissue. The brei is filtered through Miracloth (Calbiochem, LaJolla, Ca.) that had been rinsed with GB. Particulate matter is removed by a 10-min centrifugation at 1000 *g* (4°C); the supernatant is then centrifuged at 31,000 *g* for 20 min. The pellet, here the source of bound enzyme, is resuspended with a Dounce homogenizer using one half the original volume of RB (the supernatant from this centrifugation can serve as a source of the soluble enzyme). The resuspended pellet is centrifuged again at 31,000 *g* for 20 min and resuspended in 0.1 mℓ RB/g of original tissue. This is the bound RNA-dependent RNA polymerase.

***b. Unbound Enzyme***

The soluble enzyme found in the 31,000 *g* supernatant (above) may be concentrated either by acid precipitation (pH 4.3[115]) or by ammonium sulfate precipitation at 50% saturation.[93] Enzyme activity with similar properties may also be released from bound enzyme preparations by detergents [92,94] or by washing the bound preparation with a buffer devoid of $Mg^{++}$.[95]

*2. Preparation of Reaction Mixtures*

[$^3$H] UTP is pipetted into a plastic scintillation vial and the solvent (50% ethanol which interferes with the reaction) is removed *in vacuo.* The amount of tritium added is a function of the number of reactions to be performed, normally adding 2.5 μCi per reaction. After drying, to the vial add (per reaction):

| Reagent | Volume per reaction | Final concentration or amount per 0.1 mℓ |
|---|---|---|
| $MgCl_2$ | 10 μℓ | 5m*M* |
| NTPs | 2.5 μℓ | .05 μmol each |
| Actinomycin D | 2.5 μℓ | 40 μg |

| | | |
|---|---|---|
| DTT | 7.5 μℓ | 7.5 m$M$ |
| $(NH_4)_2SO_4$ | 2 μℓ | 20 m$M$ |
| Tris buffer | 10 μℓ | 100 m$M$ |
| RNA | 5 μℓ | 6 μg |

$H_2O$ to bring volume to 50 μℓ per reaction. Hold on ice until ready to proceed

*Note:* These concentrations of reagents have been optimized for TMV RNA-dependent enzyme. They should be optimized in other situations, particularly $Mg^{++}$ and pH. RNA is not required for bound enzyme reactions.

### 3. *The Reaction*

With a micropipet put 50 μℓ of the reaction mixture into small silicone treated (dichlorodimethyl silane) test tubes or shell vials. These tubes should be wide enough to allow the tip of the pipet to reach bottom.

Place the rack of tubes in a water bath at 33°C. At 30-sec intervals, place 50 μℓ of enzyme solution into successive tubes and agitate gently to mix. After the appropriate incubation time, (usually 20 to 30 min for bound enzyme or 60 min for a soluble enzyme assay) remove 50 μℓ from each tube in sequence (at 30-sec intervals) and pipet onto 2.3 cm 3MM filter paper discs (Whatman) impaled on a straight pin or on precut discs in a sheet.[124] When there is no free liquid on the surface of the disc, put it into a solution of ice-cold 5% trichloracetic acid (TCA) containing 1% sodium pyrophosphate and 0.02% uracil. Wash the discs at least six times as described by Byfield and Sherbaum.[125] Radioactivity on the discs after solubilization with a reagent such as NCS (Amersham, Arlington Heights, Ill.) is determined as described.[71]

Zero time controls should also be included. These are especially important for bound enzyme assays where they can represent a significant proportion of the total incorporated radioactivity. Zero time controls, when subtracted from the experimental values, also automatically correct for background and for residual counts left in the disc washing procedure. Zero time samples are obtained by using cold reaction mixtures and enzyme solutions: 50 μℓ of enzyme is pipeted into the reaction mixture, the contents are mixed quickly, and 50 μℓ are immediately withdrawn and placed on the 3MM filter paper disc, which is then placed in the TCA solution and incorporated radioactivity determined as above. With practice, this operation takes less than 20 sec.

## V. SOME CHARACTERISTICS OF PLANT VIRAL RNAs

Plant viral RNAs have many of the characteristics of other mRNAs, in particular some are capped at the 5′-end with a methylated base joined by its 5′-hydroxyl group through a triphosphate bridge and an inverted 5′-5′ linkage to the next sugar. It has been suggested (Shatkin[126]) that the cap "protects the mRNA at its terminus against attack by phosphatases and other nucleases and promotes mRNA function at the level of initiation of translation".

In the monopartite plant virus group, only TMV RNA and TYMV RNA have thus far been shown to be capped;[8,127,128] whereas, both TNV and STNV, both of which may serve as effective in vitro mRNAs[18,129] are not.[80,130,129] Some of the smaller subgenomic RNAs which serve as mRNAs in viruses of this group (see Section VI) are also capped. With TYMV this has been shown chemically;[18] in TMV, translation from these smaller RNAs is inhibited by $m^7GP$,[4] which has been shown to indicate capping.[130] A further discussion of capping and its importance in translation of viral RNA is considered in Volume II, Section III by Davies.

Many plant virus RNAs also have the property of accepting amino acids at their 3′-ends, as is discussed in Volume II, Section III by Lane and in Section VI. 2 of this chapter.

## VI. ORGANIZATION OF GENETIC INFORMATION ON THE VIRAL RNAs

As indicated in the Introduction of this chapter, the RNAs of the monopartite plant viruses, by definition, contain all of their genetic information in a single strand of RNA. With TMV[34,52,131,132] and TYMV,[18,19,133] however, the in vitro translation of some of the genes on these RNAs—the coat protein gene in particular—is not directed by the complete virion RNA, even though those genes are present there. Smaller RNAs which are subsets of the virion RNA serve as mRNAs in vitro, and their association with polyribosomes suggests that they also serve as mRNAs in vivo. [69,134] Full-length PVX RNA, when translated in an in vitro wheat germ system failed to elicit coat protein[135] indicating it also probably has a subset mRNA for coat protein. An apparent exception to this phenomenon may be TNV, where coat protein synthesis in the in vitro wheat germ system is translated from the intact viral RNA.[129] No other viruses in the monopartite virus group have been investigated in detail. It seems characteristic of plant viral RNAs, whether they be from monopartite or multipartite viruses (with TNV as the only exception noted so far), that the coat protein is translated from small, monocistronic mRNAs, rather than polycistronic mRNAs, typical of eukaryotic mRNAs. (see Volume II, Section III by Davies).

Particularly with TMV, the subgenomic mRNAs have been useful in determining the position of the genes on the virion RNA; this information, coupled with sequencing and other data, has allowed a partial characterization of the organization of the virion RNAs of TMV and TYMV, as given below.

### A. Tobacco Mosaic Virus

#### *1. Genes and Their Organization on the RNA*

Judging from the products of in vitro translation in various systems, TMV RNA codes for a minimum of three proteins and presumably contains at least three genes whose sequences occupy > 75% of the total genome; some regions apparently are not translated (see Figure 4). These genes, starting from the 5′-end of the RNA, code for proteins of molecular weights of ∿ 130,000,[34,52] ∿ 30,000 ($I_2$ product,[52,136]) and the 17,500 mol wt coat protein, which is the only one of the three with a known function. In different laboratories, the 130,000 protein has been assigned molecular weights of 130,000 to 140,000.[105] On occasion, a larger (165,000 mol wt) protein is observed both by in vitro translation, and in infected protoplasts;[117,118] on polyacrylamide gels the $I_2$ band is frequently split. It has been suggested that there may be overlapping genes in the region of both the 130,000 mol wt protein[34,136,182] and the $I_2$ product.[136] Thus, TMV could contain five genes instead of three (Figure 4). Such overlapping sequences are known in other systems. For example, in Qβ bacteriophage, the A′ protein is a "read through" product of the coat protein gene; the A′ and coat proteins share common N-terminal sequences.[137] With TMV, the 165,000 and the 130,000 proteins contain many tryptic peptides in common,[139,182] as do the two bands of the $I_2$ product[139] indicating overlaps. It has recently been determined that the two large proteins share a common initiation site.[182] Thus, it is conceivable that more genetic information is available to the virus than would be predicted from the size of RNA alone. However, a confirmation of this postulate must await an assignment of function to the supernumerary proteins which have been detected.

*2. Sequence Determination of Various Regions of the RNA*

Mandeles[140] was the first to sequence significant sized pieces of TMV RNA, obtained after partial digestion of viral RNA with RNase $T_1$. One of these, an oligonucleotide termed "Ω," 70 nucleotides long, was thought to be near the 3′-end of the RNA. This conclusion was based on the then-believed assumption that sodium dodecyl sulfate (SDS) removed protein from the end of the virus[141] which exposed the 3′-end of the RNA. Oligonucleotide Ω was found in the exposed part of the RNA. It is now known that in the common strain SDS and alkali remove protein preferentially from the 5′-end of the RNA.[142,143] Thus, Ω is near the 5′-end of the RNA, and recently it has been shown to be the leader sequence of the RNA.[144] Oligonucleotide Ω starts immediately after the cap, the sequence being $m^7$gpppGΩ . . . As Ω contains no guanosine residues, the first initiation sequence (presumably for the 165,000 and possibly the 130,000 dalton protein) (Figure 4) could start at position 69 with an AUG.[183] (The 5′ terminal sequence of TMV RNA is now known to position 236.[183]) This leader sequence is considerably longer than that of RNA 4 of brome mosaic virus (BMV), the monocistronic mRNA for coat protein where the initiating AUG codon is only 10 nucleotides from the 5′-terminus.[145]

The sequences of the 71 and 74 nucleotides at the 3′-OH termini of the common and GTAMV strains of TMV RNA respectively have been determined.[146,147] The sequences terminate in CCA-OH, which is consistent with their capacity to accept amino acids at the 3′-terminus[148] in the manner characteristic of tRNAs — which also terminate in that sequence. The structures internal to the 3′-OH end do not contain any unusual bases (as do tRNAs) nor can the secondary structures be rationalized to a configuration closely resembling that of tRNAs. There are several triplets in the common strain sequence which could serve as termination signals to stop the translation of a protein, most probably the coat protein, which is coded for near the 3′-end of the RNA (see Figure 4). However, the sequences preceeding these termination signals cannot be reconciled with the amino acid sequence of the carboxyl end of the coat protein, demonstrating that the coat protein gene terminates more than 71 nucleotides from the 3′-OH end of the RNA.

A number of fragments of TMV-RNA, isolated after partial digestion with $T_1$ RNase, have been shown to contain sequences coding for the coat protein of TMV; several taken together constitute a continuous stretch of 232 nucleotides of known sequence which code for amino acids in positions 53 to 130 of the coat protein.[149,150] These studies have revealed considerable degeneracy in the genetic code used to specify TMV coat protein (see Table 3 in Volume II, Section III by Davies).

The overall base sequences of many TMV strains are apparently very different because their RNAs do not compete in hybridization-competition experiments,[175,176] calling into question their relationships as strains.[176]

There has been considerable interest over the past 2 decades in the mechanism of in vitro reconstitution of TMV from its constituent parts.[172] Until recently, it was believed that the first interaction between the RNA and coat protein occurred at the 5′-end of the RNA, and polymerization proceeded from that point.[153,154] Based on new evidence (below) this is not correct; the region for nucleation of assembly, in which coat protein is recognized by RNA, is on the 5′-side of the coat gene[155] (see Figure 4). Encapsidation begins at this point and then proceeds in both directions — first towards the 5′- and then to the 3′-terminus of the RNA.[156,157,158] This mode of reconstitution should result in an intermediate-stage rodlet with two protruding RNA strands (one of which is

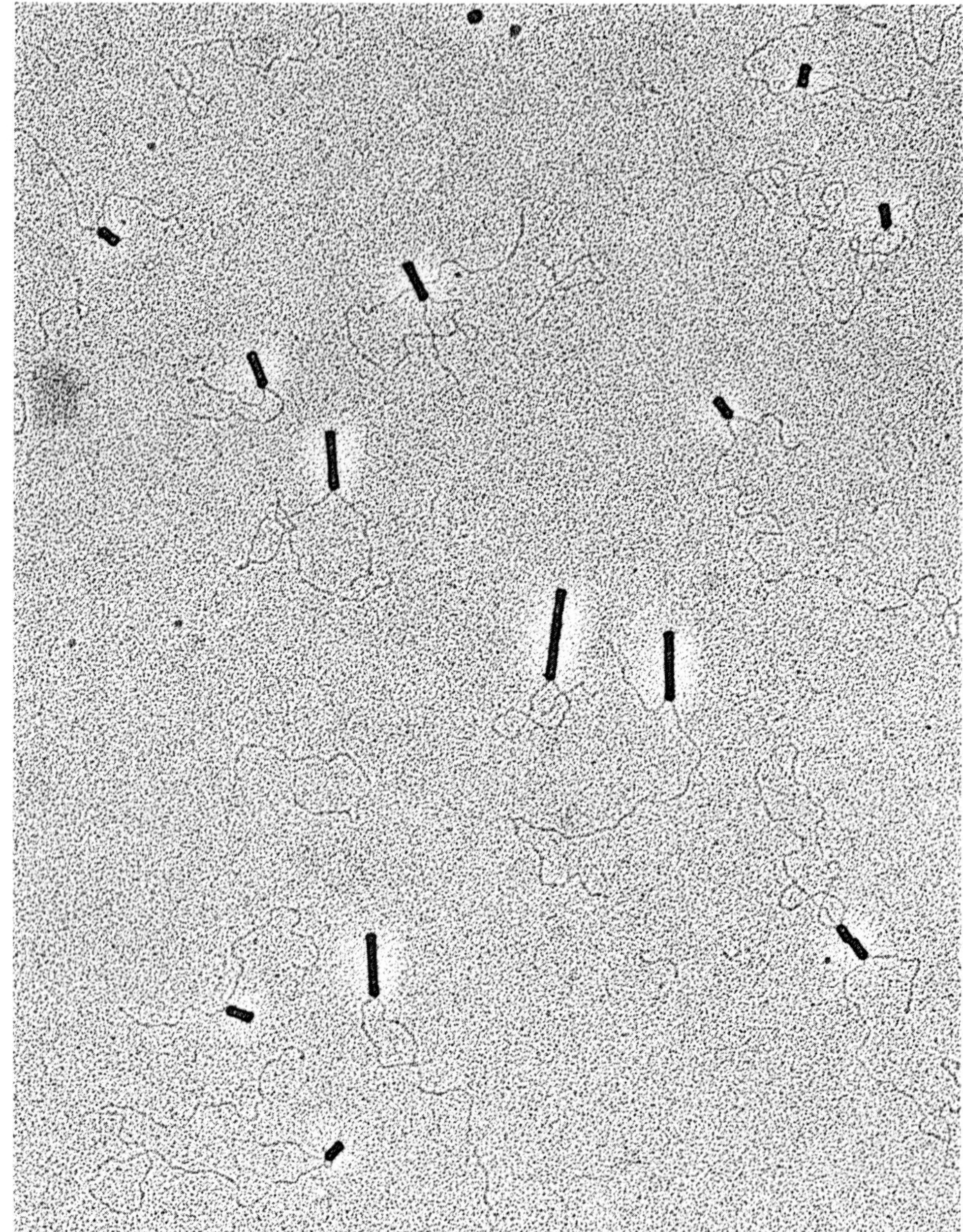

FIGURE 5. Electron micrograph (×72,000) of partially reconstituted rods of TMV, showing the two tails of RNA protruding from the particles, indicating that the reconstitution has initiated internally on the RNA rather than at one end. The longer tail of RNA represents the 5′-end, which is thought to thread through the hole in the rod.[157,159] (From Lebeurier, G., Nicolaieff, A., and Richards, K. E., *Proc. Natl. Acad. Sci. USA.*, 74, 149, 1977. With permission.)

probably looped down the central hole),[157,159] and indeed, such particles have been visualized in the electron microscope (see Figure 5).

The RNA fragments containing the nucleation sequence for assembly were isolated by reconstitution of either intact TMV RNA[160] or a partial nuclease digest of the

RNA[161] with limiting amounts of coat protein discs. The sequences which were covered with coat protein were nuclease resistant and thus could be recovered. A series of fragments were obtained of up to 550 nucleotides in length, but the shortest ones reflected a minimal core of approximately 100 bases which contained the putative nucleation region. The position of this fragment on the TMV RNA molecule was ascertained by producing TMV rods of varying length by alkali stripping followed by nuclease digestion. The RNAs in these rods represented various sized portions of the RNA, but all had the common 3′-end. Determination of how long the RNA had to be to include the fragment with the nucleation sequence localized its position to between 900 and 1350 bases from the 3′-end of the RNA.[155] By electron microscopic serology it appeared to start about 39 nm (830 bases) from the end.[158] This places it within, or on the 3′-side of the gene for the $I_2$ product (Figure 4).

The putative nucleation site is rich in purines, particularly guanine, with many repeated triplets and a purine in every third position. The sequence apparently contains several base-paired regions.[156,161]

*3. Subgenomic Messenger RNAs*

As has been alluded to earlier, and is discussed in detail in Volume II, Section III by Davies, a number of plant viral RNAs engender monocistronic mRNAs from the polycistronic viral RNAs apparently as a mechanism of controlling the synthesis of viral-encoded proteins. With monopartite viruses, this phenomenon was observed first with the common (U1) strain of TMV: A small piece of RNA, representing approximately 1/6 of the RNA including the 3′-end was found in phenol extracts of tissues[72] and on polyribosomes.[69] It has also been observed on polyribosomes in tissues infected with the $C_c$ and U2 strains.[134] This RNA of molecular weight reported variously between 280,000 and 350,000 and termed the LMC, can serve as the coat protein mRNA in vitro.[34,134,162] Based on a consideration of the location of the nucleation site for assembly of the common strain given in Section VI. A. 2, LMC lacks that site, explaining why it is not encapsidated (see Figure 4). In another strain, ($C_c$TMV) the cowpea strain of TMV, the requirements for assembly must be different, as the LMC is found as an encapsidated short rod in the virion population (Figure 6)[52,132,163] Thus, the recognition signals on $C_c$TMV RNA must be localized closer to the 3′-end of the RNA than they are in the common strain. Interestingly, in mixed infections the LOC-size RNA of the $C_c$ strain will recognize the coat protein of the U2 strain, as short rods are formed containing the LMCs of both strains.[164]

Virion populations of three strains of TMV ($U_1$, K, and $C_c$) also contain rods of an intermediate size, one class of which contains the mRNA for the 29,000 to 30,000-mol wt $I_2$ protein.[4,52,136] The proportion of these rods in any one of the three strains is too small to stand out as a unique class upon sucrose gradient centrifugation (Figure 7, top), but when the RNAs are extracted from the nucleoprotein rods taken from the slower sedimenting portion of the gradient, they resolve into several distinct components (Figure 7, bottom).

Thus, subgenomic mRNAs are constituents of several — and probably all — strains of TMV. Some of the mRNAs are encapsidated and represent some of the less-than-full-length rods seen in the virion population, while at least one other, the LMC, is usually not encapsidated and may only be found as a free RNA in vivo.[4]

**B. Turnip Yellow Mosaic Virus**

TYMV RNA is extremely rich in cytosine residues (Table 1) and probably contains significant-sized sequences of poly(C). This characteristic has made it possible to synthesize short DNA sequences complementary to TYMV RNA using reverse transcrip-

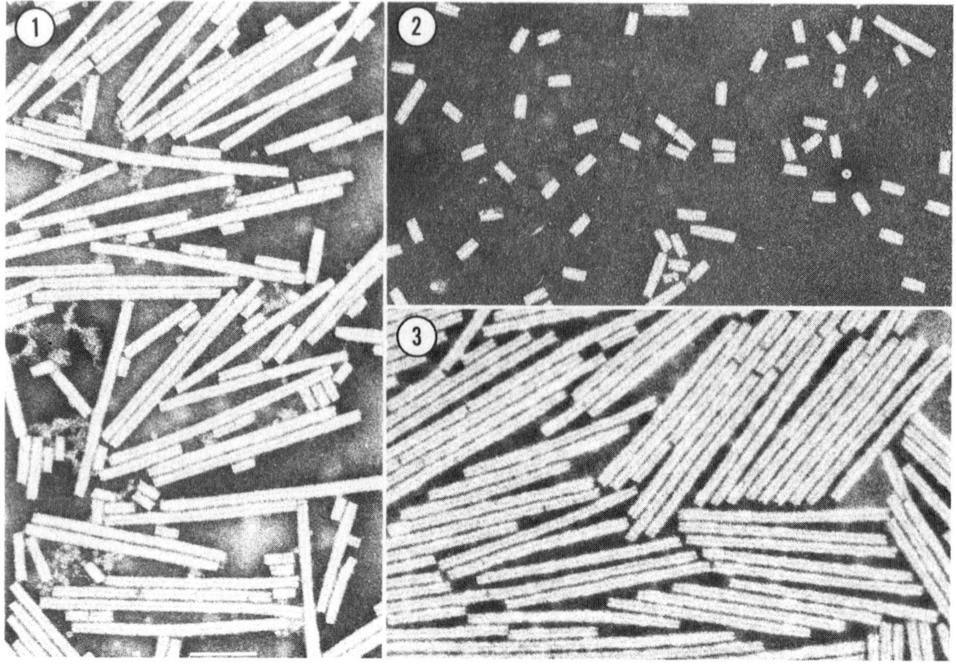


FIGURE 6. Rod sizes seen in preparations of the common and the cowpea strains of tobacco mosaic virus. (1) Leaf dip preparation of a leaf infected with the cowpea strain of TMV; preparation contains a significant proportion of short rods. (2) Short rods isolated from the cowpea strain by sucrose density gradient centrifugation. Judging from the uniform and small size of the RNAs extracted from this preparation, the longer rods seen here are dimer and trimer aggregates of the short rods. (3) Partially purified preparation of the common strain of TMV. Note the virtual absence of the very short rods seen in panels 1 and 2. Some of the intermediate size rods are considered to represent size classes encapsidating subset mRNAs, rather than breakage of virus during preparation.[4] Bar = 100 nm. (Micrographs courtesy of H.W. Israel.)

tase and oligo $(dG)_6$ as a primer.[165] The sequence of the 3′-OH extremity of TYMV RNA has been determined in two laboratories;[168,169] a similar analysis has been done for EMV.[167] There were no minor bases detected in the sequence of 159 nucleotides analyzed in TYMV RNA,[168] although with another technology minor bases may have been found in total RNA.[178]

It is well known that the 3′-OH end of the virus can be aminoacylated by valine in a manner similar to a true tRNA (see Volume II, Section III by Davies). Indeed, it has been reported that TYMV valyl RNA can serve as an amino acid doner during in vitro protein synthesis.[177] When the 3′-OH sequences are drawn, they can be folded into a cloverleaf structure superficially resembling a true tRNA, but without the minor bases, and with significantly more bases than tRNAs. The terminal sequences of TYMV and the related EMV are similar, although they show enough substitutions to suggest that one virus has not evolved from the other.[168] In the virion, the RNAs lack the terminal A residue, which is apparently added before aminoacylation can occur.[173]

The 3′-OH RNA sequences and the known sequence of the coat protein of TYMV have allowed an unequivocal localization of the coat protein sequences on the RNA. It is near the 3′-terminus, terminating 109 nucleotides from the end. There are five nonsense codons (UAA and UAG) within the terminal 108 nucleotides; they are placed at irregular intervals but they are in phase with the coding for the coat protein. The terminal 108 nucleotides are not translated.[168,169]

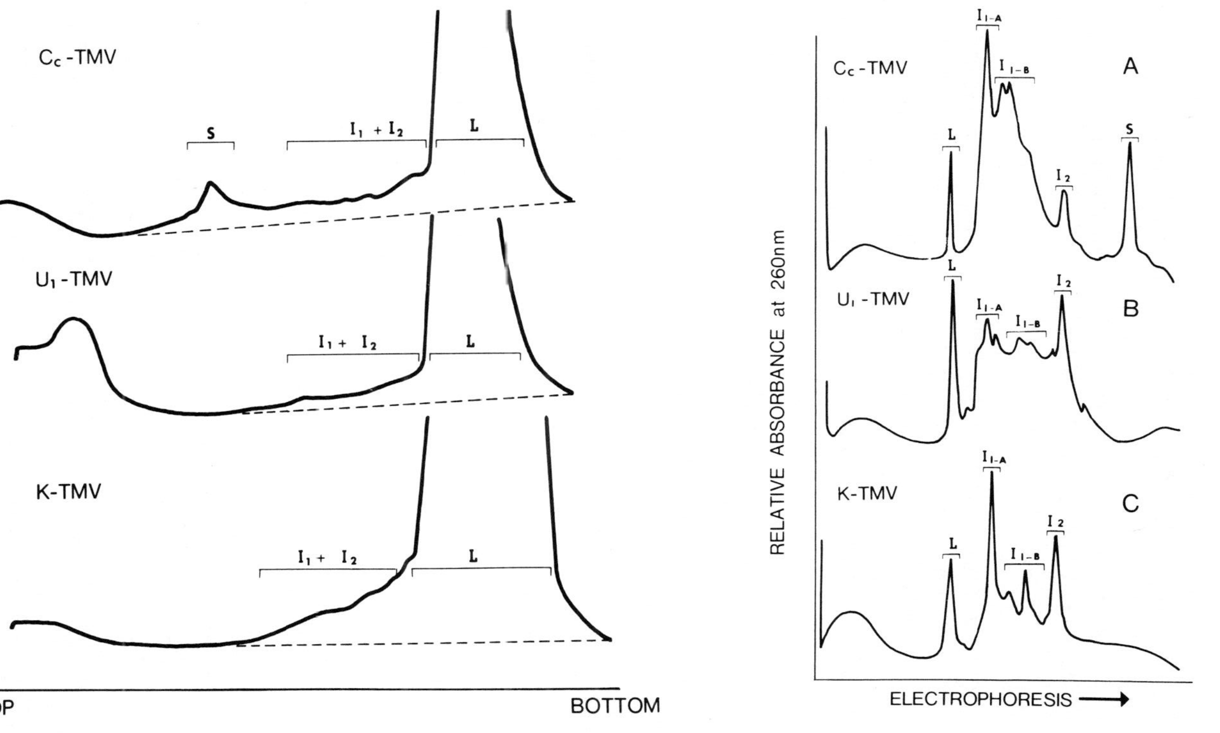


FIGURE 7. Analysis of the RNAs from less-than-full-length rods present in preparations of the common ($U_1$), $C_c$, and K strains of TMV. Top: Sucrose density gradient analysis of viral preparations. The majority of the rods in the preparations are the ∿ 300 nm L rods. Only $C_c$ TMV has short (S) rods. Bottom: Polyacrylamide gel electrophoretic pattern of the RNAs isolated from that portion of the gradients which sedimented more slowly than the L rods in the upper figure. Several classes of RNAs may be seen that are common to all strains, with the exception of the S RNA which is found only in the $C_c$ strain. S RNA is the coat protein mRNA,[52,132] while $I_2$ RNA for each strain codes for a protein of 29,000-30,000 mol wt. (Figure 4) in a wheat germ in vitro translation system; the $I_{1-A}$ and $I_{1-B}$ RNA translations show the same 29,000—30,000 mol wt proteins, but these RNAs are considered to be a heterogeneous population of molecules, some having lost the 3′- and other their 5 -ends.[4] (From Beachy, R. N. and Zaitlin, M. *Virology,* 81, 160, 1977. With permission.)

When in the virion, the RNA is complexed with spermidine.[179,180] If the virion is treated under alkaline conditions the RNA fragments, but it can reassociate to form a species with a greater sedimenation coefficient than that of the native RNA.[166]

Intact genomic TYMV RNA of molecular weight of about $1.9 \times 10^6$, like TMV RNA, does not generate coat protein upon in vitro translation. The coat protein mRNA is found in a virion associated subgenomic RNA[18,19,133] of molecular weight reported to vary from 200,000[133] to 300,000.[18] Heating the virion RNA yields five size classes of RNA, which, judging from their capacities to accept valine, may all have a common 3′-end.[19] The analogy to less-than-full-length TMV RNAs[4] is very strong. It has recently been observed that the distribution of the various classes of RNAs varies among the density classes[171] of TYMV virions.[20,170] (see Figure 8). Translation of the large RNA yields a series of polypeptides of up to 165,000 mol wt.[18]

### C. Other Viruses

Very little is known about the organization of the genomes of other viruses in the monopartite class. We can guess that many eventually may be shown to have replication strategies similar to TMV and TYMV. For example, when PVX RNA is used as the mRNA in the wheat germ in vitro system, no coat protein is translated,[135] suggesting than an LMC-like subgenomic mRNA exists to control coat protein synthesis. TNV seems to be an exception to this pattern, as coat protein is translated from the full-length RNA.[129]

We also have no information on the number, nature, and significance of viral translational products, other than coat protein, from any other monopartite viruses, although it is probable that potyvirus RNAs (Table 1) code for the proteins of the intracellular inclusion bodies characteristic of that group; the inclusion body proteins from several potato virus Y (PVY) viruses, and strains are immunologically unique. Their compositions are independent of the plant species in which the virus is propagated.[174]

## ACKNOWLEDGMENT

Preparation of this chapter and the work in my laboratory reported herein has been supported in part by grants from The National Science Foundation.

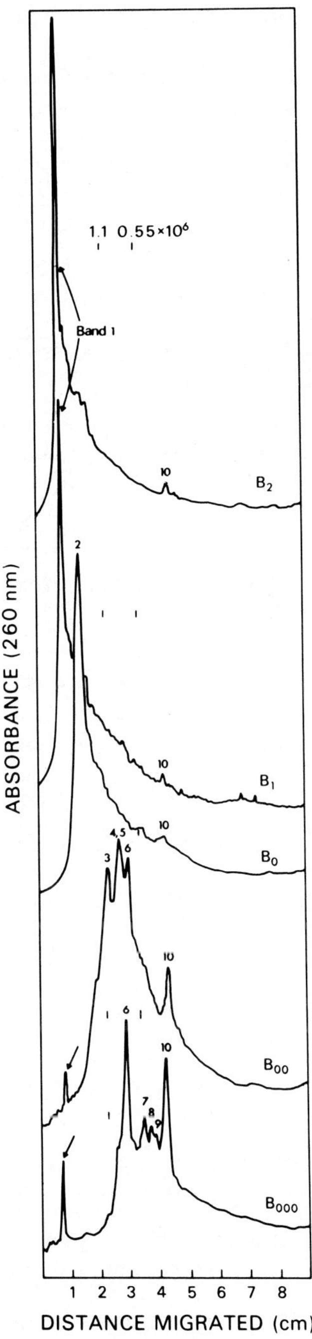


FIGURE 8. RNAs isolated from nucleoproteins of TYMV that had been fractionated on cesium chloride density gradients. The designations $B_2$, $B_1$, $B_o$, etc., are those of Matthews[171] and are shown in the order of decreasing density from top to bottom. Shown are polyacrylamide gel electrophoresis patterns of the RNAs of each nucleoprotein fraction. The major RNA species are denoted by numerals 1 to 10. The high molecular weight bands in $B_{oo}$ and $B_{ooo}$ (denoted by arrows) are due to contaminating DNA. RNA band 10 is the coat protein mRNA.[170] (Courtesy of Higgins, T. J. V., Whitfeld, P. R., and Matthews, R. E. F., *Virology*, 84, 153, 1978. (With permission.)

# REFERENCES

1. **Fenner, F.**, Classification and nomenclature of viruses, *S. Karger, Basel*, 116, 1976.
2. **Edwardson, J. R.**,Some properties of the potato virus Y group, *Fl. Agric. Exp. St. Monogr. Ser. No. 4*, 398 p, 1974.
3. **Zaitlin, M. and Israel, H. W.**, Tobacco mosaic virus (type strain), *Commonwealth Agricultural Bureau/Association of Applied Biologists Descriptions of Plant Viruses, No. 151*, 1975.
4. **Beachy, R. N. and Zaitlin, M.**, Characterization and *in vitro* translation of the RNAs from less-than-full-length, virus-related, nucleoprotein rods present in tobacco mosaic virus preparations, *Virology*, 81, 160, 1977.
5. **Bercks, R.**, Potato virus X, *Commonwealth Agricultural Bureau/Association of Applied Biologists Descriptions of Plant Viruses, No. 4*, 1970.
6. **Koenig, R.**, Nucleic acids in the potato virus X group and in some other plant viruses: comparison of the molecular weights by electrophoresis in acrylamide-agarose composite gels, *J. Gen. Virol.*, 10, 111, 1971.
7. **Purcifull, D. E. and Hiebert, E.**, Papaya mosaic virus, *Commonwealth Agricultural Bureau/Association of Applied Biologists Descriptions of Plant Viruses, No. 56*, 1971.
8. **Frowd, J. A. and Tremaine, J. H.**, Physical, chemical, and serological properties of cymbidium mosaic virus, *Phytopathology*, 67, 43, 1977.
9. **Makkouk, K. M. and Gumpf, D. J.**, Isolation and properties of potato virus Y ribonucleic acid, *Phytopathology*, 64, 1115, 1974.
10. **Makkouk, K. M. and Gumpf, D. J.**, Characterization of the protein and nucleic acid of potato virus Y strains isolated from pepper, *Virology*, 63, 336, 1975.
11. **Hill, J. H. and Benner, H. I.**, Properties of potyvirus RNAs: turnip mosaic, tobacco etch, and maize dwarf mosaic viruses, *Virology*, 75, 419, 1976.
12. **Damirdagh, I. S. and Shepherd, R. J.**, Some of the chemical properties of the tobacco etch virus and its protein and nucleic acid components, *Virology*, 40, 84, 1970.
13. **Hill, J. H. and Shepherd, R. J.**, Biochemical properties of turnip mosaic virus, *Virology*, 47, 807, 1972.
14. **Carpenter, J. M., Kassanis, B., and White, R. F.**, The protein and nucleic acid of beet yellows virus, *Virology*, 77, 101, 1977.
15. **Bar-Joseph, M. and Hull, R.**, Purification and partial characterization of sugar beet yellows virus, *Virology*, 62, 552, 1974.
16. **Bar-Joseph, M., Hull, R., and Lane, L. C.**, Biophysical and biochemical characterization of apple chlorotic leafspot virus, *Virology*, 62, 563, 1974.
17. **Wetter, C.**, Carnation latent virus, *Commonwealth Agricultural Bureau/Association of Applied Biologists Descriptions of Plant Viruses, No. 61*, 1971.
18. **Klein, C., Fritsch, C., Briand, J. P., Richards, K. E., Jonard, G., and Hirth, L.**, Physical and functional heterogeneity in TYMV RNA: evidence for the existence of an independent messenger coding for coat protein, *Nucleic Acids Res.*, 3, 3043, 1976.
19. **Pleij, C. W. A., Neeleman, A., van Vloten-Doting, L., and Bosch, L.**, Translation of turnip yellow mosaic virus RNA *in vitro*: a closed and an open coat protein cistron, *Proc. Natl. Acad. Sci. USA.*, 73, 4437, 1976.
20. **Pleij, C. W. A., Mellema, J. R., Noort, A., and Bosch, L.**, The occurrence of the coat protein messenger RNA in the minor components of turnip yellow mosaic virus, *FEBS Lett.*, 80, 19, 1977.
21. **Matthews, R. E. F.**, Turnip yellow mosaic virus, *Commonwealth Agricultural Bureau/Association of Applied Biologists Descriptions of Plant Viruses, No. 2*, 1970.
22. **Kaper, J. M. and Waterworth, H. E.**, Comparison of molecular weights of single-stranded viral RNAs by two empirical methods, *Virology*, 51, 183, 1973.
23. **Gibbs, A. J and Harrison, B. D.**, Eggplant mosaic virus, *Commonwealth Agricultural Bureau/Association of Applied Biologists Descriptions of Plant Viruses, No. 124*, 1973.
24. **Bouley, J. P., Briand, J. P., Genevaux, M., Pinck, M., and Witz, J.**, The structure of eggplant mosaic virus: evidence for the presence of low molecular weight RNA in top component, *Virology*, 69, 775, 1976.
25. **Martelli, G. P., Quacquerelli, A., and Russo, M.**, Tomato bushy stunt virus, *Commonwealth Agricultural Bureau/Association of Applied Biologists Descriptions of Plant Viruses, No. 69*, 1971.
26. **Nelson, M. R. and Tremaine, J. H.**, Physiochemical and serological properties of a virus from saguaro cactus, *Virology*, 65, 309, 1975.
27. **Brakke, M. K. and Rochow, W. F.**, Ribonucleic acid of barley yellow dwarf virus, *Virology*, 61, 240, 1974.

28. **Lesnaw, J. A. and Reichmann, M. E.,** The structure of tobacco necrosis virus. I. The protein subunit and the nature of the nucleic acid, *Virology,* 39, 729, 1969.

28a. **Hollings, M., Stone, O. M., and Barton, R. J.,** Pathology, soil transmission and characterization of cymbidium ringspot, a virus from cymbidium orchids and white clover (*Trifolium repens*), *Ann. Appl. Biol.,* 85, 233, 1977.

29. **Waterworth, H. E., Lawson, R. H., and Monroe, R. L.,** Purification and properties of Hibiscus chlorotic ringspot virus, *Phytopathology,* 66, 570, 1976.

30. **Francki, R. I. B. and Randles, J. W.,** Some properties of lettuce necrotic yellows virus RNA and its *in vitro* transcription by virion-associated trancriptase, *Virology,* 54, 359, 1973.

31. **Reeder, G. S., Knudson, D. L., and MacLoed, R.,** The ribonucleic acid of potato yellow dwarf virus, *Virology,* 50, 301, 1972.

32. **Jackson, A. O. and Christie, S. R.,** Purification and some physicochemical properties of sonchus yellow net virus, *Virology,* 77, 344, 1977.

33. **Francki, R. I. B.,** Plant rhabdoviruses, *Adv. Virus Res.,* 18, 257, 1973.

34. **Hunter, T. R., Hunt, T., Knowland, J., and Zimmern, D.,** Messenger RNA for the coat protein of tobacco mosaic virus, *Nature (London),* 260, 759, 1976.

35. **Siegel, A.,** Pseudovirions of tobacco mosaic virus, *Virology,* 46, 50, 1971.

36. **Steere, R. L.,** Purification, in *Plant Virology,* Corbett, M. K. and Sisler, H. D., Eds., University of Florida Press, Gainesville, 1964, Chap. 10.

37. **Francki, R. I. B.,** Purification of Viruses, in *Principles and Techniques in Plant Virology,* Kado, C. I. and Agrawal, H. O., Eds., Van Nostrand Reinhold, New York, 1972, chap. 11.

38. **Kassanis, B., Carpenter, J. M., White, R. F., and Woods, R. D.,** Purification and some properties of beet yellows virus, *Virology,* 77, 95, 1977.

39. **Fraser, R. S. S.,** Extraction and assay of TMV RNA, *Virology,* 45, 804, 1971.

40. **Kaper, J. M. and Alting Siberg, R.**. Degradation of turnip yellow mosaic virus by freezing and thawing *in vitro*: a new method for studies on the internal organization of the viral components and for isolating native RNA, *Virology,* 38, 407, 1969.

41. **Ralph, R. K. and Bergquist, P. L.,** Separation of viruses into components, in *Methods in Virology 2,* Maramorosch, K. and Koprowski, H., Eds., Academic Press, New York, 1967, chap 16.

42. **Kirby, K. S.,** Isolation of nucleic acids with phenolic solvents, *Methods Enzymol.,* 12B, 87, 1968.

43. **Mandeles, S. and Bruening, G.,** Tobacco mosaic virus ribonucleic acid, *Biochem. Prep.,* 12, 111, 1968.

44. **Fraenkel-Conrat, H., Singer, B., and Tsugita, A.,** Purification of viral RNA by means of bentonite, *Virology,* 14, 54, 1961.

45. **Solymosy, F., Fedorsák, I. Gulyás, A., Farkas, G. L., and Ehrenberg, L.,** A new method based on the use of diethyl pyrocarbonate as a nuclease inhibitor for the extraction of undegraded nucleic acid from plant tissues, *Eur. J. Biochem.,* 5, 520, 1968.

46. **Bagi, G. Gulyás, A., and Solymosy, F.,** A differential extraction method for the isolation of host nucleic acids from TMV-infected tobacco plants, *Virology,* 42, 662, 1970.

47. **Diener, T. O. and Schneider, I. R.,** Virus degradation and nucleic acid release in single-phase phenol systems, *Arch. Biochem. Biophys.,* 124, 401, 1968.

48. **Brakke, M. K. and Van Pelt, N.,** Properties of infectious ribonucleic acid from wheat streak mosaic virus, *Virology,* 42, 699, 1970.

49. **Boatman, S. and Kaper, J. M.,** Molecular organization and stabilizing forces of simple RNA viruses IV. Selective interference with protein-RNA interactions by use of sodium dodecyl sulfate *Virology,* 70, 1, 1976.

50. **Bruening, G. E.,** Virus degradation and nucleic acid isolation, in *Principles and Techniques in Plant Virology,* Kado, C. I. and Agrawal, H. O., Eds., Van Nostrand Reinhold, New York, chap. 16, 1972.

51. **El Manna, M. and Bruening, G.,** Polyadenylate sequences in the ribonucleic acids of cowpea mosaic virus, *Virology,* 56, 198, 1973.

52. **Bruening, G., Beachy, R. N., Scalla, R., and Zaitlin, M.,** *In vitro* and *in vivo* translation of the ribonucleic acids of a cowpea strain of tobacco mosaic virus, *Virology,* 71, 498, 1976.

52a . **Wilcockson, J. and Hull, R.,** The rapid isolation of plant virus RNAs using sodium perchlorate, *J. Gen. Virol.,* 23, 107, 1974.

53. **Loening, U. E.,** The fractionation of high-molecular-weight ribonucleic acid by polyacrylamide-gel electrophoresis, *Biochem. J.,* 102, 251, 1967.

54. **Peacock, A. C. and Dingman, C. W.,** Molecular weight estimation and separation of ribonucleic acid by electrophoresis in agarose-acrylamide composite gels, *Biochemistry,* 7, 668, 1968.

55. **Jackson, A. O., Mitchell, D. M., and Siegel, A.,** Replication of tobacco mosaic virus I. Isolation and characterization of double-stranded forms of ribonucleic acid, *Virology,* 45, 182, 1971.

56. **Adesnik, M.,** Polyacrylamide gel electrophoresis of viral RNA, in *Methods in Virology,* Vol. 5, Maramorosch, K. and Koprowski, H., Eds., Academic Press, New York, 1971, chap. 4.
57. **Thach, S. S. and Thach, R. E.,** Mechanism of viral replication. I. Structure of replication complexes of R17 bacteriophage, *J. Mol. Biol.,* 81, 367, 1973.
58. **Dawson, W. O., German, T. L., and Schlegel, D. E.,** Homogenization-resistant and -susceptible components of tobacco mosaic virus replicative form RNA, *J. Gen. Virol.,* 32, 205, 1976.
59. **Burdon, R. H., Billeter, M. A., Weissmann, C., Warner, R. C., Ochoa, S., and Knight, C. A.,** Replication of viral RNA, V. Presence of a virus-specific double-stranded RNA in leaves infected with tobacco mosaic virus, *Proc. Natl. Acad. Sci. USA,* 52, 768, 1964.
60. **Shipp, W. and Haselkorn, R.,** Double-stranded RNA from tobacco leaves infected with TMV, *Proc. Natl. Acad. Sci. USA,* 52, 401, 1964.
61. **Mandel, H. G., Matthews, R. E. F., Matus, A., and Ralph, R. K.,** Replicative form of plant viral RNA, *Biochem. Biophys. Res. Commun.,* 16, 604, 1964.
62. **Montagnier, L. and Sanders, K. F.,** Replicative form of encephalomyocarditis virus ribonucleic acid, *Nature (London),* 199, 664, 1963.
63. **Nilsson-Tillgren, T.,** Studies on the biosynthesis of TMV III. Isolation and characterization of the replicative form and the replicative intermediate RNA, *Mol. Gen. Genet.,* 109, 246, 1970.
64. **Keilland-Brandt, M. C. and Nilsson-Tillgren, T.,** Studies on the biosynthesis of TMV IV. Some properties of double-stranded TMV RNA, *Mol. Gen. Genet.,* 121, 219, 1973.
65. **Franklin, R. M.,** Purification and properties of the replicative intermediate of the RNA bacteriophage R17, *Proc. Natl. Acad. Sci. USA.,* 55, 1504, 1966.
66. **Aoki, S. and Takebe, I.,** Replication of tobacco mosaic virus RNA in tobacco mesophyll protoplasts inoculated *in vitro, Virology,* 65, 343, 1975.
67. **Ralph, R. K. and Wojcik, S. J.,** Double-stranded tobacco mosaic virus RNA, *Virology,* 37, 276, 1969.
68. **Nilsson-Tillgren, T., Kielland-Brandt, M. C., and Bekke, B.,** Studies on the biosynthesis of tobacco mosaic virus VI. On the subcellular localization of double-stranded viral RNA, *Mol. Gen. Genet.,* 128, 157, 1974.
69. **Beachy, R. N. and Zaitlin, M.,** Replication of tobacco mosaic virus VI. Replicative intermediate and TMV-RNA-related RNAs associated with polyribosomes, *Virology,* 63, 84, 1975.
70. **Bozarth, R. F. and Harley, E. H.,** The electrophoretic mobility of double-stranded RNA in polyacrylamide gels as a function of molecular weight, *Biochim. Biophys. Acta,* 432, 329, 1976.
71. **Jackson, A. O., Zaitlin, M., Siegel, A., and Francki, R. I. B.,** Replication of tobacco mosaic virus III. Viral RNA metabolism in separated leaf cells, *Virology,* 48, 655, 1972.
72. **Siegel, A., Zaitlin, M., and Duda, C. T.,** Replication of tobacco mosaic virus IV. Further characterization of viral related RNAs, *Virology,* 53, 75, 1973.
73. **Kielland-Brandt, M. C. and Nilsson-Tillgren, T.,** Studies on the biosynthesis of TMV V. Determination of TMV RNA and its complementary RNA at different times after infection, *Mol. Gen. Genet.,* 121, 229, 1973.
74. **Ralph, R. K., Matthews, R. E. F., Matus, A. I., and Mandel, H. G.,** Isolation and properties of double-stranded viral RNA from virus-infected plants, *J. Mol. Biol.,* 11, 202, 1965.
75. **Bockstahler, L. E.,** Biophysical studies on double-stranded RNA from turnip yellow mosaic virus-infected plants, *Mol. Gen. Genet.,* 100, 337, 1967.
76. **Pinck, L., Hirth, L., and Bernardi, G.,** Isolation of replicative RNA from alfalfa mosaic virus-infected plants by chromatography on hydroxyapatite columns, *Biochem. Biophys. Res. Commun.,* 31, 481, 1968.
77. **Mouchès, C., Bové, J. M., and Bové, C.,** Turnip yellow mosaic virus-RNA replicase: partial purification of the enzyme from the solubilized enzyme-template complex, *Virology,* 58, 409, 1974.
78. **Bedbrook, J. R. and Matthews, R. E. F.,** Location, rate and asymmetry of ds-RNA synthesis during replication of TYMV in Chinese cabbage, *Ann. Microbiol. (Institute Pasteur)* 127 A, 55, 1976.
79. **Klein, A. and Reichmann, M. E.,** Isolation and characterization of two species of double-stranded RNA from tobacco leaves doubly infected with tobacco necrosis and satellite tobacco necrosis viruses, *Virology,* 42, 269, 1970.
80. **Leung, D. W., Gilbert, C. W., Smith, R. E., Saavage, N. L., and Clark, J. M., Jr.,** Translation of satellite tobacco necrosis virus ribonucleic acid by an *in vitro* system from wheat germ, *Biochemistry,* 15, 4943, 1976.
81. **Kassanis, B.,** Properties and behaviour of a virus depending for its multiplication on another, *J. Gen. Microbiol.,* 27, 477, 1962.
82. **Shoulder, A., Darby, G., and Minson, T.,** RNA-RNA hybridisation using $^{125}$I-labeled RNA from tobacco necrosis virus and its satellite, *Nature (London),* 251, 733, 1974.
83. **Bishop, J. M. and Koch, G.,** Infectious replicative intermediate of poliovirus: purification and characterization, *Virology,* 37, 521, 1969.

84. **Engelhardt, D. L.**, Assay for secondary structure in ribonucleic acid, *J. Virol.*, 9, 903, 1972.
85. **Robertson, H. D., Webster, R. E., and Zinder, N. D.**, Purification and properties of ribonuclease III from *Escherichia coli*, *J. Biol. Chem.*, 243, 82, 1968.
86. **Pinck, L.**, Chromatographie des acides nucléiques de tabac sain et infecté par divers virus sur colonne d'hydroxyapatite, *Bull. de la Soc. de Chimie Biol.*, 52, 843, 1970.
87. **Kothari, R. M. and Shankar, V.**, RNA fractionation on hydroxyapatite columns, *J. Chromatogr.*, 98, 449, 1974.
88. **Rezaian, M. A. and Francki, R. I. B.**, Replication of tobacco ringspot virus I. Detection of a low molecular weight double-stranded RNA from infected plants, *Virology*, 56, 238, 1973.
89. **van Griensven, L. J. L. D., van Kammen, A., and Rezelman, G.**, Characterization of the double-stranded RNA isolated from cowpea mosaic virus-infected *Vigna* leaves, *J. Gen. Virol.*, 18, 359, 1973.
90. **Petrovic, S. L., Sumonja, B. D., and Vasiljevic, R. B.**, Fractionation of nucleic acids from *Penicillium crysogenum* and associated ribonucleic acid viruses by selective exclusion and retention in agarose gels, *Biochem. J.*, 139, 157, 1974.
91. **Reddi, K. K. and Anjaneyalu, Y. V.**, Studies on the formation of tobacco mosaic virus ribonucleic acid. I. Nonparticipation of deoxyribonucleic acid, *Biochim. Biophys. Acta*, 72, 33, 1963.
92. **Zaitlin, M., Duda, C. T., and Petti, M. A.**, Replication of tobacco mosaic virus V. Properties of the bound and solubilized replicase, *Virology*, 53, 300, 1973.
93. **Romaine, C. P. and Zaitlin, M.**, RNA-dependent RNA polymerases in uninfected and tobacco mosaic virus-infected tobacco leaves: viral-induced stimulation of host RNA polymerase activity, *Virology*, 86, 241, 1978.
94. **Fraenkel-Conrat, H.**, RNA polymerase from tobacco necrosis virus infected and uninfected tobacco. Purification of the membrane-associated enzyme, *Virology*, 72, 23, 1976.
95. **Zabel, P., Jongen-Neven, I., and van Kamman, A.**, *In vitro* replication of cowpea mosaic virus RNA II. Solubilization of membrane-bound replicase and the partial purification of the solubilized enzyme, *J. Virol.*, 17, 679, 1976.
96. **LeRoy, C., Stussi-Garaud, C., and Hirth, L.**, RNA-dependent RNA polymerases in uninfected and in alfalfa mosaic virus infected tobacco plants, *Virology*, 82, 48, 1977.
97. **Astier-Manifacier, S. and Cornuet, P.**, Isolation of turnip yellow mosaic virus RNA replicase and asymmetrical synthesis of polynucleotides identical to TYMV-RNA, *Biochem. and Biophys. Res. Commun.*, 18, 283, 1965.
98. **Bové, J. M.**, RNA bicaténaire spécifique du virus de la mosaique du navet: expérience de dilution spécifique, *C. R. Soc. Biol.*, 161, 537, 1967.
99. **Ralph, R. K. and Wojcik, S. J.**, Synthesis of double-stranded viral RNA by cell-free extracts from turnip yellow mosaic virus-infected leaves, *Biochim. Biophys. Acta*, 119, 347, 1966.
100. **Mouchès, C., Bové, C., Barreau, C., and Bové, J. M.**, TYMV RNA replicase: formation of a complex between the purified enzyme and TYMV RNA, *Ann. Microbiol., (Institute Pasteur)* 127A, 75, 1976.
101. **Laflèche, D. and Bové, J. M.**, Virus de la mosaïque jaune du Navet: site cellulaire de la replication du RNA viral, *Physiol. Vég.*, 9, 487, 1971.
102. **Bandurski, R. S. and Maheshwari, S. C.**, Nucleotide incorporation into nucleic acid by tobacco leaf homogenates, *Plant Physiol.*, 37, 556, 1962.
103. **Semal, J., Spencer, D., Kim, Y. T., Moyer, R. H., Smith, R. A., and Wildman, S. G.**, A comparison of the properties of RNA-synthesizing systems in cell-free extracts of healthy and TMV-infected tobacco leaves, *Virology*, 24, 155, 1964.
104. **Bradley, D. W. and Zaitlin, M.**, Replication of tobacco mosaic virus II. The *in vitro* synthesis of high molecular weight virus-specific RNAs, *Virology*, 45, 192, 1971.
105. **Zaitlin, M., Beachy, R. N., Bruening, G., Romaine, C. P., and Scalla, R.**, Translation of tobacco mosaic virus RNA, in *Animal Virology*, Baltimore, D., Huang, A., and Fox, C. F., Eds., Academic Press, New York, 1976, 567.
106. **Brishammar, S. and Juntti, N.**, Partial purification of soluble TMV replicase, *Virology*, 59, 245, 1974.
107. **Sela, I. and Hauschner, A.**, Isolation and characterization of a TMV-RNA dependent enzyme from TMV-infected tobacco leaves, *Virology*, 64, 284, 1975.
108. **White, J. L. and Murakishi, H. H.**, *In vitro* replication of tobacco mosaic virus RNA in tobacco callus cultures: solubilization of membrane-bound replicase and partial purification, *J. Virol.*, 21, 484, 1977.
109. **Stussi-Garaud, C., Lemius, J., and Fraenkel-Conrat, H.**, RNA polymerase from tobacco necrosis virus infected and uninfected tobacco II. Properties of the bound and soluble polymerases, and the nature of their products, *Virology*, 81, 224, 1977.

110. **Baltimore, D., Huang, A. S., and Stampfer, M.,** Ribonucleic acid synthesis in vesicular stomatitis virus II. An RNA polymerase in the virion, *Proc. Natl. Acad. Sci. USA.*, 66, 572, 1970.
111. **Francki, R. I. B. and Randles, J. W.,** RNA-dependent RNA polymerase associated with particles of lettuce necrotic yellows virus, *Virology*, 47, 270, 1972.
112. **Randles, J. W. and Francki, R. I. B.,** Infectious nucleocapsid particles of lettuce necrotic yellows virus with RNA-dependent RNA polymerase activity, *Virology*, 50, 297, 1972.
113. **Gilliland, J. M. and Symons, R. H.,** Properties of a plant virus-induced RNA polymerase in cucumbers infected with cucumber mosaic virus, *Virology*, 36, 232, 1968.
114. **Astier-Manifacier, S. and Cornuet, P.,** RNA-dependent RNA polymerase in Chinese cabbage, *Biochim. Biophys. Acta*, 232, 484, 1971.
115. **Duda, C. T., Zaitlin, M., and Siegel, A.,** *In vitro* synthesis of double-stranded RNA by an enzyme system isolated from tobacco leaves, *Biochim. Biophys. Acta*, 319, 62, 1973.
116. **Bol, J. F., Clerx-van Haaster, C. M., and Weening, C. J.,** Host and virus specific RNA polymerases in alfalfa mosaic virus infected tobacco, *Ann. Microbiol., (Institute Pasteur)*, 127A, 183, 1976.
117. **Sakai, F. and Takebe, I.,** Protein synthesis in tobacco mesophyll protoplasts induced by tobacco mosaic virus infection, *Virology*, 62, 426, 1974.
118. **Paterson, R. and Knight, C. A.,** Protein synthesis in tobacco protoplasts infected with tobacco mosaic virus, *Virology*, 64, 10, 1975.
119. **Dawson, W. O. and Schlegel, D. E.,** The sequence of inhibition of tobacco mosaic virus synthesis by actinomycin D, 2-thiouracil, and cycloheximide in a synchronous infection, *Phytopathology*, 66, 177, 1976.
120. **Otsuki, Y., Takebe, I., Honda, Y., Kajita, S., and Matsui, C.,** Infection of tobacco mesophyll protoplasts by potato virus X, *J. Gen. Virol.*, 22, 375, 1974.
121. **Lewandowski, L. J., Kimball, P. C., and Knight, C. A.,** Separation of the infectious ribonucleic acid of potato spindle tuber virus from double-stranded ribonucleic acid of plant tissue extracts, *J. Virol.*, 8, 809, 1971.
122. **Downey, K. M., Byrnes, J. J., Jurmark, B. S., and So, A. G.,** Reticulocyte RNA-dependent RNA polymerase, *Proc. Natl. Acad. Sci. USA.*, 70, 3400, 1973.
123. **Boyd, C. D. and Fitschen, W.,** Direct evidence that ribosome bound RNA-dependent RNA polymerase does not play a role in globin messenger RNA replication, *Nucleic Acids Res.*, 4, 461, 1977.
124. **McLeester, R. C. and Hall, T. C.,** Simplification of amino acid incorporation and other assays using filter paper techniques, *Anal. Biochem.*, 79, 627, 1977.
125. **Byfield, J. E. and Scherbaum, O. H.,** A rapid radioassay technique for cellular suspensions, *Anal. Biochem.*, 17, 434, 1966.
126. **Shatkin, A. J.,** Capping of eucaryotic mRNAs, *Cell*, 9, 645, 1976.
127. **Zimmern, D.,** The 5′-end group of tobacco mosaic virus RNA is $m^7$G5′-ppp5′Gp, *Nucleic Acids Res.*, 2, 1189, 1975.
128. **Keith, J. and Fraenkel-Conrat, H.,** Tobacco mosaic virus RNA carries 5′-terminal triphosphorylated guanosine blocked by 5′-linked 7-methylguanosine, *FEBS Lett.*, 57, 31, 1975.
129. **Salvato, M. S. and Fraenkel-Conrat, H.,** Translation of tobacco necrosis virus and its satellite in a cell-free wheat germ system, *Proc. Natl. Acad. Sci. USA*, 74, 2288, 1977.
130. **Roman, R., Brooker, J. D., Seal, S. N., and Marcus, A.,** Inhibition of the transition of a 40S ribosome-Met-tRNA$_i^{Met}$ complex to an 80S ribosome-Met-tRNA$_i^{Met}$ complex by 7-methylguanosine-5′ phosphate, *Nature (London)*, 260, 359, 1976.
131. **Knowland, J.,** Protein synthesis directed by the RNA from a plant virus in a normal animal cell, *Genetics*, 78, 383, 1974.
132. **Higgins, T. J. V., Goodwin, P. B., and Whitfeld, P. R.,** Occurrence of short particles in beans infected with the cowpea strain of TMV II. Evidence that short particles contain the cistron for coat-protein, *Virology*, 71, 486, 1976.
133. **Ricard, B., Barreau, C., Renaudin, H., Mouches, C., and Bové, J. M.,** Messenger properties of TYMV-RNA, *Virology*, 79, 231, 1977.
134. **Skotnicki, A., Gibbs, A., and Shaw, D. C.,** *In vitro* translation of polyribosome-associated RNAs from tobamovirus-infected plants, *Intervirology*, 7, 256, 1976.
135. **Ricciardi, R. P., Goodman, R. M., and Gottlieb, D.,** Translation of PVX RNA *in vitro* by wheat germ I. Characterization of the reaction and product size, *Virology*, 85, 310, 1978.
136. **Beachy, R. N., Zaitlin, M., Bruening, G., and Israel, H. W.,** A genetic map for the cowpea strain of TMV, *Virology*, 73, 498, 1976.
137. **Hindennach, I. and Jockusch, H.,** Peptide mapping of phage $Q\beta$ proteins using cell-free synthesis, *Virology*, 60, 327, 1974.
138. **Sanger, F., Air, G. M., Barrell, B. G., Brown, N. L., Coulson, A. R., Fiddes, J. C., Hutchison III, C. A., Slocombe, P. M., and Smith, M.,** Nucleotide sequence of bacteriophage $\phi$X174 DNA, *Nature (London)*, 265, 687, 1977.

139. **Hunter, T.**, personal communication, 1977.
140. **Mandeles, S.**, Location of unique sequences in tobacco mosaic virus ribonucleic acid, *J. Biol. Chem.*, 243, 3671, 1968.
141. **May, D. S. and Knight, C. A.**, Polar stripping of protein subunits from tobacco mosaic virus, *Virology*, 25, 502, 1965.
142. **Wilson, T. M. A., Perham, R. N., Finch, J. T., and Butler, P. J. G.**, Polarity of the RNA in the tobacco mosaic virus particle and the direction of protein stripping in sodium dodecyl sulfate, *FEBS Lett.*, 64, 285, 1976.
143. **Ohno, T. and Okada, Y.**, Polarity of stripping of tobacco mosaic virus by alkali and sodium dodecyl sulfate, *Virology*, 76, 429, 1977.
144. **Richards, K., Guilley, H., Jonard, G., and Keith, G.**, Leader sequence of 71 nucleotides devoid of G in tobacco mosaic virus RNA, *Nature (London)*, 267, 548, 1977.
145. **Dasgupta, R., Shih, D. S., Saris, C., and Kaesberg, P.**, Nucleotide sequence of a viral RNA fragment that binds to eukaryotic ribosomes, *Nature (London)*, 256, 624, 1975.
146. **Guilley, H., Jonard, G., and Hirth, L.**, Sequence of 71 nucleotides at the 3′-end of tobacco mosaic virus RNA, *Proc. Natl. Acad. Sci. USA*, 72, 864, 1975.
147. **Lamy, D., Jonard, G., Guilley, H., and Hirth, L.**, Comparison between the 3′OH end RNA sequence of two strains of tobacco mosaic virus (TMV) which may be aminoacylated, *FEBS Lett.*, 60, 202, 1975.
148. **Hall, T. C. and Wepprich, R. K.**, Functional possibilities for aminoacylation of viral RNA in transcription and translation, *Ann. Microbiol. (Institute Pasteur)*, 127A, 143, 1976.
149. **Guilley, H., Jonard, G., Richards, K. E., and Hirth, L.**, Sequence of a specifically encapsidated RNA fragment originating from the tobacco-mosaic-virus coat-protein cistron, *Eur. J. Biochem.*, 54, 135, 1975.
150. **Guilley, H., Jonard, G., Richards, K. E., and Hirth, L.**, Observations concerning the sequence of two additional specifically encapsidated RNA fragments originating from the tobacco-mosaic-virus coat-protein cistron, *Eur. J. Biochem.*, 54, 145, 1975.
151. **Laflèche, D., Bové, C., Dupont, G., Mouchès, C., Astier, T., Garnier, M., and Bové, J. M.**, Site of viral RNA replication in the cells of higher plants: TYMV-RNA synthesis on the chloroplast outer membrane system, *FEBS Proc. Meet.*, 27, 43, 1972.
152. **Garnier, M. and Bové, J. M.**, Autoradiographical studies on RNA synthesis in healthy and TYMV infected Chinese cabbage: effect of actinomycin D, *Ann. Microbiol. (Institute Pasteur)*, 127A, 69, 1976.
153. **Butler, P. J. G. and Klug, A.**, Assembly of the particle of tobacco mosaic virus from RNA and disks of protein, *Nature (London) New Biol.*, 229, 47, 1971.
154. **Thouvenel, J. C., Guilley, H., Stussi, C., and Hirth, L.**, Evidence for polar reconstitution of TMV, *FEBS Lett.*, 16, 204, 1971.
155. **Zimmern, D. and Wilson, T. M. A.**, Location of the origin for viral reassembly on tobacco mosaic virus RNA and its relation to stable fragment, *FEBS Lett.*, 71, 294, 1976.
156. **Zimmern, D.**, The nucleotide sequence at the origin for assembly on tobacco mosaic virus RNA, *Cell*, 11, 463, 1977.
157. **Lebeurier, G., Nicolaieff, A., and Richards, K. E.**, Inside-out model for self-assembly of tobacco mosaic virus, *Proc. Natl. Acad. Sci. USA.*, 74, 149, 1977.
158. **Otsuki, Y., Takebe, I., Ohno, T., Fukuda, M., and Okada, Y.**, Reconstitution of tobacco mosaic virus rods occurs bidirectionally from an internal initiation region: demonstration by electron microscopic serology, *Proc. Natl. Acad, Sci. USA.*, 74, 1913, 1977.
159. **Butler, P. J. G., Finch, J. T., and Zimmern, D.**, Configuration of tobacco mosaic virus RNA during virus assembly, *Nature (London)*, 265, 217, 1977.
160. **Zimmern, D. and Butler, P. J. G.**, The isolation of tobacco mosaic virus RNA fragments containing the origin for viral assembly, *Cell*, 11, 455, 1977.
161. **Jonard, G., Richards, K. E., Guilley, H., and Hirth, L.**, Sequence from the assembly nucleation region of TMV RNA, *Cell*, 11, 483, 1977.
162. **Siegel, A., Hari, V., Montgomery, I., and Kolacz, K.**, A messenger RNA for capsid protein isolated from tobacco mosaic virus-infected tissue, *Virology*, 73, 363, 1976.
163. **Morris, T. J.**, Two nucleoprotein components associated with the cowpea strain of TMV, *Proc. Am. Phytopathol. Soc., (Abstr.,)* 1, 83, 1974.
164. **Skotnicki, A., Gibbs, A., and Shaw, D. C.**, Mixed infection with two tobamoviruses: the formation of particles containing the coat protein messenger RNAs of either virus, *Intervirology*, 7, 328, 1976.
165. **Demeure, M., Kummert, J., Portetelle, D., and Semal, J.**, Properties of the DNA product of reverse transcription of turnip yellow mosaic virus RNA preparations, *Virology*, 81, 449, 1977.
166. **Pleij, C. W. A., Eecen, H. G., Bosch, L., and Mandel, M.**, The formation of fast-sedimenting turnip yellow mosaic virus RNA: structural rearrangement inside the capsid, *Virology*, 76, 781, 1977.

167. **Briand, J. P., Richards, K. E., Bouley, J. P., Witz, J., and Hirth, L.,** Structure of the amino-acid accepting 3′-end of high-molecular-weight eggplant mosaic virus RNA, *Proc. Natl. Acad. Sci. USA.*, 73, 737, 1976.
168. **Briand, J. P., Jonard, G., Guilley, H., Richards, K., and Hirth, L.,** Nucleotide sequence (n = 159) of the amino-acid-accepting 3′-OH extremity of turnip-yellow-mosaic-virus RNA and the last portion of its coat-protein cistron, *Eur. J. Biochem.*, 72, 453, 1977.
169. **Silberklang, M., Prochiantz, A., Haenni, A. L., and Rajbhandary, U. L.,** Studies on the sequence of the 3′-terminal region of turnip-yellow-mosaic-virus RNA, *Eur. J. Biochem.*, 72, 465, 1977.
170. **Higgins, T. J. V., Whitfeld, P. R., and Matthews, R. E. F.,** Size distribution and *in vitro* translation of the RNAs isolated from turnip yellow mosaic virus nucleoproteins, *Virology*, 84, 153, 1978.
171. **Matthews, R. E. F.,** Properties of nucleoprotein fractions isolated from turnip yellow mosaic virus preparations, *Virology*, 12, 521, 1960.
172. **Fraenkel-Conrat, H.,** Reconstitution of viruses, *Annu. Rev. Microbiol.*, 24, 463' 1970.
173. **Litvak, S., Carré, D. S., and Chapeville, F.,** TYMV RNA as a substrate of the tRNA nucleotidyltransferase, *FEBS Lett.*, 11, 316, 1970.
174. **Purcifull, D. E., Hiebert, E., and McDonald, J. G.,** Immunochemical specificity of cytoplasmic inclusions induced by viruses in the potato Y group, *Virology*, 55, 275, 1973.
175. **Vandewalle, M. J. and Siegel, A.,** A study of nucleotide sequence homology between strains of tobacco mosaic virus, *Virology*, 73, 413, 1976.
176. **Zaitlin, M., Beachy, R. N., and Bruening, G.,** Lack of molecular hybridization between RNAs of two strains of TMV: a reconsideration of the criteria for strain relationships, *Virology*, 82, 237, 1977.
177. **Haenni, A. L., Prochiantz, A., Bernard, O., and Chapeville, F.,** TYMV valyl-RNA as an amino-acid donor in protein biosynthesis, *Nature (London), New Biol.*, 241, 166, 1973.
178. **Niblett, C. L.,** personal communication, 1977.
179. **Beer, S. V. and Kosuge, T.,** Spermidine and spermine-polyamine components of turnip yellow mosaic virus, *Virology*, 40, 930, 1970.
180. **Nickerson, K. W. and Lane, L. C.,** Polyamine content of several RNA plant viruses, *Virology*, 81, 455, 1977.
181. **Diaz-Ruiz, J. R. and Kaper, J. M.,** Isolation of viral double-stranded RNAs using a LiCl fractionation procedure, *Prep. Biochem.*, 8, 1, 1978.
182. **Pelham, H. R. B.,** Leaky UAG termination codon in tobacco mosaic virus RNA, *Nature (London)*, 272, 469, 1978.
183. **Jonard, G., Richards, K., Mohier, E., and Gerlinger, P.,** Nucleotide sequence at the 5′ extremity of tobacco-mosaic-virus RNA. II. The coding region (nucleotides 69-236), *Eur. J. Biochem.*, 84, 521, 1978.

# THE NUCLEIC ACIDS OF MULTIPARTITE, DEFECTIVE, AND SATELLITE PLANT VIRUSES

**L. C. Lane**

## TABLE OF CONTENTS

## I. INTRODUCTION

Plant viruses with multipartite genomes can be divided into two classes: those in which a single virion contains all the genome segments and those in which genome segments are separately packaged. The latter class is unique to plant viruses. For want of a better term, the latter class will be referred to as "divided genome viruses", although the term itself does not necessarily imply a distinction between the two classes.

The multipartite genome viruses are particularly amenable to genetic analysis. By constructing appropriate hybrid strains with genome segments from different origins, one can correlate genome segments with the properties (phenotypes) that they determine.

## II. DIVIDED GENOME VIRUSES

### A. Introduction

The divided genome viruses are those which require more than one nucleic acid component for infection and in which these nucleic acids are packaged separately. Divided genome viruses were first recognized in the mid-1960s. Roughly, a third of the known plant virus groups consist of divided genome viruses. Table 1 lists and summarizes properties of the known groups of divided genome viruses.

The divided genome is a convenient strategy to allow independent translation of a number of different proteins from a single genome, since ribosomes of higher organisms do not normally initiate synthesis of more than a single protein from a given RNA chain. In the case of spherical viruses, the divided genome allows efficient packaging of large amounts of genetic information into small units. The divided genome allows high-frequency recombination between related virus strains. The divided genome also confers, at least under some circumstances, resistance to *in situ* inactivation of nucleic acid infectivity. For example, a dose of UV light which inactivates a monopartite genome will inactivate only half of a bipartite genome of equal size. Any or all of these factors may exert selective pressure favoring the evolution of these viruses. All known divided genome viruses are transmitted by mechanisms that minimize physical dilution of the virus. Perhaps it is only among plant viruses that transmission mechanisms are suitable to overcome the selective disadvantage that dilution imposes on multiple component viruses.

There is still confusion in the literature about what comprises the genome of a virus. The genome of a virus is that part which determines inheritance. Viral genomes are invariably comprised of nucleic acid, but not all of the nucleic acid is necessarily part of the genome. To be part of the genome, a nucleic acid must determine the inheritance of changes in its own nucleotide sequence. As far as we know, if an RNA determines the inheritance of part of its own sequence, then it determines the inheritance of its entire sequence, except for enzymatic modifications. We know of no cases where a genomic RNA sequence is spliced to a sequence which is genetically determined by another source.

Experimentally, the simplest way to define a viral genome is to determine which nucleic acids are required for infection. This is accomplished by separating nucleic acid components and measuring specific infectivities of mixtures of components. In theory, combinations which contain the complete genome will be highly infectious, and all others will be noninfectious. In practice, it is often difficult to obtain biologically pure components. In many cases, the existence of a divided genome is surmised from a difference of less than a factor of ten in specific infectivity between the complete genome and preparations from which a genome segment has been deleted. For example, genome segments of broad bean mottle virus[2] and several of the nepoviruses[3,4] have been difficult to purify. A philosophical objection to this aproach has been raised by the discovery of a class of nucleic acids, namely, the smallest RNAs of alfalfa mosaic virus and the ilarviruses, which are required for infection, but are subgenomic; i.e., they are derived from larger genomic RNAs, but have no effect on inheritance, even of their own sequence. To be considered a part of the genome, a nucleic acid must not only contribute to infection, but also contribute genetically to the progeny. It is particularly important to establish genetic roles for the shorter RNAs that are most likely to be subgenomic. For several viruses, such as the hordeiviruses,[5] pea enation mosaic virus,[6] and carnation ringspot virus,[7] the smaller RNAs do not yet have established genetic roles.

### B. Nomenclature of RNA Components

Divided genome viral RNAs are usually designated by number. By convention, RNA 1 is the largest component, and the others are numbered sequentially in order of decreasing size.

### C. Properties of the Nucleic Acids

Table 2 shows molecular weights of divided genome RNAs. These have been deter-

TABLE 1

**Classification and Properties of Divided Genome Plant Viruses[1]**

| Group<br>Typical viruses (No. in CMI/AAB; descriptions of plant viruses) | Morphology (vector) | Genome components and molecular weights | Nucleoprotein components (coat protein molecular weight) |
|---|---|---|---|
| Bromoviruses | | | |
| Brome mosaic virus, BMV (180)<br>Broad bean mottle virus, BBMV (101)<br>Cowpea chlorotic mottle virus, CCMV (49) | Isometric (beetles) | $1.1 \times 10^6$ plus $1.0 \times 10^6$ plus $0.8 \times 10^6$ plus $0.3 \times 10^6$ subgenomic | 85S, 22% RNA (20,000) |
| Comoviruses | | | |
| Cowpea mosaic virus, CPMV (47)<br>Bean pod mottle virus, BPMV (108)<br>Broad bean stain virus, BBSV (29)<br>Radish mosaic virus, RMV (121)<br>Red clover mottle virus, RCMV (74)<br>Squash mosaic virus, SMV (43)<br>True broad bean mosaic virus TBBMV (20) | Isometric (beetles) | $2.3 \times 10^6$ plus $1.4—1.5 \times 10^6$ | Bottom 115S, 34% RNA<br>Middle 95S, 23—28% RNA<br>Top 55S, no RNA (42,000 plus 22,000) |
| Possible comovirus | | | |
| Broad bean wilt virus, BBWV (81) | (Aphids) | | |
| Cucumoviruses | | | |
| Cucumber mosaic virus, CMV (1)<br>Peanut stunt virus, PSV (92)<br>Tomato aspermy virus, TAV (79) | Isometric (aphids) | $1.1 \times 10^6$ plus $1.0 \times 10^6$ plus $0.7 \times 10^6$ plus $0.3 \times 10^6$ subgenomic | 98S, 18% RNA (24,000) |
| Ilarviruses | | | |
| Citrus leaf rugose virus, CLRV (164)<br>Citrus Variegation virus, CVV<br>Elm mottle virus, EMV (139)<br>Prunus necrotic ringspot virus, PNRSV (5)<br>Tobacco streak virus, TSV (44)<br>Tulare apple mosaic virus, TAMV (42) | Isometric irregular (none) | $1.3 \times 10^6$ plus $1.1 \times 10^6$ plus $0.8 \times 10^6$ plus $0.4 \times 10^6$ subgenomic | Bottom, middle, top, 80 to 110S, all 14% RNA (18,000) |
| Nepoviruses | | | |
| Arabis mosaic virus, AMV (16)<br>Cherry leaf roll virus, CLRV (80)<br>Grapevine fanleaf virus, GFV (28)<br>Raspberry ringspot virus, RRV (6)<br>Strawberry latent ringspot virus, SLRV (126)<br>Tobacco ringspot virus, TRSV (17)<br>Tomato black ring virus, TBRV (38)<br>Tomato ringspot virus, TomRSV (18) | Isometric (nematodes) | $2.3 \times 10^6$ plus $1.2—2.1 \times 10^6$ | Bottom, 120—130S, 43% RNA<br>Middle, 90—120S, 27—40% RNA<br>Top 50S, no RNA (55,000) |
| Pea enation mosaic virus, PEMV (25) | Isometric (aphids) | $1.7 \times 10^6$ plus $1.4 \times 10^6$ | Bottom, 112S, 29% RNA<br>Middle (top) 99S, 29% RNA (22,000) |
| Alfalfa mosaic virus, AMV (46) | Bacilliform (aphids) | $1.1 \times 10^6$ plus $0.7 \times 10^6$ plus $0.6 \times 10^6$ plus $0.3 \times 10^6$ subgenomic | Bottom 99S<br>Middle 89S<br>Topa 73S<br>Topb 63S<br>all 16% RNA (24,500) |
| Hordeiviruses | | | |
| Barley stripe mosaic virus, BSMV (68) | Rod-shaped | $1.4 \times 10^6$ plus $1.2 \times 10^6$ plus often $1.1 \times 10^6$ | 148nm, 128nm, 112nm<br>165—200S, 4% RNA (22,000) |

TABLE 1 (continued)

**Classification and Properties of Divided Genome Plant Viruses**[1]

| Group typical viruses (No. in CMI/AAB; descriptions of plant viruses) | Morphology (vector) | Genome components and molecular weights | Nucleoprotein components (coat protein molecular weight) |
|---|---|---|---|
| Tobamoviruses | | | |
| Beet necrotic yellow vein virus, BNYVV (144) | | $2 \times 10^6$ plus other components | 300nm, 190S, 5% RNA (17,500) |
| Soil-borne wheat mosaic virus, SBWMV (77) | Rod-shaped (fungi) | | |
| Tobraviruses | | | |
| Tobacco rattle virus, TRV (12) | Rod-shaped | $2.3 \times 10^6$ plus | 180—210nm, 300S |
| Pea early browning virus, PEBV (120) | (nematodes) | $0.6—1.3 \times 10^6$ | plus 50—105nm, 155—240S, 5% RNA (22,000) |
| Carnation ring spot virus, CRSV (21) | Isometric (nematodes) | $1.5 \times 10^6$ plus $0.5 \times 10^6$ | 135S, 20% RNA (38,000) |

mined in a variety of ways, including empirical gel electrophoretic and sedimentation velocity methods. More rigorous hydrodynamic methods also have been employed, either directly on RNAs or on virions to give RNA molecular weight indirectly. The latter two methods have been compared with alfalfa mosaic virus and disagree, for unknown reasons, by as much as 15% for the larger RNAs.[8] Comparison of a variety of molecular weight determinations indicates that single-stranded viral nucleic acids tend to be smaller than indicated by empirical gel electrophoretic methods, even methods which employ denaturing agents.[18] At present, molecular weight determinations are not sufficiently accurate to allow precise comparisons of values from different laboratories.

### D. Base Compositions

Table 3 shows base compositions of divided genome viral RNAs. Base compositions of the RNAs are not sufficiently distinct to be of great taxonomic value. Different RNA components (for example, bottom and top RNAs of alfalfa mosaic virus) differ slightly in base composition.[34] Recently developed methods, such as oligonucleotide mapping and nucleic acid hybridization, are far more sensitive for comparing RNA components.

### E. 5′ Termini

Many, but not all, plant virus RNAs have 7-methyl guanosine caps (see chapters by Hall and Davies) at their 5′ termini (Table 4). In the cucumoviruses, bromoviruses, and alfalfa mosaic virus, all RNA components are capped. With tobacco rattle virus, the short RNA is capped, but apparently not the longer RNA.[55] Cowpea mosaic virus RNA is neither capped with 7-methyl guanosine nor phosphorylated.[56] It may have an unusual 5′ terminus, perhaps similar to that of poliovirus RNA which is blocked by a small protein.[57] So far, only brome mosaic virus (BMV) RNA 4 and alfalfa mosaic virus (AMV) RNA 4 have been sequenced extensively at the 5′ ends (Table 5).[58,59] The initiator codon begins with the tenth nucleotide in BMV RNA 4 and the 37th nucleotide in AMV RNA 4. The precistronic sequences of RNAs are rich in A and U and may participate in initiation by base-pairing with ribosomal RNA. A small fragment containing the 5′ terminal 22 nucleotides of BMV RNA 4, and which can form no stable base-paired structures, binds efficiently to wheat embryo ribosomes.[58] The 5′ terminus

TABLE 2

**Selected Molecular Weights of Divided Genome Viral RNAs (× $10^{-6}$)**

| Virus | RNA molecular weights | Ref. |
|---|---|---|
| Alfalfa mosaic | 1.04, 0.73, 0.62, 0.25 | 8 |
| Barley stripe mosaic | 1.5, 1.35, 1.2 | 5 |
| Bean pod mottle | 2.5, 1.5 | 9 |
| Beet necrotic yellow vein | 2.3, 1.8, 0.7, 0.6 | 10 |
| Broad bean | | |
| mottle | 1.10, 1.03, 0.90, 0.36 | 2 |
| stain | 2.64, 1.62 | 11 |
| wilt | 2.0, 1.5—1.6 | 12 |
| Brome mosaic | 1.09, 0.99, 0.75, 0.28 | 13 |
| Carnation ringspot | 1.5, 0.5 | 7 |
| Cherry leaf roll | 2.3, 2.1 | 14 |
| Citrus leaf rugose | 1.1, 1.0, 0.7, 0.3 | 15 |
| Cowpea | | |
| chlorotic mottle | 1.15, 1.10, 0.85, 0.32 | 16 |
| mosaic | 2.02, 1.37 | 17 |
| Cucumber mosaic | 1.01, 0.89, 0.65, 0.33 | 18 |
| Elm mottle | 1.3, 1.15, 0.83, 0.39, 0.30 | 19 |
| Grapevine fanleaf | 2.4, 1.4 | 20 |
| Pea early browning | 2.5, 1.3 | 21 |
| enation mosaic | 1.74, 1.44, 0.28 | 22 |
| Prunus necrotic ringspot | 1.30, 0.89, 0.67, 0.27 | 23 |
| Radish mosaic | 2.2, 1.3 | 24 |
| Raspberry ringspot | 2.4, 1.4 | 25 |
| Red clover mottle | Same as CPMV | 26 |
| Soil-borne wheat mosaic | 1.84, 0.95 | 27 |
| Strawberry latent ringspot | 2.6, 1.6 | 28 |
| Tobacco | | |
| rattle | 2.5 & 0.7, 1.0 or 1.3 | 21 |
| ringspot | 2.3, 1.4 | 29 |
| streak | 1.3, 1.06, 0.78, 0.72, 0.36 | 30 |
| Tomato | | |
| aspermy | 1.26, 1.10, 0.90, 0.43 | 31 |
| black ring | 2.5, 1.5 | 32 |
| ringspot | 2.3 (two similar components) | 33 |
| True broad bean mosaic | 2.71, 1.75 | 11 |

of AMV RNA 4 has at least two binding sites for *Escherica coli* ribosomes, one of which utilizes AUG as an initiator to incorporate methionine and the other of which can utilize UUU in the precistronic region to incorporate acetylphenylalanine.[60]

### F. 3′ Termini

Table 4 shows 3′ termini. As a rule, 3′ termini are identical in the different components of a virus. Several classes of termini are apparent so far: those which are similar to the 3′ termini of tRNA, those which are polyadenylated, and those which fit neither category. Terminal poly-A segments are common among eukaryotic messenger RNAs. They have been postulated to stabilize the message against nucleolytic degradation.[66]

### G. Amino Acid Acceptor Activity

Charging of transfer RNAs with amino acids is an intermediate step in protein syn-

TABLE 3

**Selected Base Compositions of Divided Genome Viruses**

| Virus | G | A | C | U | Ref. |
|---|---|---|---|---|---|
| Alfalfa mosaic | 23 | 26 | 21 | 30 | 34 |
| Barley stripe mosaic | 20 | 31 | 19 | 30 | 35 |
| Bean pod mottle | 20 | 32 | 16 | 31 | 9 |
| Broad bean | | | | | |
| mottle | 25 | 27 | 19 | 29 | 36 |
| stain | 23 | 27 | 19 | 32 | 37 |
| wilt | 26 | 30 | 17.5 | 26.5 | 12 |
| Brome mosaic | 28 | 27 | 21 | 24 | 38 |
| Carnation ringspot | 26 | 27 | 23 | 24 | 39 |
| Cherry leaf roll | | | | | |
| bottom | 27 | 22 | 22 | 29 | 14 |
| middle | 27 | 21 | 23 | 30 | 14 |
| Cowpea | | | | | |
| chlorotic mottle | 26 | 25 | 20 | 28 | 40 |
| mosaic | | | | | |
| bottom | 23 | 28.5 | 17 | 31.5 | 41 |
| middle | 21 | 28.5 | 19 | 31.5 | 41 |
| Cucumber mosaic | 23 | 24 | 23 | 29 | 42 |
| Pea enation mosaic | 26 | 24 | 24 | 26 | 43 |
| Peanut stunt | 24 | 26 | 21 | 29 | 44 |
| Prunus necrotic ringspot | 27 | 25 | 21 | 27 | 45 |
| Red clover mottle | 20 | 29 | 20 | 30 | 46 |
| Squash mosaic | 23 | 32 | 16 | 30 | 47 |
| Tobacco | | | | | |
| rattle | 25 | 29 | 17 | 29 | 48 |
| ringspot | 25 | 23 | 22 | 30 | 49 |
| Tomato | | | | | |
| aspermy | 24 | 26 | 21 | 29 | 50 |
| ringspot | 26 | 23 | 22 | 29 | 51 |
| True broad bean mosaic | 20 | 29 | 20 | 30 | 46 |
| Tulare apple mosaic | 24 | 24 | 21 | 31 | 45 |

TABLE 4

**Terminal Sequences of Divided Genome Viral RNAs**

| Virus | 5′ termini | Ref. |
|---|---|---|
| Alfalfa mosaic virus | $m^7G$ 5′ ppp 5′ Gp all components | 52 |
| Brome mosaic virus | $m^7G$ 5′ ppp 5′ Gp all components | 53 |
| Cucumber mosaic virus | $m^7G$ 5′ ppp 5′ Np all components | 54 |
| Tobacco rattle virus | $m^7G$ 5′ ppp 5′ Ap only the short RNA | 55 |

| Virus | 3′ termini | Ref. |
|---|---|---|
| Brome mosaic virus | GACCA-OH all components | 61 |
| Bean pod mottle virus | Poly A both components | 62 |
| Cucumber mosaic virus | A-OH all components | 63 |
| Cowpea | | |
| chlorotic mottle virus | GACCA-OH all components | 61 |
| mosaic virus | Poly A both components | 64 |
| Red clover mottle virus | Poly A large component | 26 |
| Tobacco rattle virus | C-OH both components | 65 |

TABLE 5

**Extended Termini of Divided Genome Viruses**

**3′ Terminus of BMV RNA 4[72]**

161
A
C A C G C A G A C C U C U U A C A A G A
140
G U G U C U A G G U G C C U U U G A G A
120
G U U A C U C U U U G C U C U C U U C G
100
G A A G A A C C C U U A G G G G U U C G
80
U G C A U G G G C U U G C A U A G C A A
60
G U C U U A G A A U G C G U A C C G G G
40
U G U A C A G U U G A A A A A C A C U G
20
U A A A U C U C U A A A A G A G A C C A-OH

In RNA 2, base 46 is G. In RNA 1, base 46 is G and base 45 is U.

**5′ Terminus of BMV RNA 4[58]**

| | | | | | | | | | | | |
|---|---|---|---|---|---|---|---|---|---|---|---|
| | | | | | Met | Ser | Thr | Ser | Gly | Thr | Gly |
| 7MeG$^{5'}$ | p p p | $^{5'}$G U A | U U A | A U A | A U G | U C G | A C U | U C A | G G A | A C U | G G U |
| | | | | | 10 | | | | | | 30 |
| Lys | Met | Thr | Arg | Ala | Gln | Arg | Arg | | | | |
| A A G | A U G | A C U | C G C | G C G | C A G | C G U | C G— | | | | |
| | | | 40 | | | 60 | | | | | |

**5′ Terminus of AMV RNA 4[59]**

| | | | | | | | | | | | |
|---|---|---|---|---|---|---|---|---|---|---|---|
| 7Me G$^{5'}$ | p p p | $^{5'}$G U U | U U U | A U U | U U U | A U U | U U U | C U U | U C A | A A U | A C U |
| | | | 10 | | | | | 20 | | | 30 |
| | | met | ser | ser | ser | gln | lys | lys | ala | gly | gly |
| U C C | A U C | A U G | A G U | U C U | U C A | C A A | A A G | AAA | G C U | G G U | G G G |
| | | | 40 | | | 50 | | | 60 | | |
| lys | ala | gly | | | | | | | | | |
| AAA | G C U | G G— | | | | | | | | | |
| | 70 | | | | | | | | | | |

thesis. The charging reaction must be highly specific to maintain fidelity in protein synthesis. The nucleic acid components of cucumber mosaic virus and the bromoviruses can be charged with tyrosine.[67,68] Tyrosine-aminoacyl transfer RNA synthetases from wheat and bean esterify tyrosine to the 3′ terminus of the nucleic acid molecule. Apparently, the 3′ termini of the viral RNAs are similar, at least in part, to plant tyrosine tRNAs.

Since transfer RNAs recognize a variety of molecules, including messenger RNA, aminoacyl tRNA synthetases, peptide chain elongation factors, nucleotidyl transferases, and probably the 5S RNA of the large ribosomal subunit, it is of interest to know

TABLE 6

**Replicative Forms of Divided Genome Viruses**

| Virus | Molecular weights ($\times 10^{-6}$) | Ref. |
|---|---|---|
| Alfalfa mosaic | | 76, 77 |
| Barley stripe mosaic | | 78 |
| Brome mosaic | | 79, 80 |
| Cowpea mosaic | | 81, 82 |
| Cucumber mosaic | 2.02, 1.78, 1.36, 0.66 | 18 |
| Pea enation mosaic | 3.3, 2.2, 0.72 | 83 |
| Tobacco ringspot | Heterogeneous | 84, 85 |

how many of these recognition sites are present in viral RNA. Aminoacylated brome mosaic virus RNA binds to polypeptide chain elongation factor (EF)1 of wheat, and uncharged viral RNA does not bind.[69] The binding of BMV RNA to EF 1-GTP releases GTP, in contrast to the binding of aminoacyl tRNAs which yields a ternary complex. This argues that the binding of the viral RNA to EF 1 does not play a role in translation.[69] In fact, in the wheat germ protein synthesizing system, tyrosine is not transferred from tyrosyl BMV RNA into nascent protein.[70] When the 3′ terminus of BMV RNA is chemically modified so that it can no longer accept amino acids, the RNA is still fully active as a message in vitro.[71] The sequence of the 3′ terminus contains nothing comparable to the 5′-T$\psi$CG-3′ loop, which is thought to be essential for the binding of tRNA to the ribosome (Table 5).[72] Thus, the tRNA-like region of BMV RNA does not function as a tRNA, and it is unlikely that it plays a role in protein synthesis in vivo.

If the tRNA-like structure is not a tRNA, what does it do? The similarity of 3′ termini in all BMV RNA components[72] indicates involvement of the termini in a process common to all four RNA components, such as RNA synthesis. Two chain elongation factors of *E. coli* are integral subunits of bacteriophage RNA replicases.[73, 74] It is possible that chain elongation factor 1 serves a similar role in plant virus replicases. The 3′ terminus also could serve as a primer, analagous to tryptophan tRNA, which serves as a primer to initiate synthesis of Rous sarcoma viral RNA.[72]

Brome mosaic virus RNAs are substrates for *E. coli* and yeast nucleotidyl transferases[75] which replace lost adenylate and cytidylate residues at the 3′ termini of transfer RNAs. The biological role of nucleotidyl transferases in host cells is not fully understood. A current hypothesis is that they protect tRNAs from exonucleases.

## H. RNA Synthesis

### *1. Introduction*

Enzymes replicate nucleic acid strands by using them as templates to polymerize nucleotides into strands which are complementary in terms of Watson-Crick base pairing. Enzymes then use the complementary strands as templates to produce copies of the original strand. The RNA of divided genome viruses is the translatable, and therefore by convention, the plus strand (see chapter by Davies). A complementary or minus strand should be an intermediate in viral replication. The simplest evidence for a minus strand intermediate is the presence of double-stranded RNA, often referred to as replicative form (RF) in infected tissue. It is not clear whether replicative forms exist as such in vivo or whether they arise from annealing of plus and minus strands during extraction. Replicative forms can be isolated from plants infected with a variety of divided genome viruses (Table 6). There is no documented case where replicative form cannot be isolated from cells infected by a plus-stranded RNA virus. It must be borne in mind, however, that the yield of these forms under optimal conditions is roughly 0.1% of the total nucleic acid of the infected tissue (i.e., about 0.0001% by weight of the tissue). Zaitlin's chapter discusses isolation of replicative forms.

In virtually all cases, there are replicative forms containing full-length copies of all of the genome segments. The possible exception is tobacco ringspot virus (TRSV), where smaller polydisperse replicative forms have been found.[84] However, a recent report[85] indicates larger replicative forms.

### *2. Subcellular Location*

This has been studied in detail in only a few cases, but indirect evidence implicates the cytoplasm, and in particular, cytoplasmic membranes, as the usual site of RNA replication. In the case of cowpea mosaic virus (CPMV)-infected cells, the replicative form (double-stranded RNA) is associated with a membranous fraction, which has a buoyant density in sucrose slightly less than that of chloroplast membranes.[86] In broad bean mottle virus (BBMV)-infected cells, tritiated uridine is incorporated into RNA predominantly in the cytoplasm.[87] The exception to the rule of cytoplasmic RNA replication is pea enation mosaic virus (PEMV), where the RNA replicates in the nucleus. The ability to incorporate tritiated uridine in the presence of actinomycin D, reactivity with antibody to double-stranded RNA, and material hybridizing with labeled PEMV-plus strands are all found in the nucleus.[86]

### *3. The Time Course*

Efficient and synchronous infection is required to investigate the time course of RNA replication. This is usually accomplished by infection of isolated plant protoplasts. Synchrony can also be achieved by temperature shifts of systemically infected plants.[88] In all systems studied so far, RNA synthesis can be detected after a few hours, and it precedes the appearance of intact virus (BMV,[89] cowpea chlorotic mottle virus (CCMV)[89,90], cucumber mosaic virus (CMV),[91] and tobacco rattle virus (TRV)[92]). Individual RNA components have been studied only with BMV, CCMV, and CMV. RNA 3 accumulates to a higher level than the other viral RNAs.[89-91] Pulse labeling studies with infected plants showed virion RNA 3 to have an appreciably lower specific activity than the other RNAs, apparently due to the larger pool size of this component.[40,80,93] More RNA 3 is synthesized than can be encapsidated.[89,91] In systemically infected plants synchronized by temperature shifts,[90] the genomic RNAs of CCMV are synthesized synchronously throughout most of the growth cycle. RNA 4 is synthesized at low levels early in infection, but toward the end of the growth cycle, its synthesis rate is comparable to that of RNA 3.[90] This is consistent with its presumptive role as a coat protein messenger late in the infection cycle.[94] Replicative forms (BMV,[89] CCMV,[89] and CMV[91]) can be found in infected protoplasts. Shortly after inoculation, they represent a larger fraction of total viral RNA than at late times, consistent with their hypothesized roles as intermediates in RNA synthesis.[89]

### *4. Subgenomic RNAs*

Alfalfa mosaic virus, the bromoviruses, the cucumoviruses, and the ilarviruses contain, in addition to the genomic RNAs, a small RNA, RNA 4, which has no genetic function and which is a subsequence of RNA 3, and a messenger for coat protein synthesis. This RNA has been termed "subgenomic", implying not only that it is a fragment of the genome, but also that it is subservient to the genomic RNAs. This RNA could be derived from RNA 3 either by nuclease cleavage or by partial transcription. So far, we cannot distinguish between these. Once the component is derived, it may propagate, either by repeating the initial event or by independently replicating. Continuous nuclease cleavage is ruled out because pulse-labeled RNA 4 has a higher specific activity than RNA 3, precisely the opposite of what the continuous nuclease cleavage model predicts[80,93] Continuous partial replication, utilizing RNA 3-minus strand as a template, predicts that replication of RNA 4 should interfere with replication of RNA 3, but such interference is not apparent.[90]

The independent replication model predicts the existence of a replicative form corresponding to RNA component 4. This has been found for BMV[79] and CMV,[18,91] but on the other hand, it has been reported as conspicuously absent in BMV,[80,89] AMV,[77] and CCMV.[89,90] Its atypical properties, resulting from its small size, its low level, and relative poor sensitivity of its detection, could explain the difficulty of its isolation. On the other hand, its isolation could be explained by hybridization of RNA 4-plus strand to RNA 3-minus strand followed by ribonuclease digestion. The finding of replicative forms corresponding to subgenomic RNAs for several other viruses[256] is in accord with the independent replication model.

Additional subgenomic RNAs are sometimes found in small amounts in divided genome viruses. An RNA of 0.5 million mol wt often is found in small amounts associated with brome mosaic virus. In one instance, a virus RNA preparation contained almost 20% by weight of this component.[257] This component is readily lost by cloning.[257] When infected tissue is radioactively labeled, large amounts of radioactivity appear in a similar component.[80,89] It has been suggested[80] that this small component may be a messenger RNA derived from one of the larger RNAs. A replicative form corresponding to this component had been reported.[13] A rapidly labeled small RNA has been detected in broad bean mottle virus (BBMV)-infected tissue,[95,96] but the electrophoretic mobility of this component suggests that it may, in fact, be RNA 4.[96]

Two smaller RNAs associated with alfalfa mosaic virus are found in relatively large proportions in polysomal (messenger) RNA,[97] implying that they may be messengers derived from larger RNA components.

*5. Is the Viral RNA Polymerase Coded by the Host Genome or the Virus Genome?*

Little is known about plant virus polymerases. The RNA bacteriophage RNA replicases are much better characterized and provide a useful model system for the plant enzymes. Initially, phage RNA replication was thought to require two enzymes, one to produce plus strand and one to produce minus strand. It is now clear that a single enzyme performs both functions. The enzyme is complex, consisting of four polypeptide subunits, one coded by the virus genome and the other three coded by the host genome. In healthy cells, the host specific polypeptides are constituents of the protein synthesis machinery.[73,74] To initiate RNA synthesis with a plus strand template[98] and to synthesize RNA autocatalytically,[99] the phage enzymes require additional host factors. Autocatalytic synthesis is a striking feature of phage replicases. The product of the reaction serves as a template for the reaction and the rate of replication accelerates until the substrate supply is exhausted. The enzymes can multiply the amount of input RNA by many orders of magnitude.

So far, plant-virus-specific RNA polymerases have not been isolated in sufficient purity either to synthesize RNA autocatalytically or to identify the polypeptide subunits. Virtually all we know is that virus infection stimulates extractable polymerase activity early in infection, consistent with its postulated role in virus replication (AMV,[100,101] BBMV,[102] BMV,[103,104] CMV,[105,106] CPMV,[107] and TRSV[84,106]). Most extractable polymerase activities begin to fall, some of them quite sharply,[84,106] as virus accumulates. The cucumber mosaic virus polymerase reaches a maximum and remains roughly constant for at least a week.[105,106] Healthy plants of both Chinese cabbage[108] and tobacco[109] contain RNA-dependent RNA polymerases which may be related, respectively, to turnip yellow mosaic virus (TYMV)-induced RNA polymerase and tobacco mosaic virus and alfalfa mosaic virus RNA polymerases.

*6. Purification of Template-Dependent, RNA-Dependent RNA Polymerases*

Several RNA-dependent polymerases have been purified to the point where they are appreciably stimulated by adding template RNA (AMV,[101] BMV,[103] CMV,[110] and CPMV[111]). Unfortunately, these enzyme preparations synthesize little product, usually equivalent to a small amount of the input template. The plant virus enzymes which so far have been isolated are, therefore, unsuitable for studying many important features, such as enzyme-RNA binding sites and RNA nucleotide sequences. Most plant virus RNA polymerase activities are bound to material sedimentable at 10,000 g. The exceptions are the tobacco ringspot virus polymerase, which is soluble,[84,106] cucumber mosaic virus polymerase,[112] where roughly two thirds of the enzyme activity is soluble, and alfalfa mosaic virus,[101] where equal amounts are bound and soluble.

The soluble enzymes generally show little template specificity. The cucumber mosaic virus polymerase uses several RNAs, including *E. coli* RNA, almost as efficiently as it uses CMV RNA.[105] Poly C is the most efficient template[105] for the CMV polymerase. Phage replicases also utilize poly C.[74] The poly G polymerase activity can be partially separated from CMV polymerase during purification.[110] The most specific plant virus polymerase so far isolated is that of brome mosaic virus.[103] Brome mosaic viral RNA is the most effective template, but other RNAs also serve as templates. Several RNAs, including tobacco mosaic virus RNA, however, do not detectably stimulate the enzyme. The specificity of even the most specific plant virus polymerase is much less than that of bacteriophage RNA replicase. A curious feature of the BMV polymerase is its high (32 m*M*) and broad magnesium ion optimum. The activity is relatively unaffected by magnesium ion concentration between 18 and 40 m*M*. The high optimum arises in part from the high (roughly 3 m*M* each) nucleoside triphosphate concentration, since the RNA precursors chelate magnesium ions.

The cucumber mosaic RNA polymerase has been purified roughly 100-fold.[110] The partially purified preparation gives many protein bands on gel electrophoresis. Since several thousandfold purification will almost certainly be necessary to give homogeneity,[110] the major polypeptides of the partially purified enzyme are unlikely to be polymerase components. Several polymerases sediment on sucrose gradients as proteins of 100,000 to 150,000 mol wt (BMV,[113] CMV,[106] and TRSV[106]). The sedimentation pattern of enzyme activity, however, may represent overlapping of several different activities. This could be tested by combining fractions from different parts of the gradient to see if combined fractions are more active than the sum of the pure fractions. The sensitivity of the CMV RNA polymerase activity to dilution, indeed, suggests that it consists of more than one component.[110]

*7. Products of RNA Polymerases*

Both template-bound and template-dependent RNA polymerases incorporate radioactively labeled nucleotides, largely into double-stranded (ribonuclease-resistant) RNA. With the membrane-bound enzymes, reannealing melted radioactive product in the presence of excess unlabeled plus strand appreciably decreases the amount of label in double-stranded RNA, indicating that a large part of the incorporation is into plus-stranded viral RNAs. The membrane-bound AMV polymerase is exceptional and appears to make appreciable amounts of minus strand.[101] The membrane-bound BMV polymerase[114-116] incorporates radioactivity into replicative intermediate (partially double-stranded), replicative form and full-length single-stranded RNA. Pulse labels can be chased from double-stranded RNA and from replicative intermediate into single-stranded RNA. The membrane bound BBMV[117] and CPMV[111] polymerases incorporate labeled precursors into replicative forms and into genome-sized single-stranded RNA. With the BBMV polymerase,[117] pulse label can be chased from double-stranded RNA into single-stranded RNA.

The soluble AMV,[101] BMV,[103] and CMV[112] polymerases, using viral RNA as a template, incorporate label predominantly into double-stranded RNA of about the size of the replicative form of the smallest (subgenomic) RNA. The smallest BMV RNA is apparently a more efficient template for the BMV polymerase than the larger BMV RNAs.[118]

In summary, RNA dependent RNA polymerases increase in amount soon after plant virus infection and that the intermediates of RNA replication are similar to those in bacteriophage RNA replication. We still need to know how many enzymes and factors are involved in viral RNA snthesis and which polypeptide subunits are contributed by the host and which by the virus. We need to know how synthesis of RNA components is coordinated. We need to know if the RNA polymerases can be specifically inhibited. We need to know if the host contribution to viral RNA synthesis limits the host range of the virus. Since polymerase activity can be detected in crude juice, it should be possible to purify all the RNA synthesis factors. The purification will be tedious because the absolute amount of polymerase is very small. The concentration of plant virus polymerase is likely to be much lower than that of bacteriophage RNA replicase, because only a small fraction of the total volume of a plant cell is cytoplasm, and because infection by plant viruses is asynchronous. Proteases and ribonucleases may pose serious obstacles to isolation. If multiple components are involved in plant virus RNA synthesis, the labor of purification will increase roughly with the square of the number of components. In all likelihood, methods for stabilizing some of the factors will have to be investigated.

### I. RNA Homology Between Different Viruses

One might imagine that a given segment of a divided genome could be replaced by any of a large number of different RNA components that coded for the missing functions. It is becoming clear, however, that genome segments are complex and highly specialized, so that only similar genome segments from closely related viruses can be interchanged. In no case will a genome segment from a virus in one of the recognized plant virus groups substitute for a segment of a virus in another group. In this regard, the compatibility of genome segments can be compared with the compatibility of chromosomes in higher organisms. In fact, segments are usually not exchangeable among distantly related viruses within a group. For example, several strains of tobacco rattle virus are incompatible[119] and different viruses within the comovirus group are incompatible,[120,121] including two viruses which, on the basis of host range, are considered to be cowpea mosaic virus.[122] To a first approximation, close serological relatives are compatible, while distant serological relatives are incompatible. There are, however several cases of limited compatibility between distant serological relatives; for example, brome mosaic and cowpea chlorotic mottle viruses,[123] cucumber mosiac and tomato aspermy viruses,[124] the yellow and Campinas strains of tobacco rattle virus,[125] and the beet ringspot and potato bouquet strains of tomato black ring virus.[126] In general, not all the RNA segments are interchangeable among distantly related strains. The only segments which interchange are apparently those which carry coat-protein genes. The interchange often is unilateral; for example, RNA 3 of CCMV can substitute for RNA 3 of BMV, but not vice versa.[123] Genetic hybrids between distantly related strains grow poorly, but the cucumber mosaic tomato aspermy hybrid is exceptional and multiplies as readily as its parents.[124]

RNA homology can be measured by RNA-RNA hybridization. Radioactively labeled viral RNA will anneal with melted replicative form isolated from infected cells. The relationship of another viral RNA to the radioactively labeled viral RNA is measured by its ability to prevent the labeled RNA from reannealing with the melted replicative form.

RNA-RNA hybridization is less sensitive than other techniques for detecting distant relationships. Several distant serological relatives, which can form genetic hybrids, are totally unrelated by competitive RNA-RNA hybridization; for example, the Campinas strain and Oregon yellow strains of tobacco rattle virus[127] and cucumber mosaic and tomato aspermy viruses.[124] The absence of competitive hybridization implies considerable differences in nucleotide sequence between two RNAs. There is still insufficient nucleotide sequence data to interpret weak homologies or the absence of homology on a molecular basis.

Knowledge of nucleotide sequences of RNAs from a number of distinct, but related, viruses should help establish evolutionary relationships among viruses; however, RNA nucleotide sequencing is still tedious and such knowledge seems many years away. The only interesting sequence comparison so far has been the 3′ terminal regions of brome mosaic and cowpea chlorotic mottle viruses. The sequences are highly conserved among the four components within each virus, but differ considerably between the two viruses.[72]

### J. Do the RNA Components of a Divided Genome Virus Have Independent Sequences?

Early in vitro protein-synthesis experiments showed that the smallest (subgenomic) RNA components of brome mosaic[128] and alfalfa mosaic virus[129] RNAs contained sequences which code for coat protein. When it became apparent that RNA 3 contained the coat-protein gene, it was clear that RNA 4 must be a subsequence of RNA 3. Oligonucleotide mapping confirms this for AMV,[130] BMV,[131] and CMV.[132] Nucleic acid hybridization also confirms it for AMV[133] and CMV.[134]

The relationship among RNA components of barley stripe mosaic virus (BSMV) is not so clear. RNA-RNA hybridization suggests that the two largest RNAs have independent sequences, but that the smaller RNAs are subsequences of RNA 2.[135] The biological role of the smaller RNAs, however, remains unclear. Some strains of BSMV lack these components, but in at least one strain it appears to be required for infection.[5]

RNA-RNA hybridization indicates homology of about 600 bases between the long and short RNAs of the Campinas strain of tobacco rattle virus.[127] The homology is apparently imperfect, since the heterologous double-stranded RNA melts about 6°C lower than either homologous double-stranded RNA. Small degrees of homology can be demonstrated with this virus because the components differ greatly in size, simplifying separation and minimizing cross-contamination, which gives spurious homology.

The 3′ termini of the brome mosaic virus RNAs are homologous for at least 161 nucleotides (Table 5).[72] The sequences are identical in RNAs 3 and 4, and the RNA 1 and 2 sequences differ by 1 and 2 bases, respectively. A different sort of homology has been detected between the two largest RNAs of alfalfa mosaic virus. In the pancreatic ribonuclease digest, 2 of 17 total nonamers, 9 of 31 total octamers, and 5 of 31 total heptamers have identical sequences in the two RNAs.[130] Neither of these two RNAs, however, contains large oligonucleotides with sequences identical to those in the two shorter RNAs. With the exceptions already mentioned, the sequences of genomic RNAs of divided genome viruses are different by oligonucleotide mapping (AMV[130] and CMV[132]), and by nucleic-acid hybridization (AMV,[133] BSMV,[135] CMV,[134] CPMV,[136] TRV,[127] and TRSV[29]).

Part of the sequence homology detected by nucleic acid hybridization in the TRV

RNAs could be analogous to that detected by nucleotide sequencing in the BMV RNAs, but hybridization studies with fragments derived from the 3′ termini of TRV RNA[258] indicate that this is not the case. Coat-protein recognition sites for TRV assembly could account for additional homology.[127] The TRV sequences that are conserved between long and short RNAs are not conserved among TRV strains, since Campinas has no detectable RNA sequence homology to either PRN or Oregon yellow. If homology is involved in protein recognition, then TRV RNA should prefer its own protein to those of other strains in in vitro reconstitution.

## K. Genetics

### *1. Introduction*

Nucleic acids of RNA plant viruses serve two functions. They determine inheritance and they also serve as messengers. The latter function is discussed in the chapter by Davies. The divided genome viruses are amenable to simple genetic analysis and their genetic properties will therefore be discussed here.

Complex mechanisms, which are obscure to the layman, often determine inheritance in higher organisms. In contrast, simple mechanisms, understandable with minimal genetic training, determine inheritance in divided genome viruses.

The goal of plant virus genetics is to determine what each part of the genome contributes to virus multiplication and to host-virus interaction. With current technology, we can ask only the yet simpler question: what does each RNA component contribute to the infection? We are not yet equipped to subdivide an individual RNA component and map the genetic contribution onto particular regions of a component.

The fundamental tool of the geneticist is the genetic marker. A genetic marker is not a lump on the nucleic acid, but rather, an easily detectable characteristic of a virus which differs in a discrete way between two strains of a virus. For example, a mutation in the coat-protein gene which changes an amino acid in the coat protein of the virus can provide a convenient genetic marker. If, for example, the change affects the surface charge of the virus particle, we can detect the change by following the electrophoretic mobility of the virus. The difference between the normal (often called "wild type") and mutant strain is then a useful genetic marker. When gene products are detectably different, but coded for by different forms of the same gene, as in the case of the wild type and mutant coat proteins, the different forms of the gene are called "alleles". The mutant coat protein is therefore coded by a mutant allele and the wild-type coat protein by the wild-type allele.

One can follow the genetic marker in progeny virus after reshuffling the genomes between the two strains. The ability to reshuffle (reassort) genomes depends on the fact that a divided genome virus requires all its genome segments to infect. That is, the effect of an RNA component is all or nothing. If the component is present, the virus multiplies in the normal way. If it is absent, the virus does not multiply detectably. In this respect, the RNA components are similar to chromosomes of higher organisms. The absence of a single chromosome is, under most circumstances, lethal.

The all-or-nothing role of RNA components in infection can be exploited to obtain pure virus strains by cloning. A virus clone is a virus population derived from a single parent and is obtained by inoculating plants with low levels of virus, such that most infections arise from the minimum infectious unit. Infection by an aggregate of virus particles can produce progeny that are still a mixture of strains. Multiple cloning is, therefore, advisable to obtain a pure clone. Since plant viruses are, of course, haploid, clones will carry only a single allele of a given gene.

The behavior of an RNA component in a divided genome contrasts with the behavior of the RNA of a satellite virus. In the divided genome virus, the ratio of components in the progeny will be the same as in the parent and independent of the amount of each component actually used to infect the plant, whereas the level of a satellite virus in the progeny will depend on the amount of satellite in the inoculum, provided that this level is below saturation.

Infection requires intact RNA segments. A single break in the RNA chain is lethal. The cell, as far as we know has no method to rejoin fragments. Individual fragments cannot express themselves and depend on the integrity of the chain in the same way as infection depends on the presence of all genome segments.

### 2. *Analysis by Reassortment*

Classical genetic studies exploit recombination of DNA molecules. During mating of bacteria and gametogenesis in higher organisms, homologous regions of DNA recombine with a finite frequency. Genetic markers that are far apart in the DNA recombine with appreciable frequency, while those which are close together recombine with lower frequency. Genetic markers are mapped by measuring their relative rates of recombination with one another. Maps constructed in this way correlate with maps constructed from nucleotide sequences.

The genomes of RNA viruses are very small compared even with bacterial genomes; therefore, RNA molecules break and reunite with a very low frequency. Physical breakage and rejoining of viral RNAs have been detected only with poliovirus, where the extreme ends of the RNA (2.6 million daltons) recombine with a frequency of about 2%.[137] However, when a plant is infected by two strains of a divided genome virus, the individual RNA components reassort (provided the strains are compatible), and the progeny will contain a random mixture of virus particles. One can then isolate genetically hybrid strains by cloning the progeny. For a two-component virus, assuming equal infection by the two parental strains, roughly 25% of the progeny clones will be identical to each of the parental clones and 50% of the clones will be genetic hybrids containing an RNA from each of the two parental strains (Figure 1).

These genetic hybrids arise by recombination of RNA components. The recombination is, in fact, reassortment rather than breakage and reunion. In the animal virus literature, such strains are referred to as "recombinants", which does not distinguish them from strains which have recombined by classical breakage and reunion of nucleic acid molecules. In the plant virus literature, such strains have been called "pseudorecombinants" to distinguish them from classical recombinants. This is a cumbersome word and, in fact, the recombinants are not "pseudo". "Reassortants" describes these strains more simply and precisely.

In plant virus literature, the combination of purified RNA components to give infectious virus (enhancement of infectivity by mixing) or the mixing of defective strains to give normal strains (reassortment) has often been described as "complementation". Complementation in either of these senses differs from usage of "complementation" in bacterial genetics, where it refers to interaction between gene products, rather than genes, of two defective strains and does not generate wild-type progeny.

By characterizing reassortants, one can determine which RNA component carries a genetic marker. For example, in a two-component virus, if a reassortant has the small RNA of strain A, the large RNA of strain B, and the serological properties of strain A, one can immediately state that the small RNA contains the gene which determines serological properties (Figure 1). In most cases, this will be the coat protein gene.

There is a problem in this type of analysis. How does one know the source of the RNA components in the reassortant? There are, in fact, two ways to know this. The most straight forward and definitive method is to use parental strains whose RNA components, though homologous, are detectably different. This approach has been employed in an elegant way with influenza virus.[138] Influenza virus contains eight RNA components and there exists at least one pair of strains which, when mixed, contain

PARENTS

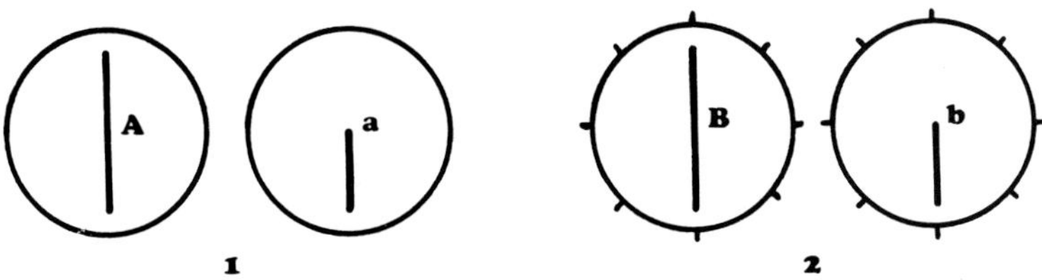


PROGENY

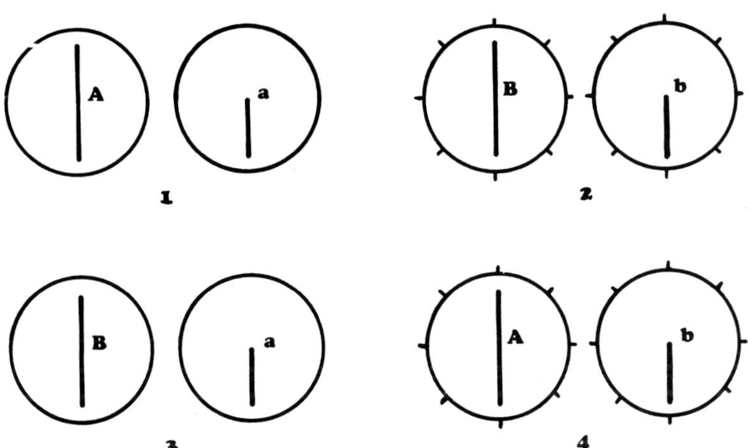


FIGURE 1. Construction of reassortants from a two-component virus (comovirus). A-large RNA of strain 1, a-small RNA of strain 1, B-large RNA of strain 2, b-small RNA of strain 2, smooth capsid-strain 1 serotype, "spiked" capsid-strain 2 serotype; 1,2-parental genotypes; 3,4-reassortant genotypes. In this example, the short RNA determines the serotype.

16 gel electrophoretically distinct RNA components. By gel electrophoresis of RNA of a reassortant, one can immediately determine which RNA component came from which parent. Furthermore, the virus-specific proteins produced by these two strains can also be distinguished by differences in gel electrophoretic mobility. By examining RNA components and virus-specific proteins from reassortant clones, one can deduce which RNA component codes for which protein. Unfortunately the physical properties of genome segments are usually quite uniform among different strains of a plant virus. However, useful size differences have been found in genome segments of the tobraviruses (Table 7), the cucumoviruses,[178,182] and in soilborne wheat mosaic virus.[201]

*3. Construction of Reassortants In Vitro*

Where RNA components are indistinguishable between strains, reassortants with RNA components of defined origin are created only by inoculating mixtures of purified RNA components of the two strains. Purification of RNA components requires considerable thought and effort as reassortants constructed from impure components are contaminated with parental genotypes. If components can be purified to the point where they are noninfectious, but become highly infectious when mixed, then the

TABLE 7

**Tobacco Rattle Virus Strains and Mutants**

| Strain | Serotype[151] | Particle lengths Short (nm) | Long (nm) | Ref. |
|---|---|---|---|---|
| Mild (Oregon)[a] | II | 81 | 195 | 146 |
| Severe (Oregon)[a] | II | 90 | 195 | 146 |
| Yellow (Oregon)[ab] | II | 100 | 195 | 146, 151 |
| German[b] | | 70 | 180 | 119 |
| Campinas (CAM, Brazil) | III | 52 | 195 | 151 |
| Campinas N8 | III | 52 | 195 | 149 |
| Campinas N10 | III | 52 | 195 | 149 |
| Campinas Yellow Symptoms | III | 52 | 195 | 147 |
| Bel (England) | I | 90 | 185 | 151 |
| PRN (Scotland)[c] | I | 66 | 183 | 151 |
| HSN (Japan) | I | — | — | 151a |
| A (Japan) | I | — | — | 151a |
| Pea early browning | | 105 | 210 | 119 |
| Wag (Netherlands)c | (I) | 58, 74 | 176 | 152 |
| SAL (USA) | (II) | 50 | 172 | 151 |

[a] Form stable hybrids.[146]
[b] Form stable hybrids.[119]
[c] Form stable hybrids.[152]

of the parental RNA components is not in doubt. However, in most cases, it is difficult or impossible to achieve such purity, and rigorous precautions are necessary to make sure that the progeny are, in fact, reassortants and not parental contaminants.

*4. Purification of RNA Components*

Since the majority of plant virus reassortants have so far been constructed by in vitro recombination of purified components, discussion of component purification is in order. The striking feature of purification is the difficulty usually encountered in obtaining noninfectious components. This is especially true of larger components. Studies of component mixtures[139] show that their infectivity is relatively independent of component ratio until components are roughly 90% pure. This means that biological purity is not related to physical purity in a simple way and that ordinary physical methods may not detect contaminants which are biologically significant.

Aggregation of smaller components is an obstacle to purifying large components. Another generally overlooked, potential problem is the sensitivity of large components to inactivation. For example, deamination sufficient to kill 90% of molecules weighing 1 million daltons will kill 99% of molecules weighing 2 million daltons.

Additional problems plague purification of components at the nucleoprotein level. A nucleoprotein particle that is pure by the criterion of sedimentation velocity may actually consist of more than one type of virus particle. For example, several nepoviruses have particles containing two small RNAs and having physical properties virtually identical to particles with a single large RNA. If particles containing the two short RNAs are present at a low level, they can easily go undetected by physical techniques. If particles contain more than one RNA, UV light, and perhaps other agents, can cross-link the two RNAs,[140] producing an RNA aggregate with physical properties virtually identical to the large RNA.

To obtain genetically purified components, one must consider all possible sources of biological contamination. One should begin with virus of high specific infectivity. One should use separation techniques with the highest possible resolving power and, where multiple cycles of separation are necessary, one should employ techniques that separate on the basis of different physical properties. For example, sucrose density gradient centrifugation of nucleoproteins followed by gel electrophoresis of isolated RNA would be an appropriate strategy for many viruses. In this example, the second separation procedure would eliminate contamination by minor particles with sedimentation coefficients identical to the particle of interest, but with atypical RNA compositions.

*5. The Supplementation Test*

Another strategy to locate genes on RNA components is particularly convenient with mutants that produce small local lesions.[141] The mutant in question is supplemented (mixed) with a purified wild-type component. Only where the supplementing component supplies the wild-type allele of the mutant gene (or, in the exceptional case, where the supplementing component carries a suppressor) can the mixture give rise to wild-type lesions. The supplementing components must be purified adequately, and all components of the virus must be tested to demonstrate that the supplementation is specific to one component.

*6. Epistasis*

In some cases, properties of a virus infection, most notably the symptoms, are determined not by a single gene, but by the interaction of two or more genes. Often, the effect of the combination of two genes is not what one might predict from the effects of the individual genes. The effect of a gene or group of genes on the expression of another is called ''epistasis''. Epistatic interactions between RNA components of a divided genome virus, however, do not necessarily result from a simple interaction between two genes. In the case of distantly related viruses, for example, epistasis can reflect incompatibility of RNA components. An example is the genetic hybrid between brome mosaic virus and cowpea chlorotic mottle virus,[123] which grows neither on barley (a systemic host of BMV) nor on cowpea (a systemic host of CCMV). One possible interpretation is that both BMV and CCMV require at least two genes to grow on their respective systemic hosts. An alternate explanation is that the RNAs of the two viruses have limited compatibility and that the hybrid does not grow well enough to infect the normal systemic hosts of its parents.

*7. Where Does One Find Genetic Markers?*

The most convenient source of genetic markers is naturally occurring virus strains. Unfortunately, the pathogenicity of plant viruses often restricts their free exchange among scientists. Uncritical use of the term ''strain'' can give a falsely optimistic estimate of the variability of a virus. Strain designations often are based on the host from which a virus is isolated. However, a strain isolated from one host may or may not differ from a strain of the same virus isolated from another host. In the future, it will be advisable to curb the relatively free use of the term ''strain''.

The other source of genetic markers is mutation. The literature describes a variety of mutagenic procedures. Perhaps the most common is nitrous acid deamination, which converts cytidine into uracil, guanine into xanthine, and adenine into hypoxanthine. A typical mutagenesis[142] involves treating two parts of purified virus or nucleic acid with one part of 4 *M* sodium nitrite and one part of 1 *M* acetate buffer, pH 4. The mixture is incubated at room temperature, and the reaction is stopped by diluting 100-fold into 1/15 *M* phosphate buffer. The product of the reaction is then cloned and individual clones are examined for altered properties. Symptoms, host range, antigens, electrophoretic mobility, and in vivo temperature sensitivity are among properties often examined. The value of mutagenesis is largely determined by the experimenter's ability to discover simple methods for selecting interesting mutants. Selection techniques are poorly developed in plant virology, but bacteriologists have devised ingenious selection procedures. For example, mutants defective in a particular enzyme can be selected by a number of strategies. One strategy is to find a substrate which this enzyme converts into a toxic product and which, under normal circumstances, kills the bacterium. The relatively few bacterial colonies which grow in the presence of this substrate will be enriched in mutants of the enzyme in question.

*8. What Genes Do Plant Viruses Have?*

A gene is a nucleic acid region which codes for and determines the inheritance of a physiologically active product, such as a protein, ribosomal RNA, or transfer RNA. Genetic material contains other functional regions, including binding sites for repressors (operators) and nucleic acid synthesizing enzymes (promoters), as well as regions which are transcribed and serve as binding sites for ribosomes and other regulators of protein synthesis. Genes normally are named for their products. The only well-documented gene products of plant viruses are the coat proteins, which are coded for by the coat-protein genes.

The symptoms of virus infection are not a gene product, but rather, the result of the action of one or more gene products. It is, therefore, incorrect to refer to a gene as a "symptom gene", but rather, the gene should be designated by its protein (or perhaps RNA) product. Though "symptom gene" is not proper terminology, genes which affect symptoms are of great interest to geneticists who would like to identify the gene products and determine how they interact with one another and with the host to produce symptoms.

In vitro protein synthesis (chapter by Davies) shows that plant viruses code for several polypeptides. We can speculate that one or more polypeptides are involved in RNA replication. It is conceivable that virus-coded polypeptides regulate host metabolism in the infected cell and assist the transport of infectious units from cell to cell and into and out of the vascular system. Unravelling the roles of virus-specific polypeptides is a major task facing virologists in the immediate future.

## L. Genetics of Individual Virus Groups

*1. Tobraviruses*

The tobraviruses are unique among the divided genome viruses in that the large RNA multiplies on its own to give a so-called unstable infection. In this state, the large RNA has no coat protein. Multiplication of the complete virus, or stable form, requires both RNA components. All other divided genome viruses require the complete complement of genomic RNAs for detectable multiplication. Earlier reviews[137, 143-145] describe the evolution of the divided genome concept with this virus. The genetics of the tobraviruses is summarized in Table 7.

In general, long and short particles of distantly related strains are incompatible.[119] However, two serologically distinct strains, GER and USA, are genetically compatible.[119] The long and short particles of these strains have characteristic lengths (Table 7). Lengths of long and short particles are inherited independently in reassortment tests, proving that the two RNA segments are genetically independent. Furthermore, the short RNA determines the serological properties of the hybrid. Hybrids constructed from three American isolates (Table 7) demonstrated that, with this set of strains, the short RNA determines the systemic symptoms on *Nicotiana clevelandii.*[146]

The independent replication of the long RNA shows that it codes for at least one replication factor. An RNA replicase is the most obvious candidate for this factor. A protein which facilitates cell-to-cell transport of the RNA is an alternative candidate for the factor. More recently, a naturally occurring yellow mutant of the Campinas (CAM) strain has been isolated.[147] As in other strains, the short RNA determines systemic symptoms of TRV reassortants. Thus, the short particle determines two properties, coat-protein serotype and systemic symptoms. Genetic studies cannot determine whether both these properties are inherited through the same gene. However, the tryptic peptide maps of CAM and CAM-yellow coat proteins are identical.[147]

Although mapping of tryptic peptides does not compare amino acid sequences rigorously, the identical maps suggest that the yellow symptoms may arise from a mutation outside the coat-protein gene. In vitro protein synthesis directed by CAM short RNA gives no evidence of a second gene product; however, PRN short RNA does code for two proteins.[148] Many plant viruses produce variants which elicit yellow symptoms, but we know neither why yellow symptoms are common nor how they are produced. The CAM-yellow mutation is a useful model system for studying the molecular basis of virus-induced yellowing.

CAM-long RNA and ORE-yellow-short RNA are comptabile, but the reverse combination does not produce stable virus.[125] Again, the short particle determines the coat protein and systemic symptoms. ORE yellow produces systemic vein yellowing and yellow patterns in Petunia, and CAM produces only local lesions. The hybrid produces only local lesions, indicating that the long particle influences petunia local-lesion phenotype.

Two mutants which are temperature-sensitive on *Chenopodium amaranticolor,* N8 and N10, map on the long RNA.[149] In *Nicotiana clevelandii,* only N8 is temperature sensitive. This could be explained if the gene product which is mutated in N10 interacts with host components and if this interaction influences thermal stability. Both mutants develop resistance to UV inactivation between 1½ to 3 hr after inoculation at the permissive (20°C) temperature.[150] The mutants fail to develop resistance to UV inactivation at the permissive temperature when they are incubated at the nonpermissive (30°C) temperature prior to UV treatment. This implies that the N8 and N10 mutations affect an early event in replication, possibly RNA synthesis.[150]

The variation in rod length, particularly in the short rods of tobacco rattle virus, suggests that some strains contain nonessential RNA.

### 2. *Comoviruses*

The comoviruses are icosahedral, consisting of a bottom component with a 2.0 million-dalton RNA, a middle component with a 1.4 million-dalton component, and often a top component devoid of RNA. There have been no reports of comovirus virions containing multiple RNAs. For many years, CPMV bottom component was thought to be intrinsically infectious; however, bottom component purified by buoyant-density centrifugation is virtually noninfectious.[121] A variety of phenotypes are inherited through the middle component (Table 8), showing that middle component is, indeed, genetically independent of bottom component.

Some of the most interesting known plant virus mutants are comovirus mutants. For example, Wood[153] (Table 8) has found mutations affecting local-lesion type and inherited through bottom component. One of the two mutants (7b) has a middle component which interacts with the bottom component of the other (3d) and prevents it from expressing the atypical lesions. This phenomenon is known as "suppression", and this is the only documented example of genetic suppression in plant viruses. The suppression mechanism is, in all likelihood, unrelated to the well-known nonsense and missense suppression mechanisms in *E. coli.* In backcross experiments, the 3d bottom component can be recovered from the 3d, 7b hybrid and still expresses the abnormal

TABLE 8

**Comovirus Strains and Mutants**

| Strain | Characteristics | Inheritance |
|---|---|---|
| Cowpea Mosaic Virus | | |
| Surinam (SB) | A yellow strain of CPMV | — |
| Sb-24[161] | Produces more top component than Sb | Middle |
| Sb-N3[162] | Produces more top component than Sb and produces local lesions instead of systemic symptoms at 22°C | Middle |
| Sb-N168[160] | Does not multiply at 30°C | Middle |
| Sb-N123[163] | Atypical local lesions; low titer | Middle |
| Sb-N163[163] | Atypical local lesions; normal titer | Middle |
| Sb-N140[163] | Atypical local lesions; low titer | Bottom |
| Sb-Bil[164] | Higher specific infectivity; enhanced S→F conversion | Bottom |
| Nigeria (Nig) | A yellow strain of CPMV | — |
| Nig 63[165] | Produces a virion with atypical electrophoretic mobility containing a precursor of the normal small polypeptide | Middle |
| CPMV-Vu | A naturally occurring CPMV strain that contains more bottom than middle component | — |
| Vu-7b[153] | More middle than bottom; atypical local lesions | Bottom |
| | Factor that suppresses atypical lesions produced by Vu-3d | Middle |
| Vu-3d[153] | Atypical local lesions | Bottom |
| Bean Pod Mottle Virus | | |
| BPMV-J10[166] | Serologically different from BPMV | Middle |
| Radish Mosaic Virus | | |
| KV (kale virus)[159] | Has specific antigens; forms aggregates | Both components required to show these properties |
| HZ (turnip virus)[159] | | |

lesion type.[153] This is strong evidence that the unusual properties of the 3d, 7b hybrid do not result from contamination.

Comovirus capsids are unusual in that they contain equimolar amounts of two polypeptides of 42,000 and 22,000 mol wt.[154] Cowpea mosaic virions have two electrophoretic components. The slow electrophoretic form can be converted to the fast form in vivo or in vitro with proteases.[155-158] The conversion involves cleavage of the small polypeptide from a 25,000-dalton protein to give a 22,000-dalton protein.[155] Slow virions contain the 25,000-dalton protein and fast virions contain the 22,000-dalton protein. Surprisingly, no virions have intermediate electrophoretic mobility, suggesting involvement of a cooperative event, perhaps a change in capsid configuration, in the cleavage mechanism.

Genetics can be expected to tell which RNA component codes for which capsid polypeptide and, perhaps, show whether virus specific proteins participate in the in vivo cleavage. Table 8 describes several genetic markers which affect the capsid. Some map on bottom and some on middle component. Reassortment tests with two radish mosaic strains, Kale virus (KV) and turnip virus (HZ), indicate that middle and bottom components each code for one of the two coat polypeptides.[159] Only virus clones containing both middle and bottom components from HZ exhibit HZ specific antigenic determinants and tend to aggregate as does the HZ strain. However, the relative impurity of radish mosaic virus strain HZ bottom component in the reassortment experiments sheds some doubt on the role of bottom component in the inheritance. So far, the

TABLE 9

**Nepovirus Genetics**

| Virus and RNA component | Property | Ref. |
|---|---|---|
| Raspberry ringspot virus | | |
| RNA — 1 | Seed transmission | 168, 169 |
| | Competitiveness in *Chenopodium quinoa* | 169 |
| | Symptom severity on *C. quinoa* | 170 |
| | Infection of Lloyd George raspberry | 170 |
| | Systemic invasion of *Phaseolus vulgaris* | 170 |
| RNA — 2 | Serological properties | 170, 171 |
| | Severe yellowing of Petunia | 171 |
| | Transmissibility by *Longidorus elongatus* | 170 |
| Tomato black-ring virus | | |
| RNA — 1 | Seed transmission | 168 |
| RNA — 2 | Serological properties | 167 |
| | Transmissibility by *L. elongatus* | 167 |
| | Local lesion type | 167 |

relationship between genetic markers and polypeptides is unclear. They could affect the structural gene for either polypeptide or they could be involved in cleavage. Isolation and characterization of mutations in the structural genes of the capsid polypeptides should clarify the genetics of the capsid.

De Jager[160] has found mutants with similar phenotypes, atypical local lesions and greatly reduced titers within the plant, but inherited independently, one through the bottom component and the other through the middle component. These mutant will be useful for reassortment tests with new mutants since, between them, they carry easily recognized genetic markers on both RNA components.

### *3. Nepoviruses*

The nepoviruses are icosahedral and contain bottom, middle, and often top components. The bottom component contains an RNA of 2.4 million daltons and, often, a class of particles with two shorter RNA molecules. Middle component contains a shorter RNA weighing from 1.4 to 2.2 million daltons. These viruses are all stable and structurally similar to the comoviruses, except that their capsids contain a single polypeptide of 55,000 mol wt. The nepoviruses require both RNAs for infection; however, the individual RNA components often are difficult to obtain biologically pure.

Genetic studies of raspberry ringspot virus and tomato black-ring virus are summarized in Table 9. The two strains of tomato black ring virus which have been studied have limited compatibility. Middle component of G12 forms a genetic hybrid with bottom component of A which multiplies slowly, but the reciprocal hybrid does not multiply detectably. Properties such as speed of lesion appearance[167] and speed of systemic movement,[126] which are influenced by both RNA components, may simply reflect the incompatibility of the RNA components.

If the nepoviruses are related to the comoviruses, then the comovirus capsid proteins probably result from cleavage of a precursor. Since RNA component 2 determines the coat protein of the nepoviruses, then both capsid peptides of the comoviruses should, likewise, be specified by RNA component 2. Siler et al.[164] have noted that cleavage of a precursor is a convenient way to generate two polypeptides in equimolar ratio. Strawberry latent ringspot virus provides further evidence of structural similarity between comoviruses and nepoviruses. It is, by most criteria, a nepovirus, but its capsid contains two polypeptides weighing 44,000 and 29,000 daltons.[28]

*4. Bromoviruses*

The bromoviruses contain four RNAs packaged into three types of particles which sediment at almost the same rate, but differ slightly in density.[13] The viruses require the three largest RNAs to infect,[13,141] but the smallest RNA, which is a subgenomic derivative of RNA 3, is not required for infection.[131] The genome of broad bean mottle virus is not so well defined. The two largest RNAs are difficult to separate from one another, but both seem to be required for infection.[2] BBMV contains only small amounts of RNA 3, and there is still no clear evidence that it is required for infection.[2]

Brome mosaic virus and cowpea chlorotic mottle virus are genetically poorly compatible; however, a reassortant containing RNA components 1 and 2 of BMV and RNA 3 of CCMV multiplies to a low level in *Chenopodium quinoa* and produces tiny local lesions on *Chenopodium hybridum*.[123] It infects neither barley nor cowpea, the normal systemic hosts of the parental viruses. The reassortant has the serotype of CCMV. The purified virus is virtually devoid of RNA components 3 and 4, but contains a small amount of a component which migrates more slowly than normal RNA 3. Studies of RNA metabolism of tripartite viruses show that RNA 3 normally predominates. The reassortant must be defective in either synthesis or packaging of RNA 3. At first glance, the latter hypothesis seems implausible, since the reassortant effectively encapsidates BMV RNAs which, otherwise, are not naturally encapsidated by CCMV protein. The defective packaging hypothesis is feasible if the reassortant cannot process RNA 3 in the normal fashion to produce components of appropriate size for packaging. Studies of the RNA metabolism of this reassortant should distinguish between these mechanisms.

Table 10 summarizes the genetics of the bromoviruses. With both BMV and CCMV, component 3 codes for the coat protein. The high frequency with which coat protein mutants have altered symptoms (in CCMV) and reduced amounts of RNA components 3 and 4 (in BMV) makes it likely that the coat protein gene affects both symptoms and the amount of encapsidated RNAs 3 and 4.

In the case of BMV, small lesion mutations map reproducibly onto RNA component 3, while in the case of CCMV, they map reproduciby onto RNA component 2.[141] It is surprising, first of all, that genes affecting lesion type should be restricted to a specific RNA component and, secondly, that this component should differ between two similar viruses.

Naturally occurring variants of CCMV have recently been found[172] and should be useful for genetic studies.

*5. Cucumoviruses*

The cucumoviruses are genetically similar to the bromoviruses. They require the three largest RNAs, but not the smallest, to infect. As with the bromoviruses, RNA 3 determines coat-protein inheritance (Table 11).[124,177,178] The inheritance of local-lesion type also has been studied (Table 11).[179,180] Mossop and Francki[178] have described two naturally occurring CMV strains which differ antigenically and in aphid transmissibility. RNA 3 determines both properties. This supports a growing body of evidence that, with many plant viruses, coat proteins play a critical role in vector specificity.

Naturally occurring CMV strains are serologically diverse,[181] indicating the potential to find many natural genetic markers. Several CMV strains appear to differ in the gel electrophoretic mobilities of their RNA components.[178,182]

The V strain of tomato aspermy virus (TAV) and the Q strain of CMV form a reassortant which contains RNA 1 and 2 of TAV and RNA 3 of CMV and which multiplies as efficiently as the parental strains,[124] despite the fact that the parental viruses are unrelated serologically or by nucleic aicd hybridization. This is perhaps the most striking example of genetic compatibility between distantly related plant viruses.

TABLE 10

**Bromovirus Mutants**

| Virus | Mutant | Properties | RNA component which determines inheritance | Ref. |
|---|---|---|---|---|
| BMV | V2 | Altered electrophoretic mobility, RNA component ratio | 3 | 13, 141 |
| | V5 | Altered electrophoretic mobility and RNA component ratio | 3 | 141 |
| | F | Local primary lesions; slow sedimenting accessory components | 2 | 141 |
| | MB1b, Mb2a, Mb4a, Mb4b | Small local lesions on *Chenopodium hybridum* | 3 | 141 |
| CCMV | Perturbed assembly | Dense nucleoprotein component; altered amino acid composition; purified protein forms unusual aggregates | Probably 3 | 173 |
| | Salt stable mutant | Very mild symptoms on cowpea; lys→arg change in coat; does not swell normally at high pH | Probably 3 | 174 |
| | Arg→cys | Noninfectious if extracted without reducing agent Arg→cys in coat protein | Probably 3 | 175 |
| | ts | Temperature-sensitive growth; altered component ratio; (lys, val)→(glu, ala) in coat; reduced specific infectivity; altered systemic symptoms | 3 | 141, 176 |
| | Mild | Small local lesions; altered systemic symptoms; temperature-sensitive coat protein | 2<br>3 | 141 |
| | MC2a | Small local lesions | 2 | 141 |
| | MC2d | Low infectivity at 32°C; small local lesions | 1<br>2 | 141 |

### *6. Alfalfa Mosaic Virus*

Alfalfa mosaic contains particles of different sizes, ranging from small spherical particles about 20 nm in diameter to bacilliform particles about 20 nm wide and 60 nm long. The RNA composition of the virus is similar to that of the bromoviruses. Alfalfa mosaic virus (AMV) requires three nucleoproteins for infection,[183] but requires all four RNAs at the RNA level;[184] however, AMV coat protein can replace RNA 4.[184] RNA 4 is not a component of the genome, but rather, an infection factor. The amount of protein required for infection is much less than required to coat the RNA. This biological activity of the protein is destroyed by mild heat treatment (5 min @ 60°c), indicating that the tertiary structure of the protein is important. The similar biological activities of RNA 4 and coat protein make sense when one considers that RNA 4 is the message for the coat protein.[129] The role of the coat protein in establishing the infection is obscure, but it could be a subunit of the viral RNA replicase.[184] The coat protein binds strongly and specifically to AMV RNA[185,186] The coat protein binding

TABLE 11

**Cucumovirus Genetics**

| Strain | Phenotype | RNA component which determines inheritance | Ref. |
|---|---|---|---|
| CMV-D | Necrotic lesions | 3 | 179 |
| CMV-TL | Large lesions | | |
| CMV-DS | Large chlorotic lesions | 2 and 3 | 180 |
| CMV-D | Necrotic lesions | | |
| Q-CMV | Aphid transmissible fast, migrating RNAs 3 and 4 | 3 | 178 |
| M-CMV | Nonaphid transmissible, slow-migrating RNAs 3 and 4 | | |
| V-TAV | Serological differences | 3 | 124 |
| Q-CMV | Symptoms, larger RNAs | | |
| CMV-D | Slow electrophoretic migration, specific antigen | 3 | 177 |
| CMV-R | Fast electrophoretic migration, specific antigen | 3 | |
| | CMV-D and CMV-R differ in symptoms. More than one RNA component contributes to these differences | | |

site of RNA 4 is near the 3′ terminus beyond the end of the coat-protein gene.[259] Reassortment (Table 12) shows that RNA 3 determines the serological properties, the nucleoprotein-component ratio, and the symptoms on tobacco, while RNA component 2 determines the symptoms on bean.[187] A series of nitrosoguanidine mutants[188] produce local lesions instead of systemic symptoms on tobacco and have altered component ratios. RNA 3 determines both characteristics. A series of naturally occurring variants, which produce local lesions on *Vigna catjang* at high temperature (34°C) show reduced growth at normal temperatures (22°C).[189] The heat resistance is inherited through RNA 1 and, surprisingly, the reduced growth at low temperature is inherited through RNA 3. The association between these two apparently independent mutations might be explained if the mutation in RNA 3 does, in fact, assist growth at 34°C or if the selection procedure has selected strains with high mutation rates.

### 7. *Ilarviruses*

The ilarviruses have roughly spherical, but irregular, particles.[191] The ilarviruses differ from AMV not only in particle morphology, but also in method of transmission. AMV is aphid-transmitted, whereas the ilarviruses are apparently transmitted through pollen.[1] There are serological relationships among some members of the group.[1] The genomes of the ilarviruses are similar to that of alfalfa mosaic virus.[192] Tobacco streak virus (TSV) is the most extensively studied ilarvirus. It requires three particles for infection[193] and, as AMV, requires coat protein or RNA 4 to activate the genome.[193,194] Surprisingly, the coat protein of TSV will activate the genome of AMV and vice versa,[193,194] despite the serological unrelatedness and the genetic incompatibility of these two viruses.[193]

Naturally occurring genetic markers have been documented for tobacco streak virus,[195,196] but technical problems have plagued inheritance studies. Purified ilarvirus RNA often has very low specific infectivity, because most ilarviruses are deficient in the subgenomic RNA which is required for infection in the absence of coat protein.[193] Ilarvirus components have been purified by sucrose density gradient centrifugation of

TABLE 12

**Alfalfa Mosaic Virus Strains and Mutants**

| Strain | Properties | RNA components which determine inheritance | Ref. |
|---|---|---|---|
| AMV 425 | Mild chlorosis on tobacco | 3 | 187, 190 |
| | Cycloheximide sensitive | 3 | |
| | Serological properties | 3 | |
| | Component ratio | 3 | |
| | Pinpoint lesions on bean | 2 | |
| YSMV (yellow spot mosaic) | Yellow necrosis on tobacco | 3 | 187, 190 |
| | Cycloheximide resistant | 3 | |
| | Serological properties | 3 | |
| | Component ratio | 3 | |
| | Systemic infection of bean | 2 | |
| AMV S | Systemic in tobacco | | |
| | Bottom and top "b" components predominant | | |
| | High yield | | |
| | Growth optimum 28°C | | |
| $-A_2$fi | Local necrotic lesions on tobacco | 3 | 188 |
| | Middle component predominates | 3 and/or 2 | |
| | No top "b" component | 3 | |
| | Low yield, growth optimum 22°C | Not tested | |
| | Differences in coat tryptic peptides | Not tested | |
| — 246 | Necrotic lesions on tobacco, systemic at high temperature | 3 | 188 |
| | Bottom component predominates | Not tested | |
| | Low yield, growth optimum 22°C | Not tested | |
| | Differences in coat tryptic peptides | Not tested | |
| —F8A | Necrotic lesions on tobacco | 3 | 188 |
| | Bottom component predominates | 3 | |
| | Low yield, growth optimum 28°C | Not tested | |
| | Differences in coat tryptic peptides | Not tested | |
| — H4B, E2, A1 | Thermoresistant in vivo 34°C | 1 | 189 |
| | Grow poorly at low temperature 22°C | 3 | |

the virions. Purified components have appreciable residual infectivity.[197] This infectivity is clearly not due to infection by single virus particles, since purified components show steep infectivity dilution curves characteristic of multiparticle infection.[197] The infectivity does not result from cross-contamination. Purified bottom component can be mixed with middle component from a strain producing atypical lesions and then repurified to the point where it is still infectious, but does not produce atypical lesions.[197] The best explanation for infectivity of purified bottom component is the presence of small amounts of "pseudobottom" component containing the smaller RNAs.[193]

Inheritance studies with TSV are intellectually taxing because component purity has been defined in terms of homogeneity in sucrose density gradient centrifugation rather than in terms of biological purity. For example, "pure" middle component probably contains a small fraction of atypical virions containing top component RNA. When "a gene is carried by both top and middle component," this probably means that it is carried by middle component and by the atypical middle component particles with top component RNA.

Fulton's inheritance studies of TSV are, for the most part, consistent with inheritance studies of the other tripartite viruses, if one assumes that the components are genetically contaminated. Bottom component probably contains RNA 1, middle component RNA 2, and top component RNA 3.[193] A reasonable guess is that centrifugally purified top component is relatively pure, but that it contains a small amount of RNA 2; that middle component contains in addition to RNA 2, some RNA 3 and small amounts of RNA 1; and that bottom component contains, in addition to RNA 1, some RNA 2 and small amounts of RNA 3. Fulton's large and small lesion alleles (Table 1[196]) would then be associated with RNA 3. The toothed and entire leaf symptom alleles (Table 3[197]) would also be carried by RNA 3. The lesion color alleles (Table 1[196]) would be carried by RNA 2. Since the antigenicity alleles segregate from these other alleles, they must be inherited through RNA 1.[195] Since RNA 3 determines antigenicity in other tripartite viruses, this scheme must be considered tentative.

The only allele carried exclusively by top component produced nonrecovered (necrotic) symptoms on tobacco.[195] Since the lesion size alleles, which are apparently carried by top component RNA, are not so exclusive, one can postulate that the nonrecovered allele resides on a satellite virus which sediments more slowly than top component. The satellite hypothesis is consistent with the apparent instability of this allele.[195]

The dwarfing characteristic requires two genes, one on RNA 1 and the other on RNA 2.[197] Fulton suggests that the two dwarfing genes may be identical and that reassortants can acquire both of these genes from middle component, one gene from each of the two parents. If this is true, then only elaborate control mechanisms can explain why strains with a single dwarfing gene do not generate progeny with two dwarfing genes. A more likely source of the apparent second middle component dwarfing gene in reassortment studies is "pseudomiddle" component containing RNA 1.

This assignment of alleles to RNA components is hypothetical. If, for example, my initial assignment of the lesion-size gene to RNA 3 is inaccurate, then all my other assignments are inaccurate. In any case, TSV appears to be a promising system for determining how individual RNA components affect the course of virus-host interaction.

*8. Pea Enation Mosaic Virus*

Pea enation mosaic virus (PEMV) is the only persistently transmitted, aphid-borne virus with a divided genome.[6] The icosahedral virus has two components that sediment at different rates, but contain similar percentages of RNA. The virus contains predominantly a single type of protein subunit, but the two components have capsids of different sizes.[6] The virus is similar to tobacco streak virus in these respects. Enhancement of infectivity on mixing components varies from 100-fold in favorable experiments to just a few fold in unfavorable experiments. This suggests that each of the components may be contaminated with an atypical nucleoprotein containing the other part of the genome. Gel electrophoresis resolves the nucleoprotein components much better than other methods. Unsatisfactory resolution may have hampered early attempts to demonstrate enhancement of infectivity by mixing components.

The coat protein is inherited through the largest RNA[6] and, in this respect, PEMV differs from other divided genome viruses, with the possible exception of tobacco streak virus and carnation ringspot virus. Gel electrophoresis resolves PEMV RNA components poorly in comparison with other viral RNAs.[6,22] Since we lack evidence that the smaller RNA carries genetic information, it is possible that PEMV does not have a divided genome, but that the smaller RNA is subgenomic and helps establish infection, as do the smallest nucleic acids of alfalfa mosaic and the ilarviruses. If the small RNA of PEMV is subgenomic, then it acts by a mechanism different from that of the small AMV RNA, since infection requires two components at either the nucleoprotein or RNA levels.

TABLE 13

**Inheritance of SBWMV Properties[201]**

| Particle | Property |
|---|---|
| Short | 1 — own particle length |
| | 2 — nature of inclusion bodies in host |
| | 3 — serological differences |
| Long | 1 — ability to infect tobacco |
| | 2 — virus concentration in spinach |

Aphid-transmissible isolates of PEMV have a small amount of a second protein in the capsid.[198,199] This protein is incorporated into the capsid in variable amounts[199] and is also present in low molecular weight form in the sap of infected plants.[198] PEMV easily loses aphid transmissibility, as well as the second protein, through repeated subculture by manual inoculation.[198] This suggests the possibility of an additional RNA component in aphid-transmissible isolates. In the future, genetic analysis should determine which RNA component codes for the transmission polypeptide and, perhaps, explain how manual transmission attenuates aphid transmissibility.

*9. Tobamoviruses*

Tobacco mosaic virus (TMV) is a classical monopartite virus. Several TMV strains contain short particles, but most of these seem to be encapsidated messenger RNAs (see Chapter by Zaitlin). Soil-borne wheat mosaic virus (SBWMV) is serologically a distant relative of TMV.[200] It consists of particles the length of TMV plus particles about half the length of TMV. The virus grows only at low temperatures, tends to aggregate when purified, and is difficult to transmit mechanically. Tsuchizaki et al.[201] found that although purified long particles were infectious, mixtures of long and short particles were still more infectious. Noninfectious short particles were easily obtained, but since short particles aggregate, noninfectious long particles could not be obtained. They constructed reassortments by a combination of in vitro and in vivo methods. They mixed purified long particles with purified short particles from another strain of SBWMV where the short particles were of a different length. They inoculated this mixture and then selected the genetically hybrid clones by choosing clones whose short particles had the same length as the purified short particles. Thus, in forming the hybrid, the short particles could be obtained biologically pure by physical methods, but long particles could be freed from their associated short particles only by selecting clones containing reassortant short particles. Table 13 shows the inheritance of genetic markers between the two strains.

Cloned beet necrotic yellow-vein virus, which has many biological similarities to SBWMV, contains four RNA components (Table 2).[10] Infectivity and genetic studies are necessary to define its genome.

It seems unlikely that a monopartite virus, TMV, and a bipartite virus, SBWMV, should be related. It would seem worthwhile to independently confirm the bipartite nature of SBWMV and, if it is confirmed, to compare the relationship of the viruses in more detail, perhaps by comparing coat protein amino acid sequences.

*10. Barley Stripe Mosaic Virus*

Barley stripe mosaic virus (BSMV) is a rod-shaped virus. Some naturally occurring strains contain two components, some contain three, and still others contain four.[5,201a] In some strains, the smaller RNA components appear or disappear during culture.[201a]

With several three-component strains, however, RNA component 3 cannot be lost on subculture.[135] In one strain containing three RNAs, all three are required for infection.[5] RNA-RNA hybridization[135] shows that several common strains are closely related. However, within a strain, RNAs 1 and 2 have largely independent sequences, while the smaller RNAs appear to be subsequences of RNA 2.

The present state of knowledge indicates that the BSMV genome consists of at least two RNA components. In the Norwich strain, which requires three components for infection, RNA 3 could be an infection factor analagous to RNA 4 of AMV and the ilarviruses. Inheritance studies are needed to more completely define the genome of BSMV and, in particular, the role of RNA 3 of the Norwich strain.

*11. Carnation Ringspot Virus*

Carnation ringspot virus (CRSV) is a small isometric virus, 3.4 nm in diameter, weighing $7.1 \times 10^6$ daltons, and containing 20.5% RNA.[202] Two RNAs, weighing $1.5 \times 10^6$ and $0.5 \times 10^6$ daltons, can be isolated from the virus. Purified RNA 1 is less infectious than a mixture of the two RNAs by a factor of up to 20[7]. The infectivity of purified large RNA, as with the infectivity of large components of other divided genome viruses, can probably be accounted for by contamination with the smaller RNA. The virus, in all likelihood, has an absolute requirement for both components.[7] CRSV-A aggregates irreversibly and is stable to disruption by sodium dodecyl sulfate (SDS), while CRSV-N aggregates reversibly with increasing temperature and is labile to SDS. The aggregation properties which are probably determined by the coat-protein gene are inherited through RNA 1.[7] As with pea enation mosaic virus, no genetic functions have yet been associated with the smaller RNA, and it may be an infection factor rather than a genetic component.

## III. TOMATO SPOTTED WILT VIRUS (TSWV)

Tomato spotted wilt virus is one of a relatively few plant viruses which contain a lipid envelope. It is also the only known plant virus transmitted specifically by thrips. The virus is a large, pleomorphic sphere.[203] Purified virus is stable only under strongly reducing conditions. The reported base composition of TSWV RNA is 38% G, 35% A, 9% C, and 19% U,[203] which is remarkable in its excess of purines. The nucleic acid has recently been isolated and consists of four single-stranded RNA components weighing 2.6, 1.9, 1.7, and 1.3 million daltons. These components are not equimolar and their ratios depend on the time of year the virus is isolated.[204] Isolated RNA is noninfectious[204] and, possibly, consists of negative strands, in which case the virion would require an RNA polymerase to infect. The structure of the nucleocapsid and the organization of the genome are still unclear.

## IV. THE PHYTOREOVIRUSES

### A. Introduction

Wound tumor virus (WTV) was among the first known double-stranded RNA-containing viruses. It has been isolated only once, from leafhoppers near Washington, D.C.,[205] and has never been found associated with plants outside the laboratory. Medical interests and ease of culture have led to great progress of our understanding of the animal reoviruses in the last decade,[206,207] and this knowledge is now being applied to the study of phytoreoviruses. Vector transmission and tumor induction are unique to the latter and are of particular interest.

TABLE 14

**Phytoreoviruses**

| Viruses with 12 RNA segments | Viruses with 10 RNA segments |
|---|---|
| Wound tumor virus (WTV) | Fiji disease virus (FDV) |
| Rice dwarf virus (RDV) | Maize rough dwarf virus (MRDV) |
| | Maize wallaby ear virus (MWEV) |
| | Rice black streaked dwarf virus (RBSDV) |

TABLE 15

**Polypeptide Composition of Wound Tumor Virus[214]**

| Polypeptide | MW | Outer shell | Inner Shell | Relative amount |
|---|---|---|---|---|
| 1 | 160 K | | X | |
| 2 | 131 K | X | | |
| 3 | 118 K | | X | Major |
| 4 | 96 K | X | | Minor |
| 5 | 58 K | | X | |
| 6 | 36 K } | | | Major — varies with preparation |
| 7 | 35 K } | | | |

## B. Biology

The phytoreoviruses have worldwide distribution. They have narrow host ranges, are leafhopper-transmitted, and are not mechanically transmissible. In the plant, they are restricted to the phloem and associated tissues. They induce proliferation of phloem tissue to produce galls or tumors. Because of the tissue proliferation, these viruses reach higher concentrations than other phloem-limited viruses. For example, WTV may reach more than 100 μg/g of plant tumor tissue.[208] The phytoreoviruses multiply within the vector and can be grown and assayed in vector tissue cultures.[209]

## C. Physical Properties

The molecular weights and chemical compositions of the phytoreoviruses, with the exception of WTV, have not been carefully investigated. WTV has a sedimentation rate of 510 S,[210,211] a molecular weight of about $70 \times 10^6$,[211] and contains 23% RNA.[211] The phytoreoviruses, like reovirus, have both inner and outer protein shells. Each virion contains a single copy of each of the genomic RNA segments. The phytoreoviruses are divisible into two groups, those with 10 RNA segments and those with 12 RNA segments (Table 14). Those with 10 RNA segments have spikes protruding at the icosahedral vertices from both inner and outer shells.[212] Those viruses with 12 RNA segments have no obvious spikes on either outer or inner shells. Though the detailed surface structures of phytoreoviruses are not known, the shells appear to have icosahedral symmetry.

Among the well-characterized phytoreoviruses, only maize rough dwarf virus and rice black-streaked dwarf virus are known to be serologically related.[213] Table 15 shows the polypeptide composition of WTV, which is more similar to that of blue tongue virus than to reovirus.[214]

TABLE 16

**RNA Compositions of Phytoreoviruses**

Molecular weights of RNA components × $10^{-6}$

| RNA | Reo (type 3)[219] | WTV[220] | RDV[220] | FDV[221] | MRDV[222] | MWEV[223] | RBSDV[222] |
|---|---|---|---|---|---|---|---|
| 1 | 2.79 | 2.90 | 3.10 | 2.90 | 2.88 | 3.10 | 2.91 |
| 2 | 2.71 | 2.40 | 2.50 | 2.50 | 2.50 | 2.60 | 2.50 |
| 3 | 2.55 | 2.20 | 2.20 | 2.48 | 2.35 | 2.60 | 2.35 |
| 4 | 1.62 | 1.80 | 1.80 | 2.48 | 2.35 | 2.48 | 2.35 |
| 5 | 1.55 | 1.78 | 1.76 | 2.12 | 2.12 | 2.28 | 2.12 |
| 6 | 1.46 | 1.10 | 1.05 | 1.85 | 1.75 | 1.72 | 1.75 |
| | 0.88 | 1.05 | 1.02 | 1.45 | 1.45 | 1.24 | 1.45 |
| | 0.75 | 0.83 | 0.78 | 1.21 | 1.25 | 1.05 | 1.25 |
| 9 | 0.65 | 0.57 | 0.70 | 1.15 | 1.18 | 1.03 | 1.18 |
| 10 | 0.61 | 0.55 | 0.67 | 1.12 | 1.08 | 0.82 | 1.10 |
| 11 | | 0.54 | 0.48 | | | | |
| 12 | | 0.32 | 0.48 | | | | |
| Sum | 15.57 | 16.04 | 16.54 | 19.26 | 18.91 | 18.92 | 18.96 |

## D. Properties of RNA Components

The nucleic acid of the phytoreoviruses is resistant to DNase and contains uracil, but not thymine. The double-stranded nature of the nucleic acid has been surmised from its interaction with dyes, from its resistance to thermal denaturation, and from its base composition. More recently, a host of other criteria have been employed, such as buoyant density and behavior during polyacrylamide gel electrophoresis. WTV contains 38% G + C[210,215] and RDV contains 44% G + C.[216] The wound tumor virus ds (double stranded) RNA components all have independent and unrelated nucleotide sequences.[217] The 3′ termini of WTV are C-OH and U-OH in equimolar ratio.[218] This suggests that one strand of a double-stranded structure terminates with C-OH and the other with U-OH, but this hypothesis has not been directly tested.[218] The 5′ termini have not been examined. In reovirus, the (+) strands in the double-stranded RNA contain a 7-methyl guanosine cap joined by a 5′ to 5′ triphosphate linkage to the rest of the sequence, while the (−) strands terminate with ppGpPy.[206,207] Reovirus RNA components are completely double-stranded with no single-stranded tails at either end.[206,207] The properties of phytoreovirus RNA termini are probably similar. Table 16 shows molecular weights of phytoreovirus RNA segments.

## E. RNA Transcriptase

Reovirus core has RNA transcriptase activity. When supplied with appropriate substrates, it synthesizes large amounts of plus strands corresponding to each of the genome segments.[206,207] This reaction requires the integrity of the reovirus core. Exogenously added double-stranded RNAs are not transcribed. Transcription is conservative, i.e., neither of the parental strands is released during the reaction. Each of the single-stranded products of the enzyme is a monocistronic messenger for protein synthesis. For reovirus to express its transcriptase activity, the outer shell must be removed.

The phytoreoviruses have similar transcriptases, but do not require removal of the outer shell to express the activity. In the presence of radioactive precursors, the WTV enzyme produces single-stranded RNAs which will hybridize with denatured virion RNA to generate radioactive ds RNAs which coelectrophorese with all 12 genome segments,[217] showing that all genome segments are transcribed. The WTV core also contains enzymes which block the 5′ termini of the single stranded products.[217a] All 12 RNA products contain 5′ terminal $^7$mG(5′)ppp (5′)Ap$^m$.[217a] Although chymotrypsin removes the outer capsid, it has little effect on the transcriptase activity.[217] Both rice dwarf virus (RDV)[224] and Fiji disease virus (FDV)[225] have transcriptases, but these have not been characterized as well as the WTV enzyme. Viruses with genetic information in the form of negative-stranded or double-stranded RNAs presumably must carry their own transcriptases into the host, which is unable to process this kind of genetic information.

### F. Role of WTV RNA Components in Tumorogenicity

WTV contains genetic information which leads to phloem proliferation during the course of infection. We know neither the number of viral genes involved nor their mode(s) of action.

### G. Role of WTV RNA Components in Virus Replication in the Plant and in the Vector

Extensive subculture of WTV in the plant host produces strains which have lost RNA components or portions of RNA components.[208] Many of the mutants are transmitted inefficiently or not at all by the vector and grow poorly or not at all on vector cell monolayers. WTV strains completely devoid of RNAs 2 and 5 grow normally on clover and produce tumors.[226] Loss of the 131,000 dalton polypeptide from the outer shell accompanies loss of RNA 2.[226] Deletion mutants of RNA 1 invariably contain a small amount of virus with wild-type RNA 1,[226] indicating that the virus requires at least a small amount of the length RNA to multiply. In a single deletion mutant of RNA 7, the deleted segment completely replaced the wild-type segment. Similar studies of virus strains cultured extensively in vector cells may be useful in defining those portions of the genome specifically required for tumorogenicity and growth in the plant.

### H. Packaging of RNA Components

The packaging of phytoreovirus genome segments is a simple example of a system which coordinates a large number of different components. The simplest way to coordinate the packaging of the RNA segments would be to prevent them from segregating. Physical studies of reovirus show that the structure of RNA termini is not likely to hold the segments together, and genetic studies show that the segments do, in fact, reassort.[206,207] The full capsid hypothesis, which argues that only the proper genomic complement fills the capsid, cannot be reconciled with either the high specificity of encapsidation or the existence of deletions. If WTV assembles in a manner similar to reovirus, the initial RNA packaging probably takes place at the level of plus strands.[206,207]

The astonishing feature of encapsidation of WTV is that some structural component must have one, and only one, copy of each of 12 different binding sites. If protein-RNA interactions are involved, this requires at one extreme a single protein with 12 different binding sites, but the same task could be accomplished by a smaller number of proteins, each with a few binding sites. These proteins would have to recognize each other with a 1 to 1 stoichiometry and exist as single copies within the virion. In either case, the protein(s) has a remarkable number of nucleic acid binding sites and is worthy of study. The RNA binding sites of the proteins would have to recognize a sizeable nucleotide sequence, roughly 10 nucleotides, to provide the necessary specificity.

A temporal scheme could involve structural proteins present at more than one copy per virion. The binding of one RNA strand would change the configuration of the assembly to create a binding site for another RNA which would, in turn, change the configuration to produce another binding site. This model predicts the possibility of obtaining conditional lethal mutants, blocked at various stages of assembly which, under nonpermissive conditions, would accumulate nucleoprotein assemblies containing less than the complete genome.

Watson-Crick base pairing between plus strands of viral RNA provides a straightforward recognition mechanism. However, reovirus plus strands produced in vitro do not aggregate under normal conditions of electrophoresis or centrifugation. Conceivably, proteins could stabilize such an interaction.

The discovery of the assembly mechanism of ds RNA viruses should give insight into how a relatively simple biological system, such as the mitotic apparatus of eucaryotes, can order a large number of different components.

## V. SATELLITE AND DEFECTIVE VIRUSES

### A. Introduction

Satellite and defective viruses cannot replicate by themselves and require another virus, usually termed the "helper" virus, to replicate. Satellite viruses are distinct in origin from defective viruses. The former have no obvious evolutionary relationship to the helper, whereas defective viruses are derivatives of the helper viras. Satellite viruses are apparently highly evolved to parasitize the helper virus genome. Satellite viruses grow so vigorously that, in some cases, it is difficult to obtain helper viruses free of satellites. Defective viruses arise from continuous subculture of viruses at high multiplicity of infection. The term "defective" describes the inability of these viruses to multiply on their own, but misleads in that to reach detectable levels, a defective virus must, in fact, multiply faster than the helper virus.

Defective viruses have been studied predominantly among animal viruses where they are widespread and where they are called "defective interfering particles".[227] Defective interfering particles of animal viruses have four essential properties. First, they are derived from the viral genome. Second, they are defective; i.e., they cannot multiply on their own. Third, they interfere with the multiplication of the helper virus. Fourth, they have a selective advantage in the presence of helper virus which allows them to reach a detectable level. Defective interfering particles are felt to play a role in persistent infections of animal viruses. There is no *a priori* reason why they could not play a similar role in plant virus infections where they could, for example, mediate cross-protection. Potential defective interfering particles among plant viruses have not been well characterized, particularly in regard to their ability to interfere with the helper.

### B. Possible Examples of Defective Interfering Particles in Plant Viruses

The plant virus which produces particles most similar to animal virus defective interfering particles is wound tumor virus (WTV). Some WTV strains lack RNAs 2 or 5 completely; however, some deletion mutants of RNA 1 contain reduced amounts of the full-length component.[226] These latter strains appear to be mixtures of normal WTV, the helper virus, along with defective particles. There are other reports of phytoreoviruses containing submolar amounts of RNA components which may result from contamination of cultures with defective interfering particles.

The smaller components of certain barley stripe mosaic virus strains can be lost by cloning.[135] RNA-RNA hybridization studies are consistent with ttese components being subsequences of RNA 2.[135] Small amounts of atypical RNA components occasionally appear in brome mosaic virus cultures, and they are lost on cloning.[228] Citrus leprosis infections contain particles which could be fragments of rhabdoviruses.[229]

TABLE 17

**Base Compositions and Terminal Sequences of Tobacco Necrosis Virus and Satellite Tobacco Necrosis Virus**

| | TNV | | | | STNV | | | |
|---|---|---|---|---|---|---|---|---|
| Base[237] composition | G | A | C | U | G | A | C | U |
| | 24 | 26 | 21 | 29 | 22 | 28 | 20 | 30 |
| 5′ terminus | pp AGU[238] | | | | pppAGU[239]<br>ppAGU[238,239] | | | |
| 3′ terminus | | | | | G A C U A C C C—OH[239]<br>G A C U A C C C—p[239] | | | |

## C. Satellite Tobacco Necrosis Virus

Satellite tobacco necrosis virus (STNV) was the first well-characterized satellite virus. This virus is a small sphere, roughly 18 nm in diameter, weighs $1.7 \times 10^6$ daltons, and contains a single RNA molecule weighing 300,000 daltons, and 60 coat-protein subunits, each weighing 22,900 daltons.[230] Although the neutron scattering data[230] suggest that STNV RNA is the same size as BMV RNA 4, comparative gel electrophoresis suggests that STNV RNA is appreciably larger.[231] The virus is serologically distinct from and contains a different coat protein than tobacco necrosis virus.[232] There are several serologically distinct strains of STNV, which are activated only by specific serotypes of TNV.[233,234] This type of specificity is analagous to the compatibility of RNA components among divided genome viruses. Base compositions of STNV and its helper virus, TNV, as well as terminal sequences are shown in Table 17. By RNA-RNA hybridization, TNV and STNV have less than 2% of their nucleotide sequences in common.[235]

Since STNV has only enough genetic information to code for its coat protein, and since coat protein indeed appears to be the only product of STNV RNA directed in vitro protein synthesis (Chapter by Davies), STNV presumably depends on TNV for additional gene products, for example, the RNA polymerase. The similarity in 5′ termini between these two viruses is consistent with the common polymerase hypothesis. STNV-infected tissue contains replicative forms corresponding to both TNV RNA and STNV RNA.[236] STNV reduces the yield of TNV, suggesting that TNV and STNV compete for at least one common intermediate.[232]

## D. Panicum Mosaic Virus Satellite

Panicum mosaic virus (PMV) has physical properties similar to those of TNV.[240,241] A 42S component is associated with PMV.[20,241] This component is not infectious by itself. Its diameter is 15 to 18 nm by electron microscopy.[241] It differs serologically from and has a smaller coat protein than PMV.[241,242] Though the purified 42 S component is noninfectious, it multiplies in the presence of added St. Augustine decline virus, which is a close serological relative of PMV. The 42 S component contains 14 S and 34 S RNA components. [240,241]

## E. Satellite-Like Interaction of Tobacco Mosaic Virus with Cereal Viruses

TMV multiplies in barley without producing symptoms.[243] In systemically infected leaves, however, it reaches only 0.1% of the level it reaches in systemically infected tobacco. In mixed inoculations with a y of a number of cereal viruses, TMV multiplies as well in barley as it multiplies by itself in tobacco.[243] The yield of barley stripe mosaic virus is unaffected by mixed inoculation with TMV. The symptoms given by mixed inoculation are the same as those given by BSMV alone. In mixed infection, many cells contain both viruses. BSMV coat protein encapsidates some TMV RNA, but TMV coat protein does not encapsidate BSMV RNA.[244]

In this system, TMV has some of the properties of a satellite virus. Similar relationships may exist in nature. For example, the panicum mosaic-panicum mosaic satellite system could be such a case if conditions can be found which support independent replication of the satellite.

### F. Satellite Tobacco Ringspot Virus

Satellite tobacco ringspot, (S-TRSV) which depends upon tobacco ringspot virus (TRSV) for its replication, was originally isolated from a laboratory culture of TRSV and proved to be a short RNA component which is encapsidated by TRSV coat protein.[245,246] This RNA is 80,000 to 120,000 mol wt.[245,247] The RNA is packaged in TRSV coat protein to give a heterogeneous group of virions containing from about 12 RNAs to about 25 RNAs.[248] The RNA does not have messenger activity.[249] Like STNV, S-TRSV interferes with the replication of its helper.[245] Local lesions produced by the combination of TRSV and its satellite are smaller than those produced by TRSV alone.[246] Oligonucleotide mapping is consistent with S-TRSV having a single sequence rather than containing a mixture of components with different sequences.[250] The relationship of the S-TRSV RNA nucleotide sequence to the TRSV RNA nucleotide sequence is not yet clear. A replicative form (double-stranded RNA), specific to infections containing S-TRSV, has been isolated which has the biological properties of S-TRSV RNA after thermal denaturation.[251] This replicative form is unusual because it is heterogeneous, containing molecules up to 20 times the mass expected for double-stranded S-TRSV RNA. Pancreatic ribonuclease at high concentration converts the large molecules into smaller components which are still infectious after thermal denaturation. Most of the properties of this replicative form are consistent with it, containing a polymeric negative strand hybridized to normal S-TRSV plus strands.[251]

### G. Cucumber Mosaic Virus Associated RNA 5

A low molecular weight RNA component (approximately 100,000 daltons) is often associated with purified cucumber mosaic virus.[252] This RNA is encapsidated in CMV coat protein; it can be separated from the genomic RNAs and possesses the properties of satellite RNA[252,253] It has been termed "cucumber mosaic virus associated RNA 5" (CARNA 5). The association of CARNA 5 with cucumber mosaic virus is responsible for a severe and economically important disease of tomatoes.[253] CARNA 5 is not a subsequence of CMV genomic RNAs,[132,260] and its sequence is independent of the host in which it is propagated.[260] The relative amount of CARNA 5 in a CMV preparation depends both on the helper strain and on the host in which the virus is propagated.[254] As with other satellites, CARNA 5 interferes with multiplication of the helper and appears to reduce selectively the amount of RNA 1 produced during infection.[252] CARNA 5 differs from S-TRSV in that it is a messenger RNA in vitro where it directs synthesis of two small peptides of unknown function.[255] The 5′ terminus of CARNA-5 is $m^7GpppNp$ and the 3′ terminus is GACCG-OH.[132]

### H. Tomato Black Ring Virus Satellite

Purified tomato black ring virus (TBRV) contains an RNA of $0.5 \times 10^6$ daltons mol wt, which replicates only in the presence of the two TBRV genomic RNAs.[32] When added to the genomic RNAs, it decreases the number of lesions they produce on *Chenopodium amaranticolor.* It is not yet clear how this RNA is packaged.

## IV. SUMMARY: PRESENT STATUS OF, AND FUTURE PROSPECTS FOR, STUDIES ON RNAs OF MULTIPARTITE, DEFECTIVE, AND SATELLITE VIRUSES

The divided genome strategy has been highly successful among plant viruses despite its apparent disadvantages in virus transmission. Possible advantages of this strategy include increased genetic flexibility, increased resistance to inactivation, more efficient packaging and more effective control of translation. Hopefully, the future will tell us which of these are the important advantages. Other questions which are unique to viruses discussed in this chapter are:

- Does reassortment of genome components generate new viruses in nature?
- Are multipartite and satellite viruses derived from monopartite viruses?
- How is the synthesis of RNA components coordinated in multipartite viruses?
- How do phytoreoviruses faithfully encapsidate 12 different components with equimolar stoichiometry?
- Do satellite and detective plant viruses play roles in virus ecology?

In the future, plant viruses will continue to be a useful source of pure messenger RNAs. Satellite viruses will continue to represent a simple model for genome stategies and divided genome viruses will offer advantages in studying plant virus genetics.

The study of the viruses discussed in this chapter can be expected to parallel that of monopartite viruses. In particular, we can expect progress in nucleotide sequencing, identification of genes and their products, and isolation of RNA replicating enzymes. We will improve our understanding of the generation of subgenomic RNAs, the role of tRNA-like 3′ termini, the role of individual virus genes in replication and in host response, the feedback mechanisms which maintain persistent infections, and the origins of plant viruses.

## ACKNOWLEDGMENTS

I would like to thank all of those who contributed reprints, preprints, and unpublished information. I would like to thank Pat Entrekin and Dr. Myron Brakke for reviewing the manuscript.

## REFERENCES

1. **Fenner, F.,** Classification and nomenclature of viruses, Second report of the International Committee on Taxonomy of Viruses, *Intervirology,* 7, 1, 1976.
2. **Hull, R.,** The multicomponent nature of broad bean mottle virus and its nucleic acid, *J. Gen. Virol.,* 17, 111, 1972.
3. B. D., Murant, A. F., and Mayo, M. A., Evidence for two functional RNA species in raspberry ringspot virus, *J. Gen. Virol.,* 16, 339, 1972.
4. **Jones, A. T. and Mayo, M. A.,** Two nucleoprotein particles of cherry leaf roll virus, *J. Gen. Virol.,* 16, 349, 1972.
5. **Lane, L. C.,** The components of barley stripe mosaic and related viruses, *Virology,* 58, 323, 1974.

6. **Hull, R. and Lane, L. C.,** The unusual nature of the components of a strain of pea enation mosaic virus, *Virology*, 55, 1, 1973.
7. **Dodds, J. A., Tremaine, J. H., and Ronald, W. P.,** Some properties of carnation ringspot virus single- and double-stranded RNA, *Virology*, 83, 322, 1977.
8. **Heitjink, R. A., Houwing, C. J., and Jaspars, E. M. J.,** Molecular weights of particles and RNAs of alfalfa mosaic virus. Number of subunits in protein capsids, *Biochemistry*, 16, 4684, 1977.
9. **Semancik, J. S.,** Bean pod mottle virus, C.M.I./A.A.B., *Descriptions of Plant Viruses*, No. 108, Commonwealth Mycological Institute and Association of Applied Biologists, Kew, Surrey, England, 1972.
10. **Putz, C.,** Composition and structure of beet necrotic yellow vein virus, *J. Gen. Virol.*, 35, 397, 1977.
11. **Jones, A. T. and Barker, H.,** Properties and relationships of broad bean stain and Echtes Ackerbohnenmosaik-Virus, *Ann. Appl. Biol.*, 83, 231, 1976.
12. **Doel, T. R.,** Comparative properties of type, nasturtium ringspot and petunia ringspot strains of broad bean wilt virus, *J. Gen. Virol.*, 26, 95, 1975.
13. **Lane, L. C. and Kaesberg, P.,** Multiple genetic components in bromegrass mosaic virus, *Nature (London) New Biol.*, 232, 40, 1971.
14. **Walkey, D. G. A., Stace-Smith, R., and Tremaine, J. H.,** Serological, physical, and chemical properties of strains of cherry leaf roll virus, *Phytopathology*, 63, 566, 1973.
15. **Garnsey, S. M. and Gonsalves, D.,** Citrus leaf rugose virus, C.M.I./A.A.B. *Descriptions of Plant Viruses*, No. 164, Commonwealth Mycological Institute and Association of Applied Biologists, Kew, Surrey, England, 1976.
16. **Bancroft, J. B.,** The significance of the multicomponent nature of cowpea chlorotic mottle virus RNA, *Virology*, 45, 830, 1971.
17. **Reijnders, L., Aalbers, A. M. J., van Kammen, A., and Thuring, R. W. J.,** Molecular weights of plant viral RNAs determined by gel electrophoresis under denaturing conditions, *Virology*, 60, 515, 1974.
18. **Kaper, J. M. and Diaz-Ruiz, J. R.,** Molecular weights of the double-stranded RNAs of cucumber mosaic virus strain S and its associated RNA 5, *Virology*, 80, 214, 1977.
19. **Jones, A. T. and Mayo, M. A.,** Purification and properties of elm mottle virus, *Ann. Appl. Biol.*, 75, 347, 1973.
20. **Quacquarelli, A., Gallitelli, D., Savino, V., and Martelli, G. P.,** Properties of grapevine fanleaf virus, *J. Gen. Virol.*, 32, 349, 1976.
21. **Cooper, J. I. and Mayo, M. A.,** Some properties of the particles of three tobravirus isolates, *J. Gen. Virol.*, 16, 285, 1972.
22. **Gonsalves, D. and Shepherd, R. J.,** Biological and physical properties of the two nucleoprotein components of pea enation mosaic virus and their associated nucleic acids, *Virology*, 48, 709, 1972.
23. **Loesch, L. S. and Fulton, R. W.,** Prunus necrotic ringspot virus as a multicomponent system, *Virology*, 68, 71, 1975.
24. **Campbell, R. N.,** Radish mosaic virus, C.M.I./A.A.B. *Descriptions of Plant Viruses*, No. 121, Commonwealth Mycological Institute and Association of Applied Biologists, Kew, Surrey, England, 1973.
25. **Murant, A. F., Mayo, M. A., Harrison, B. D., and Goold, R. A.,** Properties of virus and RNA components of raspberry ringspot virus, *J. Gen. Virol.*, 16, 327, 1972.
26. **Oxelfelt, P.,** Biological and physicochemical characteristics of three strains of red clover mottle virus, *Virology*, 74, 73, 1976.
27. **Gumpf, D. J.,** Purification and properties of soil-borne wheat mosaic virus, *Virology*, 43, 588, 1971.
28. **Mayo, M. A., Murant, A. F., Harrison, B. D., and Goold, R. A.,** Two protein and two RNA species in particles of strawberry latent ringspot virus, *J. Gen. Virol.*, 24, 29, 1974.
29. **Rezaian, M. A. and Francki, R. I. B.,** Replication of tobacco ringspot virus. II. Differences in nucleotide sequences between the viral RNA components, *Virology*, 59, 275, 1974.
30. **Clark, M. F. and Lister, R. M.,** Preparation and some properties of the nucleic acid of tobacco streak virus, *Virology*, 45, 61, 1971.
31. **Habili, N. and Francki, R. I. B.,** Comparative studies on tomato aspermy and cucumber mosaic viruses. I. Physical and chemical properties, *Virology*, 57, 392, 1974.
32. **Murant, A. F., Mayo, M. A., Harrison, B. D., and Goold, R. A.,** Evidence for two functional RNA species and a "satellite" RNA in tomato black ring virus, *J. Gen. Virol.*, 19, 275, 1973.
33. **Schneider, I. R., White, R. M., and Civerolo, E. L.,** Two nucleic acid-containing components of tomato ringspot virus, *Virology*, 57, 139, 1974.
34. **Rauws, A. G., Jaspars, E. M. J., and Veldsta, H.,** The base composition of ribonucleic acids from alfalfa mosaic virus components, *Virology*, 23, 283, 1964.
35. **Atabekov, J. G. and Novikov, V. K.,** Barley stripe mosaic virus, C.M.I./A.A.B. *Descriptions of Plant Viruses* No. 68, Commonwealth Mycological Institute and Association of Applied Biologists, Kew, Surrey, England, 1971.

36. **Yamazaki, H., Bancroft, J., and Kaesberg, P.,** Biophysical studies of broad bean mottle virus, *Proc. Natl. Acad. Sci. U.S.A.*, 47, 979, 1961.
37. **Gibbs, A. J. and Smith, H. G.,** Broad bean stain virus, C.M.I./A.A.B. *Descriptions of Plant Viruses* No. 29, Commonwealth Mycological Institute and Association of Applied Biologists, Kew, Surrey, England, 1970.
38. **Bockstahler, L. E. and Kaesberg, P.,** Isolation and properties of RNA from bromegrass mosaic virus, *J. Mol. Biol.*, 13, 127, 1965.
39. **Hollings, M. and Stone, O. M.,** Carnation ringspot virus, C.M.I./A.A.B. *Descriptions of Plant Viruses*, No. 21, Commonwealth Mycological Institute and Association of Applied Biologists, Kew, Surrey, England, 1970.
40. **Bancroft, J. B., Hiebert, E., Rees, M. W., and Markham, R.,** Properties of cowpea chlorotic mottle virus, its protein and nucleic acid, *Virology*, 34, 224, 1968.
41. **van Kammen, A. and van Griensven, L. J. L. D.,** The relationship between the components of cowpea mosaic virus. II. Further characterization of the nucleoprotein components of CPMV, *Virology*, 41, 274, 1970.
42. **Kaper, J. M., Diener, T. O., and Scott, H. A.,** Some physical and chemical properties of cucumber mosaic virus (strain Y) and of its isolated ribonucleic acid, *Virology*, 27, 54, 1965.
43. **Shepherd, R. J., Wakeman, R. J., and Ghabrial, S. A.,** Preparation and properties of the protein and nucleic acid components of pea enation mosaic virus, *Virology*, 35, 255, 1968.
44. **Mink, G. I.,** Peanut stunt virus, C.M.I./A.A.B., *Descriptions of Plant Viruses*, No. 92, Commonwealth Mycological Institute and Association of Applied Biologists, Kew, Surrey, England, 1972.
45. **Barnett, O. W. and Fulton, R. W.,** Some chemical properties of prunus necrotic ringspot and tulare apple mosaic viruses, *Virology*, 39, 556, 1969.
46. **Gibbs, A. J., Giussani-Belli, G., and Smith, H. G.,** Broad bean stain and true broad bean mosaic viruses, *Ann. Appl. Biol.*, 61, 99, 1968.
47. **Mazzone, H. M., Incardona, N. L., and Kaesberg, P.,** Biochemical and biophysical properties of squash mosaic virus and related macromolecules, *Biochim. Biophys. Acta*, 55, 164, 1962.
48. **Semancik, J. S. and Kajiyama, M. R.,** Properties and relationships among RNA species from tobacco rattle virus, *Virology*, 33, 523, 1967.
49. **Stace-Smith, R., Reichmann, M. E., and Wright, N. S.,** Purification and properties of tobacco ringspot virus and two RNA-deficient components, *Virology*, 25, 487, 1965.
50. **Hollings, M. and Stone, O. M.,** Tomato aspermy virus, C.M.I./A.A.B., *Descriptions of Plant Viruses*, No. 79, Commonwealth Mycological Institute and Association of Applied Biologists, Kew, Surrey, England, 1971.
51. **Tremaine, J. H. and Stace-Smith, R.,** Chemical compositions and biophysical properties of tomato ringspot virus, *Virology*, 35, 102, 1968.
52. **Pinck, L.,** The 5′ end groups of alfalfa mosaic virus RNAs are $m^7G^{5'}$ ppp 5′Gp, *FEBS Lett.*, 59, 24, 1975.
53. **Dasgupta, R. , Harada, F., and Kaesberg, P.,** Blocked 5′ termini in brome mosaic virus RNA, *J. Virol.*, 18, 260, 1976.
54. **Symons, R. H.,** Cucumber mosaic virus RNA contains 7-methyl guanosine at the 5′ terminus of all four RNA species, *Mol. Biol. Rep.*, 2, 277, 1975.
55. **Abou Haidar, M. and Hirth, L.,** 5′ terminal structure of tobacco rattle virus RNA: evidence for polarity of reconstitution, *Virology*, 76, 173, 1977.
56. **Klootwijk, J., Klein, I., Zabel, P., van Kammen, A.,** Cowpea mosaic virus RNAs have neither $m^7$GpppN... nor mono-, di- or triphosphates at their 5′ ends, *Cell*, 11, 73, 1977.
57. **Lee, Y. F., Nomoto, A., Detjen, B. M., and Wimmer, E.,** A protein covalently linked to poliovirus genome RNA, *Proc. Natl. Acad. Sci. U.S.A.*, 74, 59, 1977.
58. **Dasgupta, R., Shih, D. S., Saris, C., and Kaesberg, P.,** Nucleotide sequence of a viral RNA fragment that binds to eukaryotic ribosomes, *Nature (London)*, 256, 624, 1975.
59. **Koper-Zwarthoff, E. C., Lockard, R. E., Alzner-deWeerd, B., RajBhandary, U. L., and Bol, J. F.,** Nucleotide sequence of the 5′ terminus of alfalfa mosaic virus RNA 4 leading into the coat protein cistron, *Proc. Natl. Acad. Sci. U.S.A.*, 74, 5504, 1977.
60. **Castel, A., Kraal, B., Kerklaan, P. R. M., Klok, J., and Bosch, L.,** Initiation of polypeptide synthesis with various $NH_2$-blocked aminoacyltRNAs under the direction of alfalfa mosaic virus RNA 4, *Proc. Natl. Acad. Sci. U.S.A.*, 74, 5509, 1977.
61. **Bastin, M., Dasgupta, R., and Kaesberg, P.,** Similarity in structure and function of the 3′ terminal regions of the four brome mosaic viral RNAs, *J. Mol. Biol.*, 103, 737, 1976.
62. **Semancik, J. S.,** Detection of polyadenylic acid sequences in plant pathogenic RNAs, *Virology*, 62, 288, 1974.
63. **Takanami, Y. and Imaizumi, S.,** Identical 3′ termini of the four RNA species of cucumber mosaic virus, *Virology*, 77, 853, 1977.

64. **El Manna, M. M. and Bruening, G.,** Polyadenylate sequences in the ribonucleic acids of cowpea mosaic virus, *Virology,* 56, 198, 1973.
65. **Darby, G. and Minson, A. C.,** The structure of tobacco rattle virus ribonucleic acids: Nature of the 3′ terminal nucleosides, *J. Gen. Virol.,* 14, 199, 1972.
66. **Huez, G., Marbaix, G., Burny, A., Hubert, E., Leclerq, M., Cleuter, Y., Chantrenne, H., Sorecq, H., and Littauer, U. Z.,** Degradation of deadenylated rabbit alpha-globin mRNA in *Xenopus* oocytes is associated with its translation, *Nature (London),* 266, 473, 1977.
67. **Hall, T. C., Shih, D. S., and Kaesberg, P.,** Enzyme mediated binding of tyrosine to brome mosaic virus ribonucleic acid, *Biochem. J.,* 129, 969, 1972.
68. **Kohl, R. J. and Hall, T. C.,** Aminoacylation of RNA from several viruses: amino acid specificity and differential activity of pant, yeast and bacterial synthetases, *J. Gen. Virol.,* 25, 257, 1974.
69. **Bastin, M. and Hall, T. C.,** Interaction of elongation factor 1 with aminoacylated brome mosaic virus and tRNAs, *J. Virol.,* 20, 117, 1976.
70. **Chen, J. M. and Hall, T. C.,** Comparison of tyrosyl transfer ribonucleic acid and brome mosaic virus tyrosyl ribonucleic acid as amino acid donors in protein synthesis, *Biochemistry,* 12, 4570, 1973.
71. **Shih, D. S., Kaesberg, P., and Hall, T. C.,** Messenger and aminoacylation functions of brome mosaic virus RNA after chemical modification of the 3′ terminus, *Nature, (London),* 249, 353, 1974.
72. **Das Gupta, R. and Kaesberg, P.,** Sequence of an oligonucleotide derived from the 3′ end of each of the four brome mosaic viral RNAs, *Proc. Natl. Acad. Sci. U.S.A.,* 74, 4900, 1977.
73. **Federoff, N.,** Replicase of the phage f2, in *RNA Phages,* Zinder, N. D., Ed. Cold Spring Harbor Laboratory, Cold Spring Harbor, New York, 1975, 235.
74. **Kamen, R. I.,** Structure and function of the Q$\beta$ RNA replicase, in *RNA Phages,* Zinder, N. D., Ed., Cold Spring Harbor Laboratory, Cold Spring Harbor, New York, 1975, 203.
75. **Busto, P., Carriquiry, E., Tarrago-Litvak, L., Castroviejo, M., and Litvak, S.,** Interactions of plant viral RNAs and tRNA nucleotidyl transferase, *Ann. Microbiol. (Institute Pasteur),* 127A, 39, 1976.
76. **Mohier, E., Pinck, L., and Hirth, L.,** Replication of alfalfa mosaic virus RNAs, *Virology,* 58, 9, 1974.
77. **Pinck, L. and Hirth, L.,** The replicative RNA and viral RNA synthesis rate in tobacco infected with alfalfa mosaic virus, *Virology,* 49, 413, 1972.
78. **Pring, D. R.,** Barley stripe mosaic virus replicative form RNA. Preparation and characterization, *Virology,* 48, 22, 1972.
79. **Bastin, M. and Kaesberg, P.,** A possible replicative form of brome mosaic virus RNA 4, *Virology,* 72, 536, 1976.
80. **Philipps, G., Gigot, C., and Hirth, L.,** Replicative forms and viral RNA synthesis in leaves infected with brome mosaic virus, *Virology,* 60, 370, 1974.
81. **van Griensven, L. J. L. D. and van Kammen, A.,** The isolation of ribonuclease resistant RNA induced by cowpea mosaic virus: evidence for the two double-stranded RNA components, *J. Gen. Virol.,* 4, 423, 1969.
82. **van Griensven, L. J. L. D., van Kammen, A., and Rezelman, G.,** Characterization of the double-stranded RNA isolated from cowpea mosaic virus-infected *Vigna* leaves, *J. Gen. Virol.,* 18, 359, 1973.
83. **German, T. L. and de Zoeten, G. A.,** Purification and properties of the replicative forms and replicative intermediates of pea enation mosaic virus, *Virology,* 66, 192, 1975.
84. **Rezaian, M. A. and Francki, R. I. B.,** Replication of tobacco ringspot virus. I. Detection of a low molecular weight double-stranded RNA from infected plants, *Virology,* 56, 238, 1973.
85. **Schneider, I. R., White, R. M., and Thompson, S. M.,** High molecular weight double-stranded nucleic acids from tobacco ringspot virus infected plants, *Proc. Am. Phytopathol.Soc.,* 1, 82, 1974.
86. **deZoeten, G. A., Powell, C. A., Gaard, G., and German, T. L.,** *In situ* localization of pea enation mosaic virus double-stranded ribonucleic acid, *Virology,* 70, 459, 1977.
87. **deZoeten, G. A. and Schlegel, D. E.,** Nucleolar and cytoplasmic uridine $^3$H incorporation in virus infected plants, *Virology,* 32, 416, 1967.
88. **Dawson, W. O. and Schlegel, D. E.,** Synchronization of cowpea chlorotic mottle virus replication in cowpea leaves, *Intervirology,* 7, 284, 1976.
89. **Bancroft, J. B., Motoyoshi, F., Watts, J. W., and Dawson, J. R. O.,** Cowpea chlorotic mottle and brome mosaic viruses in tobacco protoplasts, in *Modification of the Information Content of Plant Cells,* Markham, R., Davies, D. R., Hopwood, D. A., and Horne, R. W., Eds., North-Holland, Amsterdam, 1975, 133.
90. **Dawson, W. O.,** Time-course of cowpea chlorotic mottle virus RNA replication. *Intervirology,* 9, 119, 1978.
91. **Takanami, Y., Kubo, S., and Imaizumi, S.,** Synthesis of single- and double-stranded cucumber mosaic virus RNAs in tobacco mesophyll protoplasts, *Virology,* 80, 376, 1977.

92. **Harrison, B. D., Kubo, S., Robinson, D. J., and Hutcheson, A. M.,** The multiplication cycle of tobacco rattle virus in tobacco mesophyll protoplasts, *J. Gen. Virol.*, 33, 237, 1976.
93. **Hiruki, C.,** Properties of single- and double-stranded ribonucleic acid from barley plants infected with bromegrass mosaic virus, *J. Virol.*, 3, 498, 1969.
94. **Shih, D. S. and Kaesberg, P.,** Translation of brome mosaic viral ribonucleic acid in a cell-free system derived from wheat embryo, *Proc. Natl. Acad. Sci. U.S.A.*, 70, 1799, 1973.
95. **Romero, J.,** RNA synthesis in broadbean leaves infected with broadbean mottle virus, *Virology*, 48, 591, 1972.
96. **Romero, J.,** Properties of a slow sedimenting RNA synthesized by broadbean tissue infected with broadbean mottle virus, *Virology*, 55, 224, 1973.
97. **Bol, J. F., Bakhuizen, C. E. G. C., and Rutgers, T.,** Composition and biosynthetic activity of polyribosomes associated with alfalfa mosaic virus infections, *Virology*, 75, 1, 1976.
98. **Robertson, H. D.,** Functions of replicating RNA in cells infected by RNA bacteriophages, in *RNA Phages*, Zinder, N. D., Ed. Cold Spring Harbor Laboratory, Cold Spring Harbor, New York, 1975, 113.
99. **Kuo, C. H., Eoyang, L., and August, J. T.,** Protein factors required for the replication of phage QB RNA *in vitro*, in *RNA Phages*, Zinder, N. D., Ed., Cold Spring Harbor Laboratory, Cold Spring Harbor, New York, 1975, 259.
100. **Bol, J. F., Clerx-van Haaster, C. M., and Weening, C. J.,** Host and virus specific RNA polymerases in alfalfa mosaic virus infected tobacco, *Ann. Microbiol. (Institute Pasteur)*, 127A, 183, 1976.
101. **LeRoy, C., Stussi-Garaud, C., and Hirth, L.,** RNA dependent RNA polymerases in uninfected and in alfalfa mosaic virus infected tobacco plants, *Virology*, 82, 48, 1977.
102. **Romero, J. and Jacquemin, J. M.,** Relation between virus-induced RNA polymerase activity and synthesis of broadbean mottle virus in broadbean, *Virology*, 45, 813, 1971.
103. **Hadidi, A. and Fraenkel-Conrat, H.,** Characterization and specificity of soluble RNA polymerase of brome mosaic virus, *Virology*, 52, 363, 1973.
104. **Semal, J. S. and Kummert, J.,** Virus-induced RNA polymerase and synthesis of bromegrass mosaic virus in barley, *J. Gen. Virol.*, 7, 173, 1970.
105. **May, J. T. and Symons, R. H.,** Specificity of the cucumber mosaic virus-induced RNA polymerase for RNA and polynucleotide templates, *Virology*, 44, 517, 1971.
106. **Peden, K. W. C., May, J. T., and Symons, R. H.,** A comparison of two plant virus-induced RNA polymerases, *Virology*, 47, 498, 1972.
107. **Zabel, P., Weenen-Swaans, H., and van Kammen, A .,** *In vitro* replication of cowpea mosaic virus RNA. I. Isolation and properties of the membrane bound replicase, *J. Virol.*, 14, 1049, 1974.
108. **Astier-Manifacier, S. and Cornuet, P.,** RNA-dependent RNA-polymerase in Chinese cabbage, *Biochim. Biophys. Acta*, 232, 484, 1971.
109. **Duda, C. T., Zaitlin, M., and Siegel, A.,** *In vitro* synthesis of double-stranded RNA by an enzyme system isolated from tobacco leaves, *Biochim. Biophys. Acta*, 319, 62, 1973.
110. **Clark, G. L., Peden, K. W. C., and Symons, R. H.,** Cucumber mosaic induced RNA polymerase. Partial purification and properties of the template free enzyme, *Virology*, 62, 434, 1974.
111. **Zabel, P., Jongen-Neven, I., and van Kammen, A.,** *In vitro* replication of cowpea mosaic virus RNA II. Solubilization of membrane-bound replicase and partial purification of the solubilized enzyme, *J. Virol.*, 17, 679, 1976.
112. **May, J. T., Gilliland, J. M., and Symons, R. T.,** Plant virus-induced RNA polymerase: properties of the enzyme partly purified from cucumber cotyledons infected with cucumber mosaic virus, *Virology*, 39, 54, 1969.
113. **Hariharasubramanian, V., Hadidi, A., Singer, B., and Fraenkel-Conrat, H.,** Possible identification of a protein in brome mosaic virus infected barley as a component of viral RNA polymerase, *Virology*, 54, 170, 1973.
114. **Kummert, J.,** *In vitro* pulse-labeling of the replicative forms of bromegrass mosaic virus RNA, *Virology*, 57, 314, 1974.
115. **Kummert, J. and Semal, J.,** Properties of single-stranded RNA synthesized by a crude RNA polymerase fraction from barley leaves infected with brome mosaic virus, *J. Gen. Virol.*, 16, 11, 1972.
116. **Kummert, J. and Semal, J.,** Polyacrylamide gel electrophoresis of the RNA products labeled *in vitro* by extracts of leaves infected with bromegrass mosaic virus, *Virology*, 60, 390, 1974.
117. **Jacquemin, J. M. and Lopez, M.,** RNAs labeled *in vitro* by polymerase from leaves infected with broadbean mottle virus, *Intervirology*, 4, 45, 1974.
118. **Hadidi, A., Hariharasubramanian, V. and Fraenkel-Conrat, H.,** Template activity of brome mosaic virus-RNA components with soluble brome mosaic virus RNA polymerase, *Intervirology*, 1, 211, 1973.
119. **Sanger, H. L.,** Functions of the two particles of tobacco rattle virus, *J. Virol.*, 3, 304, 1969.

120. **Govier, D. A.,** Complementation between middle and bottom components of broad bean stain virus and Echtes Ackerbohnenmosaik Virus, *J. Gen. Virol.*, 28, 373, 1975.
121. **van Kammen, A.,** The relationship between the components of cowpea mosaic virus I. Two ribonucleoprotein particles necessary for the infectivity of CPMV, *Virology*, 34, 312, 1968.
122. **Swaans, H. and van Kammen, A.,** Reconsideration of the distinction between the severe and yellow strains of cowpea mosaic virus, *Neth. J. Plant Pathol.*, 79, 257, 1973.
123. **Bancroft, J. B.,** A virus made from parts of the genomes of brome mosaic and cowpea chlorotic mottle viruses, *J. Gen. Virol.*, 14, 223, 1972.
124. **Habili, N. and Francki, R. I. B.,** Comparative studies on tomato aspermy and cucumber mosaic viruses. III. Further studies on relationship and construction of a virus from parts of the two viral genomes, *Virology*, 61, 443, 1974.
125. **Ghabrial, S. A. and Lister, R. M.,** Coat protein and symptom specification in tobacco rattle virus, *Virology*, 52, 1, 1973.
126. **Randles, J. W., Harrison, B. D., Murant, A. F., and Mayo, M. A.,** Packaging and biological activity of the two essential RNA species of tomato black ring virus, *J. Gen. Virol.*, 36, 187, 1977.
127. **Darby, G. and Minson, A. C.,** The structure of tobacco rattle virus ribonucleic acids: common nucleotide sequences in the RNA species, *J. Gen. Virol.*, 21, 285, 1973.
128. **Stubbs, J. D. and Kaesberg, P.,** Amino acid incorporation in an *Escherichia coli* cell-free system directed by bromegrass mosaic virus ribonucleic acid, *Virology*, 33, 385, 1967.
129. **van Ravenswaay-Claasen, J. C., van Leeuwen, A. B. J., Duijts, G. A. H., and Bosch, L.,** *In vitro* translation of alfalfa mosaic virus RNA, *J. Mol. Biol.*, 23, 535, 1967.
130. **Pinck, L. and Fauquet, C.,** Analysis of the pancreatic ribonuclease digestion products of alfalfa mosaic virus ribonucleic acid: sequence homologies between the different RNAs, *Eur. J. Biochem.*, 57, 441, 1975.
131. **Shih, D. S., Lane, L. C., and Kaesberg, P.,** Origin of the small component of brome mosaic virus, *J. Mol. Biol.*, 64, 353, 1972.
132. **Lot, H., Jonard, G., and Richards, K.,** Cucumber mosaic virus RNA 5: partial characterization and evidence for no large sequence homologies with genomic RNAs, *FEBS Lett.*, 80, 395, 1977.
133. **Bol, J. F., Brederode, F. T., Janze, G. C., and Rauh, D. K.,** Studies on sequence homology between the RNAs of alfalfa mosaic virus, *Virology*, 65, 1, 1975.
134. **Gould, A. R. and Symons, R. H.,** Determination of the sequence homology between the four RNA species of cucumber mosaic virus by hybridization analysis with complementary DNA, *Nucleic Acids Res.*, 4, 3787, 1977.
135. **Palomar, M. K., Brakke, M. K., and Jackson, A. O.,** Base sequence homology in the RNAs of barley stripe mosaic virus, *Virology*, 77, 471, 1977.
136. **van Kammen, A.,** Cowpea mosaic virus, un virus au genome divise, *Physiol. Veg.*, 9, 479, 1971.
137. **Jaspars, E. M. J.,** Plant viruses with a multipartite genome, *Adv. Virus Res.*, 19, 37, 1974.
138. **Palese, P.,** The genes of influenza virus, *Cell*, 10, 1, 1977.
139. **Bruening, G. and Agrawal, H. O.,** Infectivity of a mixture of cowpea mosaic virus ribonucleoprotein components, *Virology*, 32, 306, 1967.
140. **Mayo, M. A., Harrison, B. D., Murant, A. F., and Barker, H.,** Cross-linking of RNA induced by ultraviolet irradiation of particles of raspberry ringspot virus, *J. Gen. Virol.*, 19, 155, 1972.
141. **Bancroft, J. B. and Lane, L. C.,** Genetic analysis of cowpea chlorotic mottle and brome mosaic viruses, *J. Gen. Virol.*, 19, 381, 1973.
142. **Siegel, A.,** Studies on the induction of tobacco mosaic virus mutants with nitrous acid, *Virology*, 11, 156, 1960.
143. **Lister, R. M.,** Tobacco rattle, NETU, viruses in relation to functional heterogeneity in plant viruses, *Fed. Proc.*, 28, 1875, 1969.
144. **Sanger, H. L.,** Defective plant viruses in *Molecular Genetics*, Wittman, H. G. and Schuster, H., Eds., Springer-Verlag, Berlin, 1968, 300.
145. **van Kammen, A.,** Plant viruses with a divided genome, *Ann. Rev. Phytopathol.*, 10, 125, 1972.
146. **Lister, R. M. and Bracker, C. E.,** Defectiveness and dependence in three related strains of tobacco rattle virus, *Virology*, 37, 262, 1969.
147. **Robinson, D. J.,** A variant of tobacco rattle virus: evidence for a second gene in RNA-2, *J. Gen. Virol.*, 35, 37, 1977.
148. **Fritsch, C., Mayo, M. A., and Hirth, L.,** Further studies on the translation products of tobacco rattle virus *in vitro*, *Virology*, 77, 722, 1977.
149. **Robinson, D. J.,** Properties of two temperature-sensitive mutants of tobacco rattle virus, *J. Gen. Virol.*, 21, 499, 1973.
150. **Robinson, D. J.,** Early events in local infection of *Chenopodium amaranticolor* leaves by mutant and wild-type strains of tobacco rattle virus, *J. Gen. Virol.*, 24, 391, 1974.

151. **Harrison, B. D. and Woods, R. D.,** Serotypes and particle dimensions of tobacco rattle viruses from Europe and America, *Virology*, 28, 610, 1966.
151a. **Miki, T. and Okada, Y.,** Comparative studies on some strains of tobacco rattle virus, *Virology*, 42, 993, 1970.
152. **Lister, R. M.,** Functional relationships between virus-specific products of infection by viruses of the tobacco rattle type, *J. Gen. Virol.*, 2, 43, 1968.
153. **Wood, H. A.,** Genetic complementation between the two nucleoprotein components of cowpea mosaic virus, *Virology*, 49, 592, 1972.
154. **Wu, G.-J. and Bruening, G.,** Two proteins from cowpea mosaic virus, *Virology*, 46, 596, 1971.
155. **Geelen, J. L. M. C., van Kammen, A., and Verduin, B. J. M.,** Structure of the capsid of cowpea mosaic virus. The chemical subunit: molecular weight and number of subunits per particle, *Virology*, 49, 205, 1972.
156. **Geelen, J. L. M. C., Rezelman, G., and van Kammen, A.,** The infectivity of the two electrophoretic forms of cowpea mosaic virus, *Virology*, 51, 279, 1973.
157. **Niblett, C. L. and Semancik, J. S.,** Conversion of the electrophoretic forms of cowpea mosaic virus *in vivo* and *in vitro*, *Virology*, 38, 685, 1969.
158. **Niblett, C. L. and Semancik, J. S.,** The significance of the coat protein in infection by the electrophoretic forms of cowpea mosaic virus, *Virology*, 41, 201, 1970.
159. **Kassanis, B., White, R. F., and Woods, R. D.,** Genetic complementation between middle and bottom components of two strains of radish mosaic virus, *J. Gen. Virol.*, 20, 277, 1973.
160. **de Jager, C. P.,** Genetic analysis of cowpea mosaic virus mutants by supplementation and reassortment tests, *Virology*, 70, 151, 1976.
161. **Bruening, G.,** The inheritance of top component formation in cowpea mosaic virus, *Virology*, 37, 577, 1969.
162. **de Jager, C. P. and van Kammen, A.,** Relationship between the components of cowpea mosaic virus III. Location of genetic information for two biological functions in the middle component of CPMV, *Virology*, 41, 281, 1970.
163. **de Jager, C. P., Zabel, P., van der Beek, C. P., and van Kammen, A.,** Genetic and physiological characterization of a temperature sensitive mutant of cowpea mosaic virus, *Virology*, 76, 164, 1977.
164. **Siler, D. J., Babcock, J., and Bruening, G.,** Electrophoretic mobility and enhanced infectivity of a mutant of cowpea mosaic virus, *Virology*, 71, 560, 1976.
165. **Gopo, J. M. and Frist, R. H.,** Location of the gene specifying the smaller protein of the cowpea mosaic virus capsid, *Virology*, 79, 259, 1977.
166. **Moore, B. J. and Scott, H. A.,** Properties of a strain of bean pod mottle virus, *Phytopathology*, 61, 831, 1971.
167. **Harrison, B. D. and Murant, A. F.,** Nematode transmissibility of pseudo-recombinant isolates of tomato black ring virus, *Ann. Appl. Biol.*, 86, 209, 1977.
168. **Hanada, K. and Harrison, B. D.,** Effects of virus genotype and temperature on seed transmission of nepoviruses, *Ann. Appl. Biol.*, 85, 79, 1977.
169. **Harrison, B. D. and Hanada, K.,** Competitiveness between genotypes of raspberry ringspot virus is mainly determined by RNA 1, *J. Gen. Virol.*, 31, 455, 1976.
170. **Harrison, B. D., Murant, A. F., Mayo, M. A., and Roberts, I. M.,** Distribution of determinants for symptom production, host range and nematode transmissibility between the two RNA components of raspberry ringspot virus, *J. Gen. Virol.*, 22, 233, 1974.
171. **Harrison, B. D., Murant, A. F., and Mayo, M. A.,** Two properties of raspberry ringspot virus determined by its smaller RNA, *J. Gen. Virol.*, 17, 137, 1972.
172. **Fulton, J. P., Gamez, R., and Sott, H. A.,** Cowpea chlorotic mottle and bean yellow stipple viruses, *Phytopathology*, 65, 741, 1975.
173. **Bancroft, J. B., McDonald, J. G., and Rees, M. W.,** A mutant of cowpea chlorotic mottle virus with a perturbed assembly mechanism, *Virology*, 75, 293, 1976.
174. **Bancroft, J. B., Rees, M. W., Johnson, M. W., and Dawson, J. R. O.,** A salt stable mutant of cowpea chlorotic mottle virus, *J. Gen. Virol.*, 21, 507, 1973.
175. **Bancroft J. B., McLean, G. D., Rees, M. W., and Short, M. N.,** The effect of an arginyl to a cysteinyl replacement on the uncoating behavior of a spherical plant virus, *Virology*, 45, 707, 1971.
176. **Bancroft, J. B., Rees, M. W., Dawson, J. R. O., McLean, G. D, and Short, M. N.,** Some properties of a temperature sensitive mutant of cowpea chlorotic mottle virus, *J. Gen. Virol.*, 16, 69, 1972.
177. **Marchoux, G., Devergne, J., Marrou, J., Douine, L., and Lot, H.,** Complementation entre ARN de differentes souches du virus de la Mosaique de Concombre. Localisation sur l'ARN 3, de plusieurs proprietes, dont certaines liees a la nature de la capside, *C. R. Acad. Sci.*, 279D, 2165, 1974.
178. **Mossop, D. W. and Francki, R. I. B.,** Association of RNA 3 with aphid transmission of cucumber mosaic virus, *Virology*, 81, 177, 1977.

179. **Marchoux, G., Marrou, J., Douine, L., Lot, H., Quiot, J. B., and Clement, M.,** Complementation entre souches du Virus de la Mosaique du Concombre. Localisation d'un gene sur l'ARN-3, *C. R. Acad. Sci.*, 278D, 889, 1974.
180. **Marchoux, G., Marrou, J., and Quiot, J. B.,** Complementation entre ARN de differentes souches du virus de la Mosaique du Concombre. Mise en evidence d'une interaction entre deux ARN pour determiner un type de symptome, *C. R. Acad. Sci.*, 279D, 1943, 1974.
181. **Devergne, J. S. and Cardin, L.,** Contribution a l'Etude du virus de la Mosaique du Concombre (CMV) IV. Essai de classification de plusieurs isolats la base de leur structure antigenique, *Ann. Phytopathol.*, 5, 409, 1973.
182. **Wood, K. R. and Coutts, R. H. A.,** Preliminary studies on the RNA components of three strains of cucumber mosaic virus, *Physiol. Plant Pathol.*, 7, 139, 1975.
183. **van Vloten-Doting, L., Dingjan-Versteegh, A., and Jaspars, E. M. J.,** Three nucleoprotein components of alfalfa mosaic virus necessary for infectivity, *Virology*, 40, 419, 1970.
184. **Bol, J. F., van Vloten-Doting, L., and Jaspars, E. M. J.,** A functional equivalence of top component "a" RNA and coat protein in the initiation of infection of alfalfa mosaic virus, *Virology*, 46, 73, 1971.
185. **Bol, J. F., Kraal, B., and Brederode, F. T.,** Limited proteolysis of alfalfa mosaic virus: influence on the structural and biological function of the coat protein, *Virology*, 58, 101, 1974.
186. **van Vloten-Doting, L. and Jaspars, E. M. J.,** The uncoating of alfalfa mosaic virus by its own RNA, *Virology*, 48, 699, 1972.
187. **Dingjan-Versteegh, A., van Vloten-Doting, L., and Jaspars, E. M. J.,** Alfalfa mosaic virus hybrids constructed by exchanging nucleoprotein components, *Virology*, 49, 716, 1972.
188. **Hartmann, D., Mohier, E., Leroy, C., and Hirth, L.,** Genetic analysis of alfalfa mosaic virus mutants, *Virology*, 74, 470, 1976.
189. **Franck, A. and Hirth, L.,** Temperature-resistant strains of alfalfa mosaic virus, *Virology*, 70, 283, 1976.
190. **Dingjan-Versteegh, A., van Vloten-Doting, L., and Jaspars, E. M. J.,** Confirmation of the constitution of alfalfa mosaic hybrid genomes in backcross experiments, *Virology*, 59, 328, 1974.
191. **Lister, R. M., Ghabrial, S. A., and Saksena, K. N.,** Evidence that particle size heterogeneity is the cause of centrifugal heterogeneity in tobacco streak virus, *Virology*, 49, 290, 1972.
192. **Gonsalves, D. and Garnsey, S. M.,** Functional equivalence of an RNA component and coat protein for infectivity of citrus leaf rugose virus, *Virology*, 64, 23, 1975.
193. **van Vloten-Doting, L.,** Coat protein is required for infectivity of tobacco streak virus: biological equivalence of the coat proteins of tobacco streak and alfalfa mosaic viruses, *Virology*, 65, 215, 1975.
194. **Gonsalves, D. and Garnsey, S. M.,** Infectivity of heterologous RNA-protein mixtures from alfalfa mosaic, citrus leaf rugose, citrus varietation and tobacco streak viruses, *Virology*, 67, 319, 1975.
195. **Fulton, R. W.,** Inheritance and recombination of strain-specific characters in tobacco streak virus, *Virology*, 50, 810, 1972.
196. **Fulton, R. W.,** The role of top particles in recombination of some characters of tobacco streak virus, *Virology*, 67, 188, 1975.
197. **Fulton, R. W.,** The role of particle heterogeneity in infection by tobacco streak virus, *Virology*, 41, 288, 1970.
198. **Clarke, R. G. and Bath, J. E.,** Serological properties of aphid-transmissible and aphid-nontransmissible pea enation mosaic virus isolates, *Phytopathology*, 67, 1035, 1977.
199. **Hull, R.,** Particle differences related to aphid-transmissibility of a plant virus, *J. Gen. Virol.*, 34, 183, 1977.
200. **Powell, C. A.,** The relationship between soil-borne wheat mosaic virus and tobacco mosaic virus, *Virology*, 71, 453, 1976.
201. **Tsuchizaki, T., Hibino, H., and Saito, Y.,** The biological functions of short and long particles of soil-borne wheat mosaic virus, *Phytopathology*, 65, 523, 1975.
201a. **Jackson, A. O. and Brakke, M. K.** Multicomponent properties of barley stripe mosaic virus ribonucleic acid, *Virology*, 55, 483, 1973,
202. **Kalmakoff, J. and Tremaine, J. H.,** Some physical and chemical properties of carnation ringspot virus, *Virology*, 33, 10, 1967.
203. **Best, R. J.,** Tomato spotted wilt virus, *Adv. Virus Res.*, 13, 65, 1968.
204. **van den Hurk, J., Tas, P. W. L., and Peters, D.,** The ribonucleic acid of tomato spotted wilt virus, *J. Gen. Virol.*, 36, 81, 1977.
205. **Black, L. M.,** Wound tumor virus, C.M.I./A.A.B., *Descriptions of Plant Viruses*, No. 34, Commonwealth Mycological Institute and Association of Applied Biologists, Kew, Surrey, England, 1970.
206. **Ramig, R. F. and Fields, B. N.,** Reoviruses, in *The Molecular Biology of Animal Viruses*, Nayak, D. P., Ed., Marcell Dekker, New York, 1977, 383.

207. **Silverstein, S. C., Christman, J. K., and Acs, G.,** The reovirus replicative cycle, *Annu. Rev. Biochem.*, 45, 375, 1976.
208. **Reddy, D. V. R. and Black, L. M.,** Deletion mutations of the genome segments of wound tumor virus, *Virology*, 61, 458, 1974.
209. **Reddy, D. V. R. and Black, L. M.,** Increase of wound tumor virus in leafhoppers as assayed in vector cell monolayers, *Virology*, 50, 412, 1972.
210. **Black, L. M. and Markham, R.,** Base pairing in the ribonucleic acid of wound tumor virus, *Neth. J. Plant Pathol.*, 69, 215, 1963.
211. **Kalmakoff, J., Lewandowski, L. J., and Black, D. R.,** Comparison of the ribonucleic acid subunits of reovirus, cytoplasmic polyhedrosis virus, and wound tumor virus, *J. Virol.*, 4, 851, 1969.
212. **Milne, R. G. and Lovisolo, O.,** Maize rough dwarf and related viruses, *Adv. Virus Res.*, 21, 267, 1977.
213. **Luisoni, E., Lovisolo, O., Kitagawa, Y., and Shikata, E.,** Serological relationship between maize rough dwarf virus and rice black streaked dwarf virus, *Virology*, 52, 281, 1973.
214. **Reddy, D. V. R. and MacLeod, R.,** Polypeptide components of wound tumor virus, *Virology*, 70, 274, 1976.
215. **Gomatos, P. J. and Tamm, I.,** Animal and plant viruses with double helical RNA, *Proc. Natl. Acad. Sci. U.S.A.*, 50, 878, 1963.
216. **Miura, K., Kimura, I., and Suzuki, N.,** Double stranded ribonucleic acid from rice dwarf virus, *Virology*, 28, 571, 1966.
217. **Reddy, D. V. R., Rhodes, D. P., Lesnaw, J. A., MacLeod, R., Banerjee, A. K., and Black, L. M.,** *In vitro* transcription of wound tumor virus RNA by virion-associated RNA transcriptase, *Virology*, 80, 356, 1977.
217a. **Rhodes, D. P., Reddy, D. V. R., OacLeod, R., Black, L. M., and Banerjee, A. K.,** In vitro synthesis of RNA containing 5′-terminal structure $^7$mG(5′)ppp(5′)Ap$^m$ by purified wound tumor virus, *Virology*, 76, 554, 1977.
218. **Lewandowski, L. J. and Leppla, S. H.,** Comparison of the 3′ termini of discrete segments of the double-stranded RNA genomes of cytoplasmic polyhedrosis virus (CPV), wound tumor virus (WTV) and reovirus, *J. Virol.*, 10, 965, 1972.
219. **Martin, S. A. and Zweerink, H. J.,** Isolation and characterization of two types of bluetongue virus particles, *Virology*, 50, 495, 1972.
220. **Reddy, D. V. R., Kimura, I., and Black, L. M.,** Co-electrophoresis of dsRNA from wound tumor and rice dwarf viruses, *Virology*, 60, 293, 1974.
221. **Reddy, D. V. R., Boccardo, G., Outridge, R., Teakle, D. S., and Black, L. M.,** Electrophoretic separation of dsRNA genomic segments from Fiji disease and maize rough dwarf viruses, *Virology*, 63, 287, 1975.
222. **Reddy, D. V. R., Shikata, E., Boccardo, G., and Black, L. M.,** Coelectrophoresis of double-stranded RNA from maize rough dwarf and rice black streaked dwarf viruses, *Virology*, 67, 279, 1975.
223. **Reddy, D. V. R., Grylls, N. E., and Black, L. M.,** Electrophoretic separation of dsRNA genome segments from maize wallaby ear virus and its relationship to other phytoreoviruses, *Virology*, 73, 36, 1976.
224. **Kodama, T. and Suzuki, N.,** RNA polymerase activity in purified rice dwarf virus, *Ann. Phytopathol. Soc. Japan*, 39, 251, 1973.
225. **Ikegami, M. and Francki, R. I. B.,** RNA dependent RNA polymerase associated with subviral particles of Fiji disease virus, *Virology*, 70, 292, 1976.
226. **Reddy, D. V. R. and Black, L. M.,** Isolation and replication of mutant populations of wound tumor virions lacking certain genome segments, *Virology*, 80, 336, 1977.
227. **Huang, A. S.,** Defective interfering viruses, *Annu. Rev. Microbiol.*, 27, 101, 1973.
228. **Lane, L. C.,** The bromoviruses, *Adv. Virus Res.*, 19, 151, 1974.
229. **Kitajima, E. W., Muller, G. W., Costa, A. S., and Yuki, W.,** Short rodlike particles associated with citrus leprosis, *Virology*, 50, 254, 1972.
230. **Chauvin, C., Jacrot, B., and Witz, J.,** The structure and molecular weight of satellite tobacco necrosis virus: a neutron small-angle scattering study, *Virology*, 83, 479, 1977.
231. **Bishop, D. H. L., Claybrook, J. R., and Spiegelman, S.,** Electrophoretic separation of viral nucleic acids on polyacrylamide gels, *J. Mol. Biol.*, 26, 373, 1967.
232. **Kassanis, B.,** Properties and behavior of a virus depending for its multiplication on another, *J. Gen. Microbiol.*, 27, 477, 1962.
233. **Kassanis, B. and Phillips, M. P.,** Serological relationship of strains of tobacco necrosis virus and their ability to activate strains of satellite virus, *J. Gen. Virol.*, 9, 119, 1970.
234. **Uyemoto, J. K., Grogan, R. G., and Wakeman, J. R.,** Selective activation of satellite virus strains by strains of tobacco necrosis virus, *Virology*, 34, 410, 1968.

235. **Shoulder, A., Darby, G., and Minson, T.,** RNA-RNA hybridisation using $^{125}$I-labeled RNA from tobacco necrosis virus and its satellite, *Nature (London)*, 251, 733, 1974.
236. **Klein, A. and Reichmann, M. E.,** Isolation and characterization of two species of double stranded RNA from tobacco leaves doubly infected with tobacco necrosis and satellite tobacco necrosis viruses, *Virology*, 42, 269, 1970.
237. **Uyemoto, J. K. and Grogan, R. G.,** Chemical characterization of tobacco necrosis and satellite viruses, *Virology*, 39, 79, 1969.
238. **Lesnaw, J. A. and Reichmann, M. E.,** Identity of the 5′-terminal RNA nucleotide sequence of the satellite tobacco necorsis virus and its helper virus: possible role of the 5′-terminus in the recognition by virus-specific RNA replicase, *Proc. Natl. Acad. Sci. U.S.A.*, 66, 140, 1970.
239. **Horst, J., Fraenkel-Conrat, H., and Mandeles, S.,** Terminal heterogeneity at both ends of the satellite tobacco necrosis virus ribonucleic acid, *Biochemistry*, 10, 4748, 1971.
240. **Niblett, C. L. and Paulsen, A. Q.,** Purification and further characterization of panicum mosaic virus, *Phytopathology*, 65, 1157, 1975.
241. **Niblett, C. L., Paulsen, A. Q., and Toler, R. W.,** Panicum mosaic virus, C.M.I./A.A.B. *Descriptions of Plant Viruses*, No. 177, Commonwealth Mycological Institute and Association of Applied Biologists, Kew, Surrey, England, 1977.
242. **Buzen, F. G., Niblett, C. L., and Hooper, G. R.,** A possible satellite virus of panicum mosaic virus, Abstract 69th Annual Meeting, American Phytopathological Society, No. 231, 1977.
243. **Dodds, J. A. and Hamilton, R. I.,** The influence of barley stripe mosaic virus on the replication of tobacco mosaic virus in *Hordeum vulgare* L., *Virology*, 50, 404, 1972.
244. **Dodds, J. A. and Hamilton, R. I.,** Masking of the RNA genome of tobacco mosaic virus by the protein of barley stripe mosaic virus in doubly infected barley, *Virology*, 59, 418, 1974.
245. **Schneider, I. R.,** Characteristics of a satellite-like virus of tobacco ringspot virus, *Virology*, 45, 108, 1971.
246. **Schneider, I. R.,** Satellite-like particle of tobacco ringspot virus that resembles tobacco ringspot virus, *Science*, 166, 1627, 1969.
247. **Sogo, J. M., Schneider, I. R., and Koller, T.,** Size determination by electron microscopy of the RNA of tobacco ringspot satellite virus, *Virology*, 57, 459, 1974.
248. **Schneider, I. R., Hull, R., and Markham, R.,** Multidense satellite of tobacco ringspot virus: a regular series of components of different densities, *Virology*, 47, 320, 1972.
249. **Owens, R. A. and Schneider, I. R.,** Satellite of tobacco ringspot virus RNA lacks detectable mRNA activity, *Virology*, 80, 222, 1977.
250. **Schneider, I. R.,** Defective plant viruses, in *Virology in Agriculture* Romberger, J. A., Anderson, J. D., and Powell, R. L., Eds., Allanheld Osmun, Montclair, New Jersey, 1976, 201.
251. **Schneider, I. R. and Thompson, S. M.,** Double stranded nucleic acids found in tissue infected with the satellite of tobacco ringspot virus, *Virology*, 78, 453, 1977.
252. **Kaper, J. M., Tousignant, M. E., and Lot, H.,** A low molecular weight replicating RNA associated with a divided genome plant virus: defective or satellite RNA?, *Biochem. Biophys. Res. Commun.*, 72, 1237, 1976.
253. **Kaper, J. M. and Waterworth, H. E.,** Cucumber mosaic virus associated RNAs: causal agent for tomato necrosis, *Science*, 196, 429, 1977.
254. **Kaper, J. M. and Tousignant, M. E.,** Cucumber mosaic virus associated RNA 5: I. Role of host plant and helper strain in determining amount of associated RNA 5 with virions, *Virology*, 80, 186, 1977.
255. **Owens, R. A. and Kaper, J. M.,** Cucumber mosaic virus-associated RNA 5, II. *In vitro* translation in a wheat germ protein synthesis system, *Virology*, 80, 196, 1977.
256. **Morris, T. J.,** personal communication.
257. **Lane, L. C.,** unpublished.
258. **Minson, A. C.,** personal communication.
259. **Jaspers, E. M. J.,** personal communication.
260. **Symons, R. H.,** personal communication.

# TRANSLATION OF PLANT VIRUS RIBONUCLEIC ACIDS IN EXTRACTS FROM EUKARYOTIC CELLS

J. W. Davies

## TABLE OF CONTENTS

# I. INTRODUCTION

In the early 1960s there was an awakened interest in in vitro protein synthesis, or rather amino acid incorporation, following the finding that cell-free extracts of *Escherichia coli* could be "programmed" by synthetic polynucleotides to synthesize polypeptides.[1,2] This contributed to the cracking of the genetic code and provided a most useful, yet relatively simple, means of elucidating some of the intricate steps involved in protein biosynthesis without actually synthesizing a protein. One of the first natural ribonucleic acids tested for messenger activity in such extracts was tobacco mosaic virus (TMV) RNA.[1,3] The ease of purification and ready availability of plant viral RNA was no doubt a key factor in its choice. However, whereas the RNAs of small RNA bacteriophages proved to be efficient in vitro messengers for recognizable bacteriophage proteins,[4-6] TMV RNA did not, there being no detectable TMV-specific proteins among the in vitro products.[7] Smaller plant viral RNAs showed more promise. By 1965, synthesis of satellite of tobacco necrosis virus (STNV) coat protein was shown to occur in *E. coli* extracts, in response to STNV RNA.[8] Later, it was reported that a polypeptide showing "striking similarity" to alfalfa mosaic virus (AMV) coat protein was produced in *E. coli* extracts supplied with the smallest RNA from AMV.[9] Brome mosaic virus (BMV) RNA directed the synthesis of polypeptides whose tryptic peptide maps "resembled" that of BMV coat protein.[10] While these reports indicated the plus-strand messenger nature of the RNAs and gave clues concerning the location of the coat protein genes, *de novo* synthesis of authentic coat protein was not clearly demonstrated. Further indication of the infidelity of translation was provided in the case of STNV RNA,[11] the authors concluding that only about 5% of the total in vitro products was authentic virus coat protein. Although it has been confirmed that STNV RNA and BMV RNA can be translated into some genuine coat protein polypeptides,[12,13] the efficiency of translation is less in *E. coli* cell-free extracts than in wheat embryo extracts.[12,14]

It was the advent of the wheat embryo and other eukaryote cell-free systems which opened renewed investigations of the translation of plant viral RNAs. The initial development of the system in the laboratory of Marcus,[15,16] as with early prokaryote systems, involved TMV RNA as the natural messenger. It was subsequently shown in that laboratory and others that the translation products included a heterogeneous mixture of products and little, if any, TMV coat protein.[17-19] The apparently negative results in the search for TMV coat protein as a product of TMV RNA translation in both prokaryote and eukaryote cell-free extracts proved to be correct when the mechanism of TMV RNA translation strategy was later clarified.[20-22] With *E. coli* extracts, as with wheat embryo, it was the smaller RNAs of satellite and multicomponent plant viruses that initially showed most promise in in vitro translation. STNV[12] and BMV[23] gave an efficiency of authentic coat protein production equal to that of bacteriophage RNAs in bacterial extracts.[12,14,23] In the case of multicomponent viruses, this gave a particular application in identifying which RNAs code for which viral polypeptides.

After the initial success with wheat embryo (directly prepared from seeds), wheat germ (from commercial mills) was used instead[14,24,25] with modifications made to increase the efficiency of amino acid incorporation. Cell-free extracts from animal cells were also shown to be capable of translating plant viral RNAs with fidelity.[26-28]

Some examples of applications of plant viral RNA translation in these eukaryotic cell-free extracts from wheat embryo, wheat germ, and animal cells will be considered in this chapter. It will be seen that, apart from the feasible novelty of complete in vitro synthesis and assembly of a plant virus, cell-free translation of plant viral RNAs can provide direct proof of gene location and confirmation of the existence of viral-coded proteins implied by in vivo studies (or perhaps in some cases, indication of the possible existence of a viral-coded polypeptide before it is found in vivo). Linked with RNA structure-sequence studies, some interesting data are accumulating concerning the mechanism of initiation of translation and the molecular tactics by which the viral RNA-coded proteins are derived. Whereas in some instances the viral RNA is the messenger for the viral protein per se, there are several examples of pretranslational cleavage of the RNA, or synthesis of a subgenomic RNA, and perhaps of posttranslational cleavage of a precursor polypeptide. Some ideas concerning the possible evolutionary relationships of plant viruses are discussed. Since plant viruses must adapt to the plant cell protein synthesis machinery, the study of plant virus RNA translation also gives some indication of the mechanisms involved in the translation of plant messenger RNAs (see Section II, Volume I by Hall).

## II. PREPARATION AND USE OF CELL-FREE EXTRACTS FOR TRANSLATION OF PLANT VIRAL RNAs

### A. Wheat Germ Cell-Free Extracts

The wheat embryo or wheat germ cell-free translation system has been described in several publications[12,14,23-25] and in reviews with detailed instructions.[29,30] Various small modifications have been made, but the methods of preparation and incubation are similar. A general procedure will be described here, with notes on some of the practical considerations. Minor changes in the conditions may be necessary for different RNAs. An important point, which will be referred to later in this chapter, is that the so-called optimum conditions for maximum amino acid incorporation are not necessarily the optimum conditions for high fidelity of translation.

Wheat (*Triticum aestivum*) embryos[12,23] or commercial wheat germ[14,24,25] can be used. Variation can occur with either, depending on the source. Wheat germ is generally preferred because of its availability and ease of preparation. The germ should consist mainly of yellow embryo: brown (chaff) or white (endosperm) contamination is detrimental to the activity. Flotation in a 2:5 mixture of cyclohexane and carbon tetrachloride[13,23,69] is recommended. Embryo material which floats is suitable for the preparation of cell-free extracts. This procedure may also serve to surface-sterilize the embryos. Endosperm material is perhaps not completely removed, but it is certainly minimized. It is mainly the endosperm of the seed which contains ribonuclease which, however, is induced in aleurone cells by imbibition of water. It is important therefore to store wheat germ, before or after floating, in a vacuum desiccator, preferably at 4°.

The extraction (grinding) buffer should contain 3 to 5 m*M* magnesium acetate, 90 to 120 m*M* potassium acetate (chloride ions may be inhibitory), and 1 m*M* dithiothreitol or equivalent −SH compound, added after autoclaving the buffer solution. Buffering is not obligatory at this stage,[19] but HEPES[69] (*N*-2-hydroxyethylpiperazine-*N*-2-ethane sulfonate) or potassium bicarbonate[29,69] may be used to give a pH of about 6.5 to 6.8 before the first centrifugation. If a higher pH is used (pH 7 to 8), endogenous messenger is released,[29] resulting in a high background activity and the necessity to preincubate the extract.[24,25] Preincubation can generally be omitted using the procedure which follows.

The wheat germ is ground briefly[25] with powdered glass or sand (sterile), then a small portion of grinding buffer is added as grinding and continued for 1 to 2 min. Buffer can be added stepwise during grinding.[14,25] A final proportion of about 5 to 10 mℓ buffer per 1 g of wheat germ is recommended. There is always a low level of ribonuclease present; thus a very concentrated extract can result in degradation of the messenger RNA. DNase may be added during extraction, but is not essential.

After grinding, the extract is centrifuged at 23,000 to 30,000 g for 10 to 20 min. The supernatant should be adjusted to between pH 7.4 and 7.6, with HEPES-KOH or tris-acetate buffer (e.g., with 1/100 volume of 1 *M* buffer). A second 30,000 g centrifugation is then performed. The clear part of this supernatant is collected, avoiding the surface (pellicle) material and the cloudy substance just above the pellet. This supernatant may then be dialyzed for 12 to 24 hr,[14,69] or if speed is essential, it can be passed through a small Sephadex® G-25 column.[24,25] The elution or dialysis buffer may be similar to the grinding buffer,[69] but buffered to pH 7.5 with tris-acetate, 20 m*M*. If the extract appears slightly cloudy after either of these treatments, it can be cleared with a low-speed (5000 to 10,000 g) centrifugation. The extract should then be frozen rapidly in small aliquots in liquid nitrogen, and stored in the same, or in a deep freeze at −70° or colder. Extracts remain active for several months.

The conditions for incubation of the extract with the messenger RNA should be determined carefully. It is necessary to further buffer the incubation mixture,[14,30] HEPES being preferable to tris(hydroxymethyl)aminomethane (tris). The buffer can be added in a mixture of potassium acetate and magnesium acetate, to bring the $K^+$ and $Mg^{++}$ to about 100 and 4 m*M*, respectively.[14] If HEPES-KOH buffer is used, the $K^+$ in this, and in any other ingredient (such as $K^+$ ATP), should be taken into account. The optimum concentration for $K^+$ is between 90 and 110 m*M* for incorporation measurements, but better fidelity is attained with higher $K^+$ (115 to 130 m*M*).[30,42,69,85] Unlike prokaryote cell-free systems, $NH_4^+$ is inhibitory and thus cannot be substituted for $K^+$. The $Mg^{++}$ concentration curve is very sharp, the optimum being about 4 m*M* (in the absence of polyamines) if the ATP concentration is 2 to 2.5 m*M*.[19,59,64,69] At a lower ATP concentration of, say, 1 m*M*, the $Mg^{++}$ may show an optimum of 2.5 to 3 m*M*.[25,33] The $Mg^{++}$ optimum is also lower if spermidine or spermine is used.[25,64,69] Instances of the effect of such polyamines are given later in the chapter. The ATP ($K^+$ salt) and GTP (tri-lithium) should be neutralized with tris or KOH. $Na^+$ ions may be inhibitory. The ATP and GTP may be added together in a mixture also containing the amino acids (unlabeled) and ATP generating system, if needed. The final concentrations should be around 1 to 2.5 m*M* ATP and 0.2 to 0.4 m*M* GTP. The ATP generating system recommended is creatine phosphate, 8 to 10 m*M*, and creatine phosphokinase (ATP:creatine *N*-phosphotransferase, E.C. No. 2.7.3.2.). The latter is often used at concentrations of 10 to 50 μg/mℓ, but this depends on the specific activity of the enzyme. The optimum should be determined. The pyruvate kinase system is not recommended, as batches of this enzyme can be inhibitory, probably due to ribonuclease contamination (enzyme suspended in $(NH_4)_2SO_4$ should also be avoided). In the absence of any such enzyme, phosphoenolpyruvate (PEP) may enhance activity,[14] but this effect appears to be independent of ATP. It is possible that PEP is used in wheat germ extracts as a precursor for alanine, which is negligibly incorporated as a labeled amino acid due to the high pool of newly synthesized alanine. The unlabeled amino acids should be at a final concentration of 0.01 to 0.025 m*M* (some need to be neutralized for solubility in the stock solution). One of the most suitable amino acids for radioactive labeling is leucine, as it is probably least metabolized (e.g., by decarboxylation, deamination) in the cell-free extract. Methionine is often used since it can be obtained at very high specific activities as [$^{35}S$]-L-methionine and is thus useful for rapid analysis of products by electrophoresis gel autoradiography. Proteins are initiated with methionine (via met-$tRNA_i$) in wheat germ extracts, but this is usually removed from the nascent polypeptide. With radioactive leucine incorporation, the time curve, after the initial 5- to 10-min lag phase, is almost linear for 60 min, and incorporation continues up to 90 min incubation[14] at 30°. The kinetics depend on the RNA translated, limiting ingredients of the system, and temperature. Failure of termination and attendent polypeptide release may be one reason why the synthesis stops (unpublished observations). The synthesis may proceed longer in extracts not preincubated, as any labile components which are irreversibly inactivated during incubation will already be depleting before the messenger is added. This could be important if long RNAs are to be translated.

To measure total amino acid incorporation, samples of the incubation mixtures are pipetted onto filter paper discs, which are subsequently washed in trichloroacetic acid (TCA) at 90° (usually in the presence of unlabeled amino acid or amino acid mixture), TCA at room temperature, ethanol, and ether. The discs must be free of TCA and water, both of which quench during scintillation counting. For analysis of in vitro products on polyacrylamide gels or other techniques, the reaction can be stopped by addition of ribonuclease and ethylene diaminetetra-acetic acid (disodium) (EDTA). Before electrophoresing, the mixture is heated to 60 to 100° for a few minutes, in the presence of 1 to 2% sodium dodecyl sulfate (SDS) and mercaptoethanol. Details of these and other methods will be found in the specific references given for the results presented in this chapter.

### B. Rabbit Reticulocyte Lysates

If the precautions described above are taken, and the wheat germ system is carefully optimized for quality as well as quantity of synthesis, good fidelity can be attained with most mRNAs tested. However, with some RNAs, a tendency for the synthesis of incomplete polypeptides has been reported.[13,30] While in some cases small changes in the preparation and incubation procedures and good RNA preparation can circumvent this, it may be useful to consider an alternative protein-synthesizing extract. The rabbit reticulocyte lysate system is widely used, and since it has successfully been used to translate a number of plant viral RNAs (the results of which will be referred to in this chapter), a brief description of the preparation of lysates will be given here.

Rabbits weighing 2 to 3 kg are made anemic by subcutaneous injections of 1.25% w/v acetylphenylhydrazine (0.9 mℓ/kg) over a period of about 4 days. The animal is killed after a further 5 or 6 days and the blood collected, usually by cardiac puncture.[31] The blood is kept cold, and heparin is added as an anticoagulant. Reticulocytes are harvested by centrifugation and washed several times with saline (130 m*M* NaCl, 5 m*M* KCl, and 7.5 m*M* $MgCl_2$). Some erythrocytes may be present. The reticulocytes are lysed with ice-cold sterile distilled water (1.5 volumes per volume of cell pellet). The lysate is centrifuged at 30,000 g for 15 to 20 min and the supernatant frozen rapidly in small aliquots. This lysate is very active at in vitro globin synthesis, but also translatesviral RNA. Owing to the endogenous globin synthesis, little or no stimulation is observed in total amino acid incorporation. A method has recently been described[28] for making such lysates completely dependent on added mRNA. This involves treating the extract with micrococcal nuclease, in the presence of $Ca^{++}$, to destroy globin messenger. The nuclease activity is dependent on $Ca^{++}$ ions. Afterwards, EGTA (ethylene glycol-bis(2-amino-ethylether)-*N,N*-tetraacetic acid) is added to remove $Ca^{++}$ ions. Because $Mg^{++}$ ions cannot substitute for $Ca^{++}$, the ribonuclease is no longer active in the extract. A frozen lysate is thawed and creatine kinase, 50 μg/mℓ (about 5 to 7 units/mℓ), and hemin, 25 μ*M*, are added.[28] (The hemin stock solution is 1 m*M* in 90% v/v ethylene glycol, 20 m*M* tris-HCl, pH 8.2, and 50 m*M* KCl.) The mixture is then made to (final concentrations) 100 m*M* KCl, 5.5 m*M* $MgCl_2$, 10 m*M* creatine phosphate, approximately 0.05 m*M* each amino acid, 1 m*M* $CaCl_2$, and 80 units micrococcal nuclease per milliliter (usually about 100 μg/mℓ). After incubation at 20° for 10 min, EGTA is added to 2 m*M* concentration. This concentration does not inhibit protein synthesis. The radioactive amino acid and mRNA are then added and the mixture incubated at 30°.[28] Addition of tRNA may be necessary for this system.[26,68]

To measure the total amino acid incorporation, a small aliquot of reaction mixture is pipetted into water (for example, 5 or 10 $\mu\ell$ into 1 m$\ell$), to which is added a half volume of 1 *M* NaOH and 0.5 *M* $H_2O_2$, the latter to decolorize the sample to avoid color quenching during scintillation counting. The high alkalinity discharges labeled amino acid from tRNA, thus a hot TCA wash is not necessary. After 5 min incubation at 80°, protein is precipitated with an equal volume of cold or room temperature TCA (20%). It is then necessary to collect the precipitae by filtration, usually on glass fiber discs, which afford a higher counting efficiency than cellulose filter paper. Preparation of samples for gel electrophoresis is the same for the wheat germ system.

In general, but not always, very similar results are obtained with wheat germ extracts and reticulocyte lysates. Examples will be given in this chapter. The wheat germ system is however cheaper and easier to prepare.

### C. RNA Extraction

The purified virus (see Section III, Volume II by Zaitlin and Lane) should be disrupted by a means suitable for that virus. A frequently used but not always adequate method is to heat in the presence of a detergent such as SDS. Extraction of the RNA is achieved by the well-known phenol biphasic methods, or by ultracentrifugation in the presence of SDS and an RNase inhibitor. No one method is suitable for all viruses (see Section III, Volume II by Zaitlin). Some general considerations are given in Section III, Volume II by Zaitlin and Lane. Further details of methods can be found in the specific references to the work discussed in this chapter.

The integrity and purity of the RNA should be tested, first by centrifugation and polyacrylamide gel electrophoresis under denaturing conditions (such as in formamide). Presence of RNA fragments or hidden breaks in the RNA is a likely cause of translation into more polypeptides in vitro than occur in vivo, making interpretation of the results difficult. Second, the purity of the RNA should be controlled. Proteins as well as smaller RNA breakdown fragments can bind to the RNA and interfere with translation. Freedom from protein (especially RNase!) should be an obvious requirement, but is surprisingly often assumed and not tested. Third, multicomponent RNAs can often not be completely separated from each other by gradient centrifugation of particles before RNA extraction and centrifugation of the RNA. It is possible to elute RNA from polyacrylamide gels and then translate it, being reasonably sure that the RNA component tested is free from other components. This is clearly important when attributing products, and hence cistrons, to a given RNA by analyzing the in vitro translation products.

## III. THE RNAs OF SATELLITE VIRUSES

### A. Satellite Tobacco Necrosis Virus RNA

Satellite tobacco necrosis virus (STNV) is a small (17 nm diameter) virus containing a small ($0.4 \times 10^6$ mol wt) RNA which is replicated independently in host cells, in association with certain strains of tobacco necrosis virus (TNV, see Section VI.D), which is apparently unrelated to STNV, but essential to it as the "helper virus". Hosts include tobacco (*Nicotiana tabacum*), French bean (*Phaseolus vulgaris*), and mung bean (*Phaseolus aureus*).

The RNA of STNV was the first plant viral RNA to be translated into a recognizable viral polypeptide in both prokaryote[8,12] and eukaryote[12] extracts. Klein et al.,[12] using essentially the conditions of Marcus[15,16] for wheat embryo extracts, succeeded in translating STNV RNA into STNV coat protein. The product was identified by its size (actually a little smaller than authentic coat protein) and by tryptic digest "fingerprints". Furthermore, the polypeptide synthesized in vitro possessed the N-terminal sequence Ala-Lys, as does the natural coat protein. After short incubations with [$^{35}S$] met-tRNA$_i^{met}$, the labeled peptide [$^{35}S$] met-ala-lys was found,[32] indicating that initiation began on a methionine codon and that the methionine was later removed. This work also confirmed the earlier suggestion[8] that STNV RNA was a monocistronic messenger for STNV coat protein. Recently, following treatment of the wheat ribosome-STNV RNA ($^{125}$I-labeled) initiation complex with ribonuclease A, a 38-base oligonucleotide has been isolated, which includes nucleotide sequences which correspond to the N-terminal met-ala-lys amino acid sequence.[33] The possible sequence for these amino acids is AUG · GGC · AAA. However, the exact codon assignments and sequence have not yet been ascertained.

Several years ago, it was demonstrated that STNV RNA possessed the N-terminal sequence pppApGpUp[34] or ppApGpUp.[35] This finding became particularly significant when numerous reports began to accumulate, indicating that viral mRNAs and eukaryotic cell mRNAs are "capped" at the 5′ terminus by $m^7G^{5'}$ppp (7-methylguanosine linked via its 5′-hydroxyl group triphosphate to the next base, often $^{5'}$Gp), which was implicated in ribosome-mRNA recognition for initiation of translation. It was subsequently shown that the free nucleotide $m^7G^{5'}$p inhibits the translation of RNAs with the $m^7G^{5'}$ppp cap, such as TMV RNA and rabbit globin RNA,[36] compared to STNV RNA. The extent of inhibition of STNV RNA translation by the nucleotide $m^7G^{5'}$p depends upon the concentration of the nucleotide and concentration of RNA.[38] At concentrations which exhibit a 75 to 98% inhibition of capped RNAs, STNV RNA with no cap is inhibited up to 55%.[38] The inhibition also seems to be dependent on the $K^+$ concentration. At 90 m$M$ $K^+$ it is less than at 150 m$M$. The nucleotide $m^7G^{5'}$p also inhibited the mRNA-dependent binding of the 40S ribosome-met-tRNA$_i^{met}$ in the formation of the 80S ribosome complex with TMV RNA and AMV RNA, but not with STNV RNA.[37] An $m^7G^{5'}$ppp cap can be added to STNV RNA using the capping enzyme of vaccinia virus. STNV RNA with such a cap shows increased ribosome binding (2.5-fold), the binding activity being similar to that of TMV RNA.[38] Guanylation (without the 7-methyl group) of STNV RNA has no effect on ribosome binding. Translation of STNV RNA in wheat germ extracts does not involve the in vitro formation of a 5′-terminal $m^7G^{5'}$ppp cap.[33] If the $m^7G^{5'}$ppp cap of other RNAs is involved in initiation of translation and/or its regulation (perhaps to varying degrees depending on other aspects of the RNA structure), STNV would appear to be the exception. This suggests the possibility that plant cells may be equipped with more than one mechanism for 80S ribosome-mRNA recognition.

### B. Other Satellite Virus RNAs

Little is yet known of other satellite or defective viruses. Already, however, it is clear that the function of the satellite RNA, its relationship with the helper virus, means of pathogenicity, and involvement of translation of the RNA are different with different satellites. Unlike STNV (Section III.A) in which the RNA codes for the coat protein, the satellite-like RNA5 of cucumber mosaic virus (Section IV.D) codes, at least in vitro, for two small polypeptides (perhaps with overlapping sequence) of unknown function. The RNA of the satellite of tobacco ringspot virus (a nepovirus — see Section V.B) does not code for any polypeptide[39] in vitro and in this respect is analogous to viroids (see Section IV, Volume II by Dickson). The so-called "satellite" RNA, RNA3 ($0.5 \times 10^6$ mol wt) of another nepovirus, tomato black ring virus, seems to code for a 50,000 mol wt polypeptide of unknown function (see Section V.B).

## IV. THE RNAs OF TRIPARTITE GENOME VIRUSES

The tripartite genome viruses are those types of multipartite or divided genome viruses, the genes of which are distributed among three RNA molecules which are necessary for infection (see Section III, Volume II by Lane). This includes the bromoviruses which have three virions with identical capsids. Two virions contain one or other of the largest RNAs, while the third contains the other RNA necessary for infection (intermediate size) and a fourth RNA not required for infection (small size) yet regenerated in infected cells. The RNAs of the cucumoviruses may be regarded as analogous to this system. Members of the Ilarviruses, such as tobacco streak virus (TSV), also have some analogous characteristics. Certainly alfalfa mosaic virus (AMV) is a tripartite genome virus, the RNAs of which are encapsidated in four different sized rod-like particles.

### A. Brome Mosaic Virus RNAs

Brome mosaic virus (BMV), type member of the Bromoviruses, infects a considerable number of plants (about 180).[40a] For most studies of the type discussed here, it has been grown solely in barley (*Hordeum vulgare*). BMV also systemically infects wheat (*Triticum aestivum*); thus, for this virus, the wheat embryo or wheat germ cell-free systems are truly homologous. Most of the research into BMV RNA translation has been done with the Russian strain.

The four RNAs of brome mosaic virus have molecular weights of 1.09 (RNA 1), 0.99 (RNA 2), 0.75 (RNA 3), and $0.28 \times 10^6$ (RNA 4) as calculated from gel electrophoresis under nondenaturing conditions.[40b] When added to wheat embryo cell-free extracts under conditions similar to those used for STNV,[12] BMV RNAs resulted in the synthesis of a product with tryptic peptides and molecular weight identical to BMV coat protein.[23] The theoretical coding capacity of these RNAs was clearly not being utilized, and it was concluded that the coat protein cistron when present in its monocistronic form (shown to be RNA 4) greatly reduced the translation of the other RNAs under the conditions used. At that time, it was not known if the incubation conditions were favorable for the translation of the two larger RNAs (see below), but the inhibition or preferential translation phenomenon was shown to be real for RNA 3. This RNA had previously been shown to carry the coat protein gene, by infectivity-complementation experiments.[40b] However, in the wheat embryo cell-free extracts,[23] a polypeptide (3a) of molecular weight around 34,000 to 35,000 was obtained, which had no tryptic peptides in common with the coat protein and could not be acetylated in vitro (from acetyl co-enzyme A) as could the in vitro coat protein-like product.[23] Thus, RNA 3 has two cistrons and two initiation sites, one "open" and one "closed". Possibly a pretranslational cleavage or specific partial transcription of RNA 3 is necessary to produce RNA 4, the functional coat protein messenger.

Addition of RNA 4 with RNA 3 resulted in considerable reduction in the number of translation rounds of the 3a cistron, the coat protein cistron of RNA 4 being preferentially translated. That it was not the coat protein cistron on RNA 3 that was translated was suggested by competition experiments with small RNAs from other plant viruses. For example, when cowpea chlorotic mottle virus (CCMV) RNA 4 or STNV RNA was added in place of BMV RNA 4, CCMV or STNV coat protein was synthesized.[13,41] Surprisingly, AMV RNA 4 (see Section IV.F) is not preferentially translated in the presence of BMV RNA 3, possibly indicating some specificity in this preferential translation. These data suggest an attractive regulation model, namely that RNA 4 (late functioning coat protein gene) switches off the translation of the RNA 3a cistron (earlier functioning gene) when replication has increased the number of RNA molecules to a certain critical proportion. Coat protein itself does not cause the gene to switch off. At low concentration of the RNAs (high ribosome to mRNA ratio), presumably the situation in early infection, the switch-off, or preferential translation, is less marked. Thus, at a suboptimal concentration (about one third) for incorporation activity, all four BMV RNAs are translated (see Figure 1). It is possible that variation of the other conditions for cell-free protein synthesis could bring about a similar effect. Direct extrapolation of results in vitro to the situation in vivo may be questionable, since the pattern and proportion of products are dependent, sometimes critically, upon the in vitro ingredients such as the $K^+$ and $Mg^{++}$ ion concentrations, presence of polyamines, or source of wheat germ or embryo.

Under favorable conditions, the two large RNAs of BMV (RNA 1 and RNA 2) can be translated into large polypeptides.[41,42] The $K^+$ concentration, if increased after initiation to a supraoptimal level (180 m*M* as opposed to 95 m*M*, the optimum concentration for incorporation activity), favored the translation of RNA 2 but not RNA 1.[42] The BMV RNA concentration was also important.[42] It seems from this work that RNA 1 and RNA 2 are monocistronic messengers for polypeptides corresponding to almost the entire coding capacity of each RNA (see Figure 1). Thus, three of the four BMV RNAs are monocistronic.

It is worth emphasizing the importance of carefully analyzing the optimum conditions (see Section II.A) for translation in vitro. The divalent and monovalent cation concentrations for maximum amino acid incorporation are not identical for all plant viral RNAs (cf. References 19, 24, 33, 64, 69). A $K^+$ "shift", such as for BMV RNA2, or involvement of a polyamine (see later in this chapter) may affect the translation of one RNA but not another. Furthermore, cell-free translation conditions are often optimized for maximum amino acid incorporation (total radioactivity in trichloroacetic acid precipitable material) and then the products are analyzed. The conditions for good fidelity (i.e., correct products) may not be the same as for maximum activity. The precise conditions seem to be more exacting for the translation of long RNAs into large polypeptides. The conditions for translation of small RNAs, such as BMV RNA 4, are less limiting.

The nucleotide sequence of a small T1 ribonuclease digest fragment of BMV RNA 4 has been established, after selecting the fragment from others by its ability to bind to ribosomes under conditions suitable for protein synthesis.[43] The 5′ terminal sequence up to the methionine initiation codon was found to be: XGUAUUAAUAAUG. The 5′ terminal nucleotide was later confirmed to be $pm^7G^{5'}$ linked to the next Gp by a 5′ppp linkage. This "cap" is a feature common to all four BMV RNAs.[44] Apart from this sequence being the first plant viral RNA ribosome binding site to be revealed, these data are quite intriguing from two points of view. First, the AUG codon is quite near to the 5′ terminus, unlike bacteriophage RNAs which have a longer pretranslation sequence before the AUG, which may be involved in ribosome binding. In the absence of much scope for secondary structure for ribosome recognition, a possible role for the $m^7G^{5'}$ $ppp^{5'}Xp$ cap in ribosome binding emerged. This has yet to be unequivocally proved. In the case of reovirus, and other RNAs, such a role was supported[45,46] but is questioned as a necessity for other RNAs,[47] including those of BMV.[48] Removal of the $m^7G^{5'}$ cap results in a reduction of in vitro BMV RNA translation but not its complete abolition. The structure of BMV RNAs may be such that the $m^7G^{5'}$ cap increases the efficiency of translation.

A second interesting observation concerning the structure of the 5′ terminal region of BMV RNA 4[43,48] is that the sequence from base 6 to 12 (UAAUAAU) has the potential for forming six base pairings with the sequence $AUCAUUA_{OH}$, the 3′ terminal sequence common to the several eukaryote ribosomal 18S RNAs so far analyzed.[49]

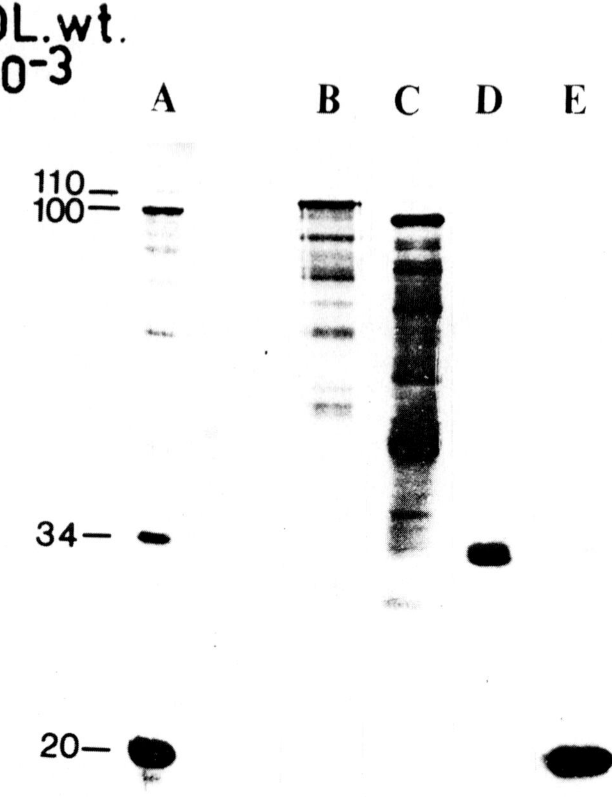


FIGURE 1. In vitro translation products of BMV RNAs. Wheat germ S-30 was prepared as described in Section II. The final incubation mixture contained per 100 μl, 50 μl of wheat germ S-30 (∿ 750 μg protein), 20 m*M* HEPES (pH 7.5 with KOH), 10 m*M* tris (pH 7.5 with acetic acid), 100 m*M* $K^+$ (mostly as acetate), 0.4 m*M* spermidine, 3 m*M* $Mg^{++}$ (acetate), 0.025 m*M* each amino acid except methionine, 5 μCi[$^{35}$S]methionine (0.002 m*M*), 2.5 m*M* ATP, 0.37 m*M* GTP, 10 m*M* creatine phosphate, 10 μg creatine kinase (0.14 unit), 0.5 m*M* dithioerythritol, and RNA as indicated. Incubation 90 min at 30°C. (A) RNAs 1, 2, 3, and 4 (unfractionated) 4 μg/100 μl; (B) RNA 1, 5 μg; (C) RNA 2, (perhaps degraded) 4 μg. Made to 150 m*M* $K^+$ after 10 min incubation; (D) RNA 3, 8 μg; (E) RNA 4, 10 μg. Aliquots (5 μℓ) were heated to 60°C with 1% SDS and 0.5% mercaptoethanol in tris-glycine pH 8.8, before applying to Laemmli[101] slab gels. BMV coat protein, BSA, ovalbumin, *E. coli* RNA polymerase subunits, β galactosidase, transferin, globulin subunits, and lactic dehydrogenase were used as molecular weight markers on 7, 10, and 12% polyacrylamide gels to calculate the sizes indicated here. This is a 10% gel.

This sequence has not yet been demonstrated in wheat germ ribosome 18S RNA, but it would not be surprising if it were present.

### B. Cowpea Chlorotic Mottle Virus RNAs

Cowpea chlorotic mottle virus (CCMV) is serologically related to BMV. Its RNAs are divided similarly to BMV among three virions. The molecular weights of RNAs 1, 2, 3, and 4, respectively, are 1.15, 1.0, 0.55, and 0.32 × $10^6$ (23S, 18S and 13S; see Section III, Volume II by Lane).

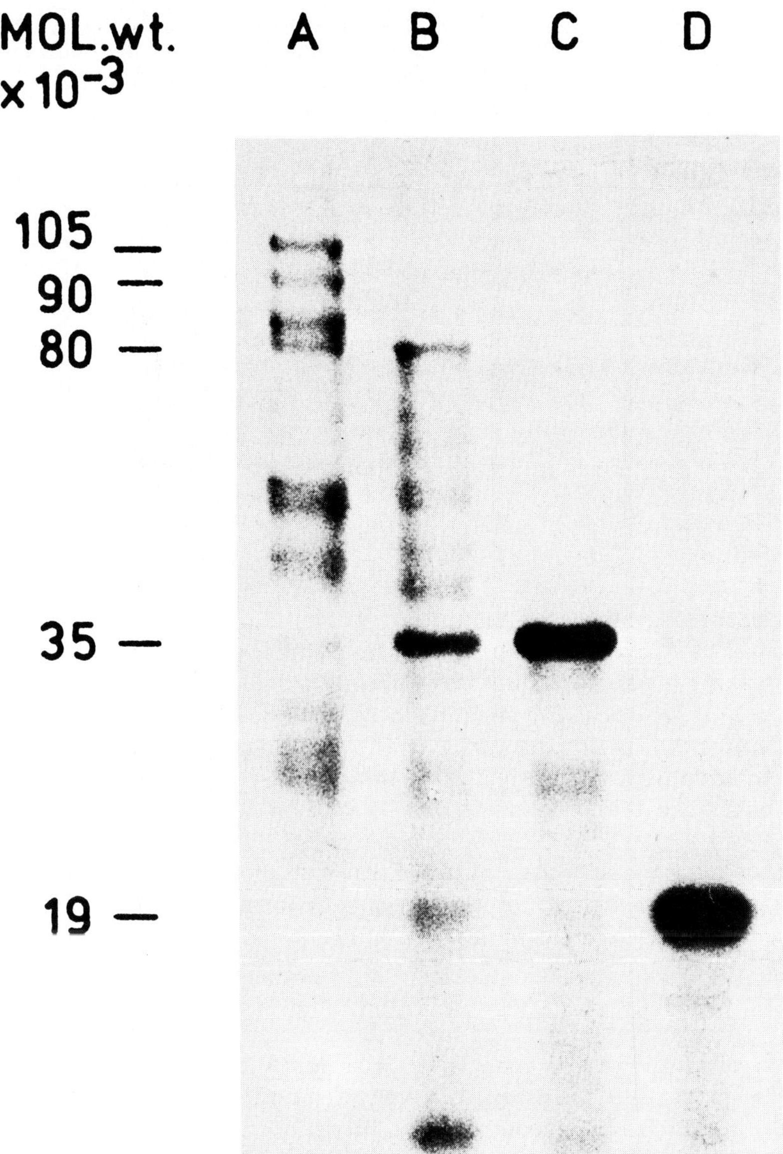


FIGURE 2. In vitro translation products of CCMV RNAs. The conditions were as described for BMV in Figure 1. CCMV RNAs were separated on polyacrylamide gels and extracted from homogenized gel particles, 4 to 5 μg per incubation. (A) RNAs 1 and 2. (B) "RNA 5". RNA from a minor gel band between RNAs 2 and 3. It may be a fragment of RNA 1 or 2 and is probably contaminated with RNA 3. (C) RNA 3. (D) RNA 4. The products were analyzed on discontinuous 10% polyacrylamide-SDS gels.[101] The approximate sizes of the products were estimated from marker proteins, including transferin, globulin subunits, β galactosidase, lactic dehydrogenase subunits, and cytochrome.

The in vitro translation of these RNAs has not been studied in much detail. Initial investigations[19] indicated similar characteristics to BMV RNA translation. Further results supporting this indication are presented in Figure 2. In wheat germ extracts, CCMV RNA 4 is translated into an approximately 19,000 mol wt polypeptide which is almost certainly coat protein. RNA 3 is translated into a 35,000 mol wt product, analogous to the BMV polypeptide 3a. RNAs 1 and 2 are translated into several products, of which there are two major polypeptides of molecular weights, approximately 105,000 and 90,000, indicating in vitro translation of these RNAs as monocistronic messengers under the conditions used. The earlier investigation under slightly different conditions showed very little of these products.[19] The significance of products of around 80,000 mol wt is not known. It is possible that they are incomplete products, at least one of which results from the translation of fragmented RNAs 1 or 2 (see Figure 2). Products around 80,000 and 50,000 mol wt also occur as in vitro products from other viral RNAs (see Figures 1, 3, 4, 6, and 7) and this may be significant.

### C. Broad Bean Mottle Virus RNAs

Little is known about the translation of BBMV RNAs, except that RNA 4 is translated into a polypeptide which comigrates on polyacrylamide (SDS electrophoresis in the same position as the BMV RNA 4 product) and is most likely BBMV coat protein (21,000 mol wt).[19] The RNA 3 products include a polypeptide of molecular weight between 30,000 and 35,000. It is likely that the translation strategy of BBMV is similar to that of the other bromoviruses.

### D. Cucumber Mosaic Virus RNAs

Cucumber mosaic virus (CMV) is the type member of the cucumoviruses. It has a wide host range which, in addition to *Cucumis sativus*, includes members of the leguminoseae (*Vigna sinensis* is a common local lesion host). For research purposes, it is most often grown in *Nicotiana tabacum*. The division of the genome is very similar to that of the bromoviruses, the small RNA 4 not being required for infection. One estimation under denaturing conditions (98% formamide) gave molecular weights for CMV RNAs 1 to 4 as 1.35, 1.16, 0.85, and 0.35 × $10^6$, respectively.[50] A fifth RNA of only 0.1 × $10^6$ daltons has been reported,[52] which appears when CMV strain S is grown in *Nicotiana tabacum* (when it exceeds the quantity of RNA 4). RNA 5 is only a minor component when CMV is grown in *Cucurbita pepo*. There is a possibility that this is a satellite RNA.[52]

The mixture of all four RNAs was shown to be translated in wheat germ[19] and wheat embryo[50] extracts into coat protein-sized polypeptide. It has been confirmed that this product is encoded in RNA 4 and is indeed coat protein.[50,51] In a recent report,[51] translation of CMV coat protein from RNA 4 was demonstrated in *Bufo marinus* oocytes. Strangely, in a rabbit reticulocyte lysate a product larger than coat protein was also observed, and this may represent the entire coding capacity of the RNA. This phenomenon, and the production of smaller minor products in this system and wheat extracts, is not understood. It may indicate in vitro read-through of a termination signal (see AMV RNA 3 translation, this chapter) and calls for a note of caution in the interpretation of in vitro results, as is indicated in a number of examples in this chapter. CMV RNA 3 produces only one polypeptide of 34,000 daltons in reticulocyte extracts, but in wheat germ it yields three products around 34,000 to 39,000 daltons, again indicating the uncertainty of extrapolating in vitro analysis of RNA genomes to the unknown in vivo situation.

The two large RNAs of CMV, isolated together from sucrose gradients, produced a wide size range of products in the wheat germ system.[50] In the later report[51] they were separated on, and extracted from, polyacrylamide gels and translated in wheat germ, wheat embryo extracts, and rabbit reticulocyte lysates. Again, a number of products were formed, including a polypeptide of 105,000 mol wt from RNA 1 and one of 120,000 from RNA 2. It is puzzling that the larger RNA produces the smaller of these two products and that neither were synthesized in oocytes.[51] The authors used a 50 m$M$ $K^+$ (KCl) concentration in the cell-free extracts, which is in contrast to 100 m$M$ $K^+$ (K acetate) and above used in most other reports described in this chapter (see Sections II.A and V.C).

Despite the uncertainty of interpretation of these results as discussed by Schwinghamer and Symons,[51] it is clear that the four RNAs of CMV can be translated in vitro into products which include 105,000, 120,000, and 34,000 mol wt polypeptides from RNAs 1, 2, and 3, and coat protein from RNA 4, indicating a translation strategy of this virus similar to that of the bromoviruses.

The CMV-associated RNA 5 has also been translated in vitro in wheat germ extracts.[53] Two polypeptides were detected, of 3800 and 5200 mol wt. The larger does not seem to be a precursor for the smaller, but it is not certain if these polypeptides are initiated independently. A $K^+$ concentration of 130 m$M$ (above the optimum for incorporation activity) favors synthesis of both products, as does the $Mg^{++}$ concentration of 2 m$M$. However, at 1.2 m$M$ $Mg^{++}$, synthesis of the smaller polypeptide is favored. The translation appears to be inhibited by 7-methylguanosine 5′-monophosphate (see above, Section III.A), suggesting that CMV RNA 5 may be capped with $m^7G^{5'}ppp$, but this is not yet confirmed.

The size of CMV RNA 5, 0.1 × $10^6$ daltons, suggests a coding capacity for a total polypeptide molecular weight of about 10,000. Thus, an RNA of this size could code for both polypeptides found in vitro without any overlapping sequences. The possibility that there could be two RNAs of identical size, each coding for one polypeptide, cannot yet be ruled out. When the mixture of all five CMV RNAs was translated, the small RNA 5 was not expressed.[53] It remains to be seen if small polypeptides of 3800 and 5200 mol wt can be detected in infected leaves, and if and how small polypeptides such as these are involved in the host-virus relationship, especially since RNA 5 may be a satellite entity responsible for a lethal necrotic disease in *Lycopersicon esculentum*. The finding of these small peptides, like the noncoat proteins of several other plant viruses, is another interesting example of the suggestion by in vitro synthesis techniques of viral coded polypeptides before their function or even existence has been demonstrated in vivo.

### E. Tobacco Streak Virus RNAs

Tobacco streak virus (TSV) is a member of the Ilar virus group, along with prunus necrotic ringspot virus and apple mosaic virus which are serologically related to each other, but not to TSV. TSV has three[54,55] heterogeneous particles all containing RNA, although a fourth type of particle has been inferred.[56] The apparent variation in number of components may depend upon which host is used. TSV is usually grown in *Nicotiana tabacum*, or in a hybrid, *N. glutinosa* × *N. clevelandii.* The RNAs show a similar electrophoretic mobility[54] to the bromovirus RNAs, there being two RNAs of around 1.0 × $10^6$ mol wt (RNAs 1 and 2), a slightly smaller RNA 3 (about 0.8 × $10^6$), and a small RNA 4 (about 0.3 × $10^6$). This complement of RNAs is also similar to AMV, with which TSV has a further striking resemblance. Like AMV, the coat protein of TSV is required for infectivity,[54] and furthermore, the coat protein of TSV can activate the infectivity of AMV RNAs, as can AMV coat protein activate TSV RNAs. It is not known why coat protein is required for infection by these viruses.

RNA 4 is a minor component, relative to RNAs 1, 2, and 3. When the mixture of all four RNAs is translated in wheat germ extracts, some 28,000 mol wt polypeptide is synthesized (Figure 3A). This product is made exclusively from RNA 4 and probably the product of the coat protein cistron. The major products from a TSV RNA mixture include the large polypeptides from RNAs 1 and 2 (see Figure 3). Only approximate

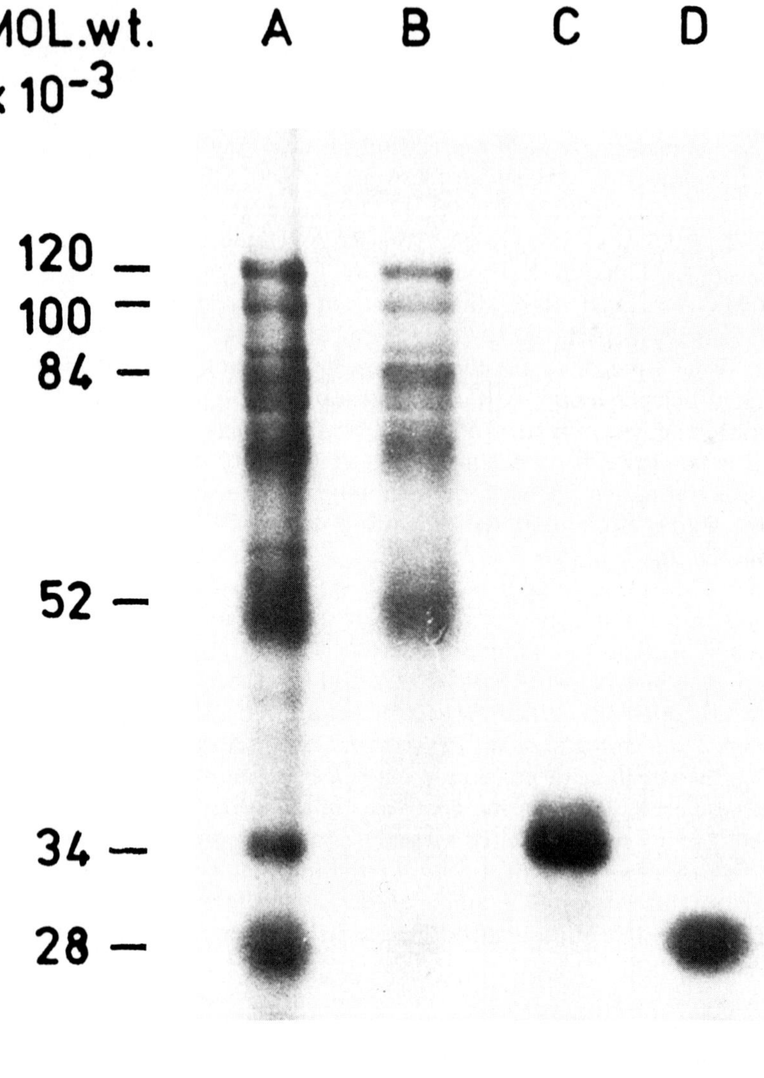


A

FIGURE 3. In vitro translation products of TSV RNAs. The conditions for wheat germ and rabbit reticulocyte extracts were as outlined in Section II. Two electrophoresic systems were used. a. Phosphate buffer (pH 7.2), 7% polyacrylamide gel[102] containing 5 *M* urea. Products of the wheat germ system (see Figure 1) are shown. (A) RNAs 1, 2, 3, and 4 (unfractionated), 2 μg/100 μℓ assay; (B) RNAs 1 and 2, 2 μg (mixture); (C) RNA 3, 5 μg; (D) RNA 4, 9 μg. b. Discontinuous tris buffer system,[101] 10% polyacrylamide gel. (A) RNA 4, 9 μg in 100 μℓ wheat germ incubation; (B) RNA 3, 5 μg in 50 μℓ rabbit reticulocyte lysate incubation; (C) RNAs 1, 2, 3, and 4 (unfractionated) 3 μg in 100 μℓ wheat germ incubation. Molecular weight approximations were as for Figure 1 (but see Reference 103).

molecular weight estimations have been made, and these differ according to the gel system used (cf. Figures 3A and B). Other products, between 50,000 and 84,000 mol wt are detected, but these may be incomplete products. A polypeptide of about 34,000 mol wt is also resolved and is shown to be from RNA 3 (Figure 3). This is probably equivalent to the 3a products of BMV and CCMV. It is not known why two products (34,000 and ∿33,000) appear to be made from RNA 3 by reticulocyte extracts.

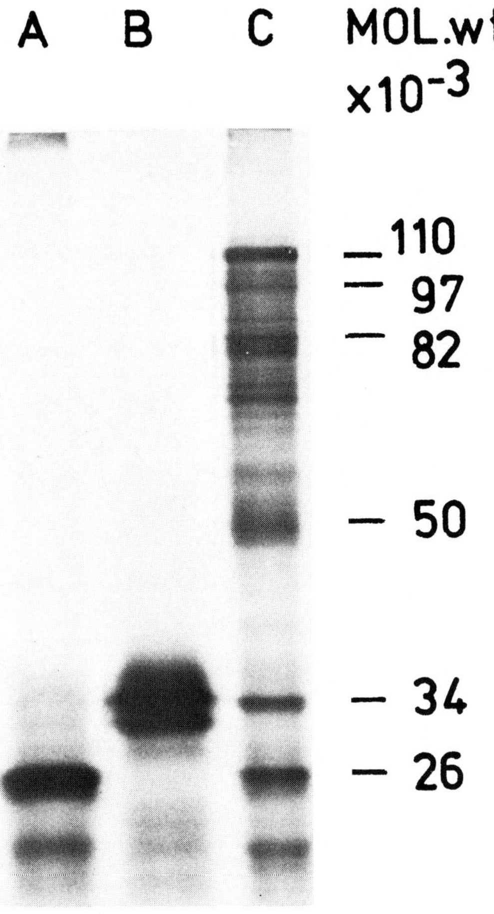


FIGURE 3B

**F. Alfalfa Mosaic Virus RNAs**

Alfalfa mosaic virus (AMV) is an example of a tripartite genome plant virus with short bacilliform virions, of different sizes, but with only one coat protein (see Section III, Volume II by Lane). AMV has a wide host range, most hosts being members of the leguminoseae. For research purposes, it is usually grown in *Nicotiana tabacum*. The genetic information is divided among three RNAs which are essential for infection, and a fourth small RNA (top component *a* RNA, or RNA 4) which is not required for infection if AMV coat protein is present.[57] The RNAs were originally designated B-RNA, M-RNA, Tb RNA, and Ta RNA according to their encapsidation in bottom, middle, and top component virions (see Section III, Volume II by Lane). By analogy with the bromoviruses they are now termed RNA 1 (24 S; $1.3 \times 10^6$ mol wt), RNA 2 (20 S; $1.0 \times 10^6$), RNA 3 (17 S; $0.7 \times 10^6$), and RNA 4 (12 S; $0.34 \times 10^6$). These figures are for the S (Strasbourg) strain. Strain 425 has similar RNAs, although they are reported after careful measurement to be smaller (1.04, 0.73, 0.62, and $0.25 \times 10^6$, respectively).[57a] The previously mentioned studies with *E. coli* cell-free extracts suggested that RNA 4 could contain a cistron for coat protein.[9,57] This has been unequivocally confirmed by translating RNA 4 in wheat germ[14,27,58] and animal cell-free extracts,[27,59] the product being identified by its size, tryptic peptide fingerprint, and also by immunoprecipitation in the animal cell extracts. Since AMV coat protein is required for

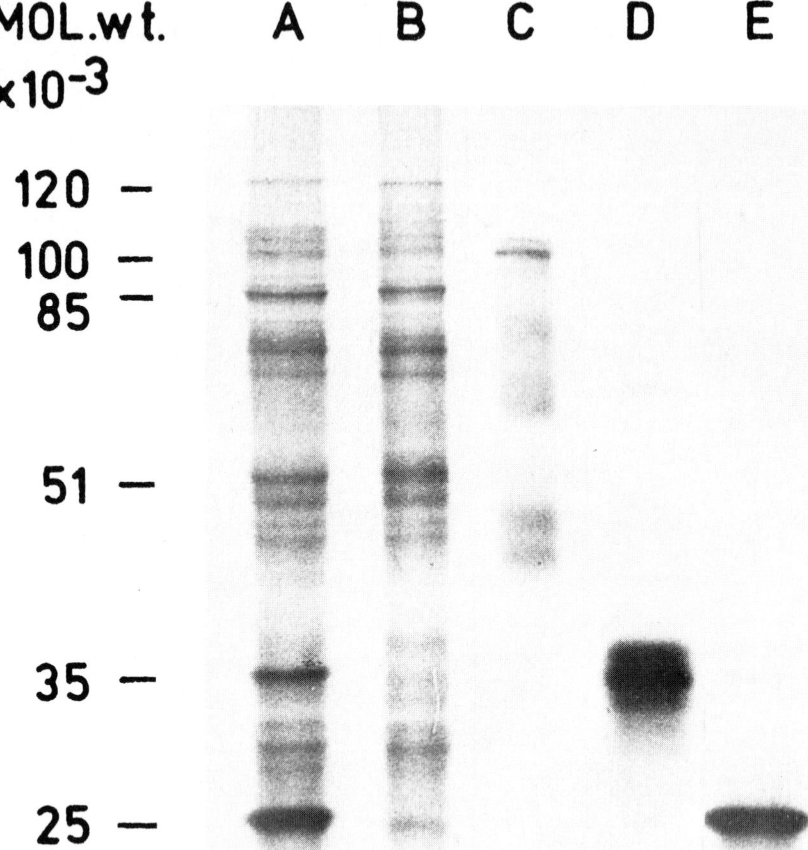


FIGURE 4. In vitro translation products of AMV RNAs. Incubation conditions as for Figure 1. AMV RNAs (strain 425) were provided by Dr. L. van Vloten-Doting. The products were analyzed on a 10% polyacrylamide discontinuous gel.[101] Marker proteins were as in Figure 1. (A) RNAs 1, 2, 3, and 4 (unfractionated), 3.4 μg/100 μℓ assay; (B) RNA 1, 3.6 μg; (C) RNA 2, 2.5 μg; (D) RNA 3, 5 μg; (E) RNA 4, 5 μg. Note that RNA 3 produces, in addition to the products around 35,000 mol wt, a very small amount of a product about 65,000 mol wt. This may be an in vitro read-through polypeptide (see Section IV.F). Note also products at 85,000 and 50 to 51,000 mol wt (cf. Figures 1 to 7).

infection, the opportunity exists for a direct biological test of an in vitro-synthesized product.

AMV RNA 3 appears to be a similar molecular entity to BMV RNA 3. The main product from S strain AMV RNA 3 in ascites extracts is a polypeptide of 35,000 mol wt.[60] A product of this size was produced in wheat germ extracts, along with a heterogeneous mixture of coproducts, when strain 425 RNA 3 was translated.[58,58a] These additional products included polypeptides of up to 65,000 mol wt (see Figure 4). Since RNA 3 is $0.62 \times 10^6$ daltons, the maximum-sized product for which it could code, if read in its entirety as a monocistronic mRNA, would be a polypeptide of about 70,000 to 75,000 mol wt. A product of about 65,000 daltons is thus equivalent to the translation of 70 to 80% of RNA 3. Thus, if this RNA has cistrons for a 3a polypeptide (35,000 mol wt) and coat protein (25,000 mol wt), then some in vitro read-through from one cistron into another must have occurred, assuming the RNA 3 is pure. Furthermore, the 65,000 mol wt product can be precipitated with anti-AMV serum, showing that it contains coat protein sequences.[58a]

Some coat protein has been detected among the strain 425 AMV RNA 3 translation products,[58b] suggesting initiation at an internal site. The translation results with AMV RNA 3 seem to differ, perhaps according to strain of virus, or more likely according to wheat germ in vitro incubation conditions. In a recent report for example,[61] only the 35,000 mol wt polypeptide is synthesized from S strain RNA 3 in a wheat germ extract (but see Figure 4 for strain 425).

When the mixture of all four AMV RNAs is translated, large polypeptides are observed (Figure 4) which are the products of RNAs 1 and 2. AMV RNAs 1 and 2 have been translated in rabbit reticulocyte,[61] ascites extracts,[27] and wheat germ extracts[61] into large products (120,000 and 95,000 daltons) corresponding to the translation of almost the entire length of the RNA, behaving as monocistronic messengers, as in the case of BMV RNAs. Similar results are shown in Figure 4. Since the molecular weight estimations are not accurate, the 100,000 product is pobably equivalent to that reported as 95,000 (clearly from RNA 2). It has been suggested[61] by analysis of initiation dipeptides synthesized in the presence of sparsomycin that AMV RNA 1 and RNA 2 each have one initiation site. If this is so, the additional products synthesized from RNAs 1 and 2 in reticulocyte[61] and wheat germ (see Figure 4) extracts must arise by differential termination. This has not been proved.

## V. THE RNAs OF BIPARTITE GENOME VIRUSES

### A. Tobacco Rattle Virus RNAs

Tobacco rattle virus (TRV), type member of the tobraviruses, infects a number of *Nicotiana* species in addition to *N. tabacum*; for RNA structure and in vitro translation experiments, it is usually grown in *N. clevandii*. The genome of TRV consists of two RNA molecules, the coat protein gene being carried on the smaller RNA,[62] designated RNA 2. The sizes of the small RNAs are considerably different for different strains. The RNA 2 of strain PRN has a molecular weight of $1.1 \times 10^6$ and that of strain CAM (Campinas) is $0.65 \times 10^6$. Both of these are larger than necessary for coat protein coding. The larger molecule, RNA 1, is $2.5 \times 10^6$ daltons and (unlike RNA 2) may not be capped with $m^7G^{5'}ppp$.[63]

Synthesis of TRV coat protein in mouse L cell extracts has been demonstrated,[26] and the wheat germ in vitro translation of RNA 2 from strains CAM and PRN has been studied.[64,65] RNA 2 from the CAM strain was translated into a product which comigrated on SDS-polyacrylamide gels with CAM coat protein, is precipitated by CAM coat protein antiserum, and can aggregate with authentic coat protein to form 36S discs. The migration and hence apparent size of TRV CAM coat protein varies with the polyacrylamide concentration, being apparently 50,000 daltons in 3% gels and 29,000 on 10% gels, whereas the actual size is believed to be 22,000 daltons.[66] Assuming the latter size, only about one third of RNA 2 is translated. If there is another cistron on this RNA, then presumably the initiation site is "closed", similarly to the coat protein cistrons of BMV and AMV RNA 3. However, with TRV RNA 2, it is the noncoat protein gene that is not expressed. Interestingly, this is so for CAM strain RNA 2 but not for PRN strain RNA 2.

It has recently been reported[65] that PRN RNA 2 is translated in wheat germ and rabbit reticulocyte extracts, into not only coat protein, but also a polypeptide of 31,000 daltons. This polypeptide is not a precursor for coat protein, but possesses a few tryptic peptides also present in coat protein, suggesting a partial overlap of amino acid sequence and hence overlap of the two genes. Both initiation sites are "open" sites, at least in vitro. The relative proportions of each of the two major products was dependent on the magnesium ion concentration. At high magnesium concentrations (supraoptimal for incorporation activity), mainly coat protein was synthesized. This again demonstrates how ionic conditions of cell-free extracts can affect the products, emphasizing the danger of extrapolating in vitro results to the in vivo situation, especially when often in vitro only one set of conditions is used and only one strain of the virus studied.

TRV RNA 1 (molecular weight $2.5 \times 10^6$) of strain PRN was translated in rabbit reticulocyte extracts, and wheat germ extracts containing spermidine, into products of molecular weights 170,000 and 140,000. The wheat germ extracts also produced a series of other products, perhaps owing to premature termination. Since the coding capacity of RNA 1 is only sufficient for a (total) molecular weight polypeptide of about 220,000, then it is likely that the two large products arise from the translation of overlapping genes, each with an "open" initiation site. In this respect, the translation strategy of TRV RNA 1 is analogous in part to the similar sized RNAs of TMV and CPMV (see below).

### B. Tomato Black Ring Virus RNAs

Tomato black ring virus (TBRV) is a member of the nepoviruses, a group which includes raspberry ringspot virus and the type member tobacco ringspot virus. It has isometric particles of three types: an empty top component (55S), a middle component (97S) containing RNA of $1.5 \times 10^6$ mol wt, and a bottom component (121S) containing a $2.5 \times 10^6$ mol wt RNA. Some small satellite-like RNA has been reported to be associated with some nepoviruses, but the relationship is not clear. TBRV is one such virus, and has a $0.5 \times 10^6$ "satellite" RNA (RNA 3).

Little is known about the in vitro translation characteristics and translation strategy of nepoviruses but, from preliminary data, the prima facie similarities between nepoviruses and comoviruses may well be upheld by translation experiments. The RNAs of TBRV are translated into large polypeptides of molecular weights of approximately 200,000 and 160,000.[65a] This is at least analogous to the in vitro translation of CPMV RNAs (see below). The satellite-like RNA (RNA 3) of TBRV is translated into a 48,000 mol wt product. The precise relationships and in vivo functions of these translation products (if they exist in infected plants) are yet to be determined.

### C. Cowpea Mosaic Virus RNAs

Cowpea mosaic virus (CPMV) is the type member of the comoviruses. Its host range includes some nonmembers of the leguminoseae, *Nicotiana tabacum* being one. It is usually grown in cowpea, *Vigna sinensis*. Both of the isometric particles, or their RNAs, are required for infection.

The genome of CPMV comprises two RNA molecules of molecular weight $2.02 \times 10^6$ (B-RNA) and $1.37 \times 10^6$ (M-RNA) and the virus capsid is composed of two proteins of approximate molecular weights 44,000 and 22,000 (25,000 for the slower migrating electrophoretic form).[67] Initial attempts to translate CPMV RNAs in a wheat germ system resulted in a heterogeneous mixture of products of a wide size range.[19] Among these products, perhaps coincidentally, were polypeptides of a similar order of size (49,000 and $\sim$27,000 mol wt on phosphate-urea-SDS polyacrylamide gels) to the coat proteins. No tests were made to ascertain amino acid sequence homology. Under the same conditions, other large RNAs, such as those of TMV and CCMV RNAs 1 and 2, also produced a wide range of heterogeneous products; thus the fidelity of large RNA translation (as opposed to that of small monocistronic messengers)[19] was doubted. Improvement of the conditions for translation of larger RNAs in wheat embryo[25,42] and rabbit reticulocyte extracts[28] prompted a reexamination of CPMV (Nigerian strain) RNA in vitro translation. The results obtained were similar in both types of extract.[68,69] The middle virion component RNA (M-RNA) was translated mainly into two large polypeptides, reported to be around 120,000 mol wt in reticulocyte extracts[28,68] and 120,000 and 140,000 in wheat germ extracts containing spermidine.[69] The size of these polypeptides has not been accurately estimated. On 7.5% gels (Figure 5), values of 130,000 and 110,000 are obtained. When coelectrophoresed, the M-RNA polypeptides synthesized in both systems comigrate, indicating similar size.[69a] This pair of polypeptides (a "doublet") accounts for nearly twice the coding capacity of M-RNA; thus it may be that the RNA is expressed in vitro as two overlapping genes. It is possible that there are two initiation sites, both "open". This has not been verified. Heterogeneity of the RNA molecule population or c posttranslational modification of the polypeptides has not been ruled out. "Doublet" products, around 220,000 mol wt, were reported for translation of the larger B-RNA in reticulocyte extracts[28] but this may be exceptional. Reticulocyte extracts more often yield only one polypeptide of this size, as do wheat germ extracts.[69,69a] This is shown in Figure 5, where on a 7.5% gel the size was estimated as approximately 200,000. Smaller products are also produced from B-RNA, perhaps as a result of premature termination or translation of B-RNA fragments. The latter may arise from hidden breaks in the RNA or nuclease degradation in the cell-free extract. A contributing factor to the ability of the wheat germ extracts to translate CPMV RNAs into large products is the addition of spermidine. At 2.5 to 3.0 m*M* $Mg^{++}$, 0.4 m*M* spermidine considerably stimulates translation. Indeed, there is virtually no translation at 2.5 m*M* $Mg^{++}$ without spermidine, as is also the case if the $K^+$ concentration is less than 70 m*M* ($K^+$ acetate).[69] The ability of CPMV RNAs to be translated into polypeptides requiring almost the entire coding capacity of the RNAs (which probably include the coat protein coding regions) suggests the intriguing possibility that in vivo these RNAs are translated into large precursor polypeptides which are subsequently processed by specific proteolytic enzymes to yield the coat proteins and other virus specific proteins. So far, however, identical large polypeptides have not been unequivocably demonstrated in vivo (but see Figure 5). If such a mechanism does occur, it would be unique among plant viruses but similar to some animal viruses, for example, poliovirus. This is a tempting comparison in view of the facts that, like poliovirus RNAs, CPMV RNAs possess poly A stretches at their 3′-ends[70] and lack the $m^7G$ cap at their 5′ termini.[71] Whether or not a polypeptide moiety exists at the 5′ terminus, as described for poliovirus RNA,[72] remains to be seen.

## VI. MONOPARTITE GENOME VIRUS RNA

Monopartite virus is a term used here for plant viruses which have their complete genome in one RNA molecule, which is the genomic form packaged into virions (see Section III, Volume II by Zaitlin). It should be observed however, that because a genome exists in such a unit does not mean it cannot have other functional forms. Evidence is now well documented that, indeed, such RNAs may form subgenomic messengers for translation. Such viruses, or at least some strains, may therefore temporarily exist in pseudo multipartite states at specific times in the replication process.

### A. Turnip Yellow Mosaic Virus RNA

Turnip yellow mosaic virus (TYMV) is the type member of the tymovirus group. It has a relatively narrow host range and is usually grown on Chinese cabbage (*Brassica pekinensis*) which is considered to be the best systemic host.

The RNA is reported to be a large single molecule ($2 \times 10^6$ mol wt). However, it may be significant that preparations of this RNA are often heterogenous. Translation of TYMV RNA has been studied using wheat germ cell-free extracts.[73] A major prod-

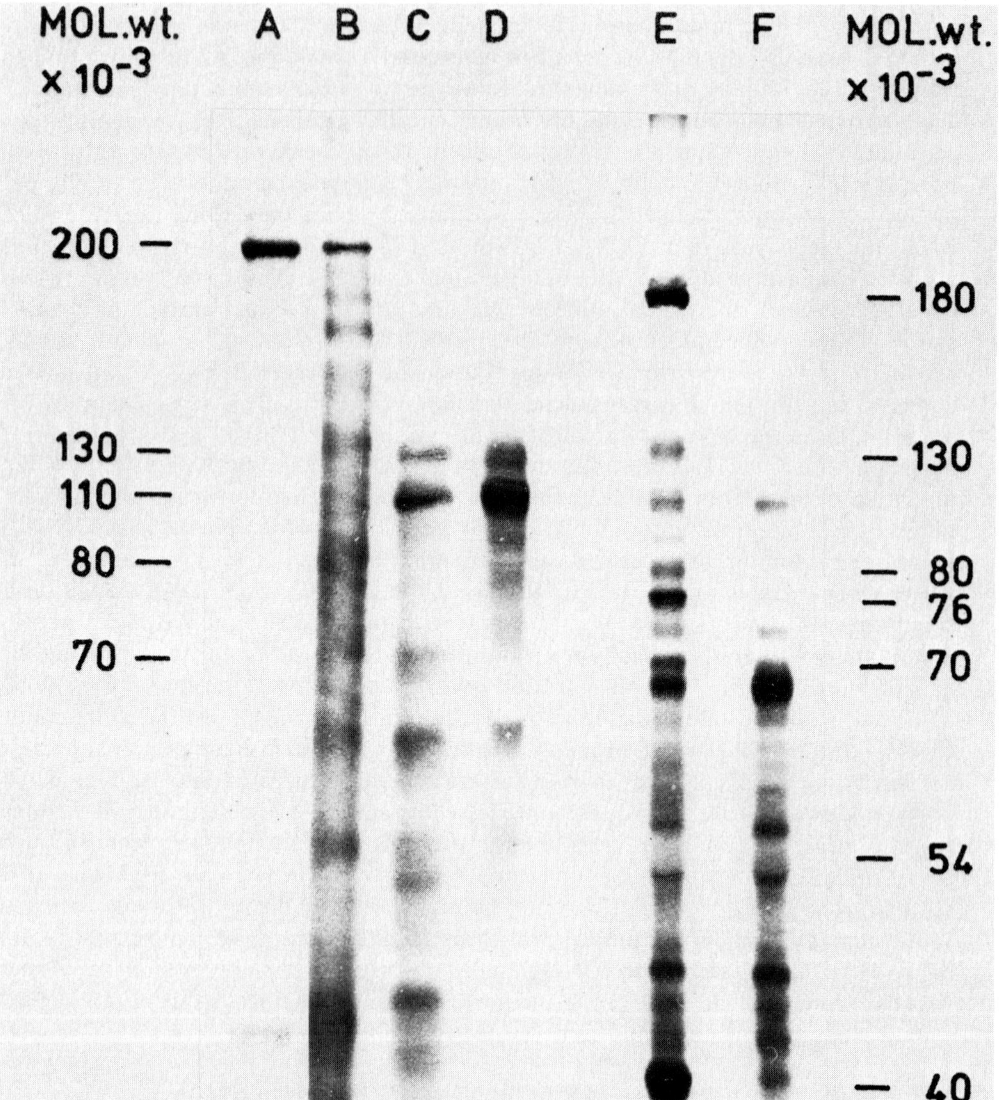


FIGURE 5. In vitro translation products of CPMV RNA. Wheat germ and rabbit reticulocyte extracts were prepared as described in Section II. The incubation conditions were also as described. CPMV RNAs were obtained from CPMV particles separated into middle and bottom components by CsCl equilibrium centrifugation. The RNAs were further purified on sucrose gradients before translation in each type of cell-free extract.[110]

For comparison, [$^{35}$S] methionine was incorporated into CPMV infected and healthy protoplasts from cowpea leaves,[111] and the soluble proteins of a 30,000 g supernatant were electrophoresed on the same type (7.5% discontinuous gels[101]) in a separate experiment. The 40,000 mol wt product is the larger of the two CPMV coat proteins. (A) B-RNA, rabbit reticulocyte lysate; (B) B-RNA, wheat germ; (C) M-RNA, wheat germ; (D) M-RNA, rabbit reticulocyte; (E) infected protoplast soluble polypeptides; (F) healthy protoplast soluble polypeptides. The 54,000 mol wt product is chloroplast Fraction I protein. The amount of smaller, heterogeneous peptides produced from B-RNA is variable, and can even be zero, with either wheat germ or reticulocyte extracts. It is not yet known if any of these polypeptides are real products (compare with infected protoplasts, E), or if all are in vitro artifacts. In wheat germ extracts, higher (up to 130 m*M*) $K^+$ concentrations than used here (100 m*M*) deplete the production of shorter polypeptides.[69]

uct of this translation was shown to be TYMV coat protein, according to its size, tryptic digest fingerprints, and specific immunoprecipitation. Other polypeptides were also detected, some of them greater than 94,000 mol wt. These were considered to be quantitatively minor products.

It was recently demonstrated[74] that whereas unfractionated TYMV RNA (total TYMV RNA, including breakdown products derived from the virus) is translated into coat protein, the large intact TYMV RNA molecules ($2 \times 10^6$ daltons) translated into little or no coat protein, yet were infective and hence contain the complete genetic information for TYMV. Other size classes of TYMV RNA were obtained after heat treatment (65°, 3 min) followed by sucrose gradient centrifugation.[74] These smaller classes of RNA (which have a reproducible sedimentation pattern) do not affect infectivity, but are readily translated in vitro. In wheat germ cell-free extracts the smallest RNA[74,74a] class yielded coat protein quite efficiently. The average molecular weight of this class of RNA is about $0.25 \times 10^6$, but it is actually a heterogenous population of molecules. It was not shown precisely which size of RNA is serving as a message for coat protein in vitro, nor if this RNA has a cap at the 5′ terminus or if the 3′ terminus has a tRNA-like structure accepting valine, as does the full-size TYMV RNA. The other selected size classes[74] were of approximate molecular weights $0.5 \times 10^6$, $1.0 \times 10^6$, and $1.3 \times 10^6$. Superficial resemblance to the multicomponent virus RNAs (see above) is clearly indicated.

A small RNA fragment of TYMV RNA can also be isolated after heating the TYMV RNA to 60° for 10 min, in the presence of 1% w/v SDS or 1 m*M* EDTA.[75] This small RNA comigrated with BMV RNA 4, and therefore has a molecular weight of 0.28 to $0.3 \times 10^6$. No other RNA species were observed other than the $2 \times 10^6$ dalton molecules. Presumably the latter consist of intact parent molecules and molecules minus the 0.3 $\times 10^6$ molecular weight fragment, which do not separate when such a large RNA is only electrophoresed a relatively short distance into a polyacrylamide gel. (There was only a small quantity of the small RNA compared to the large one.) The small RNA was shown to be a monocistronic messenger for coat protein,[75] and that it does indeed possess an $m^7G$ cap at the 5′ end. There is evidence that this small RNA occurs in the minor components of the virus.[71,71a]

The minor components are nucleoproteins designated $B_o$, $B_{oo}$ and $B_{ooo}$, which appear as faint bands of lighter density particles on cesium chloride gradients,[76,76a] whereas the main infectious particle type is more dense. This component appears as two major bands, $B_1$ and $B_2$, the latter perhaps being a CsCl gradient artifact.[76a] Which of these five components ($B_o$, $B_{oo}$, $B_{ooo}$, $B_1$, and $B_2$) carry TYMV coat protein genes? By translating RNA from each type of particle, in wheat germ extracts, Higgins et al.[76b] concluded that a similar range of polypeptides is made from all types. The small coat protein gene ($0.28 \times 10^6$ dalton RNA) was detectable in all nucleoprotein types, but was especially noticable in $B_{ooo}$ and $B_{oo}$. Matthews had earlier[76a] shown that each of the five CsCl bands is double. $B_1$ and $B_2$, thus, should be regarded as $B_{1a}$, $B_{1b}$, and $B_{2a}$, $B_{2b}$. Pleij et al.[74a] analyzed the translation products of RNAs from all particle types, purified by four cycles of CsCl centrifugation. They concluded that $B_{1a}$ and $B_{2a}$ RNAs produce negligible, perhaps no coat protein. All others produce some, especially $B_{ooa}$, $B_{oooa}$, and $B_{ooob}$. From the combined work of several laboratories[74a,76a,76b] there is no doubt that TYMV has an encapsidated small RNA that codes for coat protein: but what are the translation products of the large (intact) TYMV RNA molecule, if not coat protein?

In wheat germ extracts, the large TYMV RNA seems to be translated into a series of polypeptides. These include products of 35,000 and 37,000 mol wt[75] and others up to 170,000 including a major one of about 100,000. These larger products are more efficiently synthesized after the RNA has been heated in EDTA or dialysed against 1 *M* NaCl, presumably to remove polyamines and alter the conformation of the RNA[75] (spermidine was added to the cell-free extracts). Using the wheat extract described in this chapter and used for other viral RNA translations Figures 1 to 4), results were obtained, which agree with the published data in that there are several products, but the largest was found to be 130,000 not 170,000 and the intermediate products 37,000 and 40,000 rather than 35,000 and 37,000 (Figure 6a).

However, the wheat germ cell-free system is known to produce some artifacts or incomplete polypeptides with some RNAs, and also may not have the suppressor tRNA that allows read-through of the TMV 130,000 dalton polypeptide in reticulocyte extracts (see Section IV.C). Benicourt et al.[76d] have translated the TYMV RNA in reticulocyte lysates, and they found in addition to a 150,000 dalton product (probably equivalent to the 130,000 dalton product in wheat germ extracts), there was a synthesis of a 195,000 mol wt polypeptide, equivalent in size to the length of the large RNA minus the coat protein gene and the tRNA-like (nontranslatable) structure towards the 3′ end (see Figure 6c). Analysis of the peptides from *S. aureus* $V_8$ protease digestion suggested that the 150,000 and 190,000 polypeptides have common "in phase" sequences. It is possible that the larger was a leaky read-through of the smaller.

Following the rather interesting revelations from in vitro translation of TYMV RNAs, it is clear that further in vitro and in vivo (especially protoplast) work is required before the precise translation strategy of TYMV is known.

### B. Eggplant Mosaic Virus RNA

Eggplant mosaic virus (EMV) is also a member of the tymoviruses. It shows some serological relationship to some members of the group, but not to the type member TYMV. Like TYMV, the virus has a large RNA ($2.5 \times 10^6$ mol wt). Particles that contain this RNA are infectious.

In vitro, the large RNA is translated in wheat germ extracts, into a mixture of polypeptides which include products of approximately 22,000, 39,000, 90,000, and 130,000 (Figure 6), but little 20,000 molecular weight, which is the size of EMV coat protein. A small amount of polypeptide of this size can be detected (Figure 6), but it is not known if it is EMV coat protein. High salt (1 *M* NaCl) or EDTA treatment, as described above for TYMV RNA, made no difference. This is consistent with the apparent failure to release a small coat protein-coding RNA similar to that observed with TYMV RNA.[75a] It is suggested therefore that EMV may resemble TMV in that a nonencapsidated low molecular weight RNA coding for EMV coat protein may be produced in infected leaves. The possibility that EMV RNA in vivo serves as a polycistronic messenger is not ruled out.

### C. Tobacco Mosaic Virus RNA

Tobacco mosaic virus (TMV), the type member of the tobamovirus group, has many different strains (see Fraenkel-Conrat).[77] Much of the work on TMV RNA translation has been with the wild type U1 strain, grown in *Nicotiana tabacum*, but some recent work with a strain isolated from *Vigna sinensis* and grown in that host, *N. tabacum* or *Phaseolus vulgaris*, has revealed some interesting features.

TMV RNA (wild type, U1) has a molecular weight of $2 \times 10^6$ (see Section III, Volume II by Zaitlin) encapsidated in the well known rod-like virus particles. Using TMV RNA, it was demonstrated that a wheat germ ribosome-RNA initiation complex is formed, which requires 40S and 60S ribosome subunits (the former binding first), $tRNA_i^{met}$, ATP, GTP, and two protein factors,[78-81] indicating an initiation of protein synthesis in wheat germ parallel to prokaryotes.

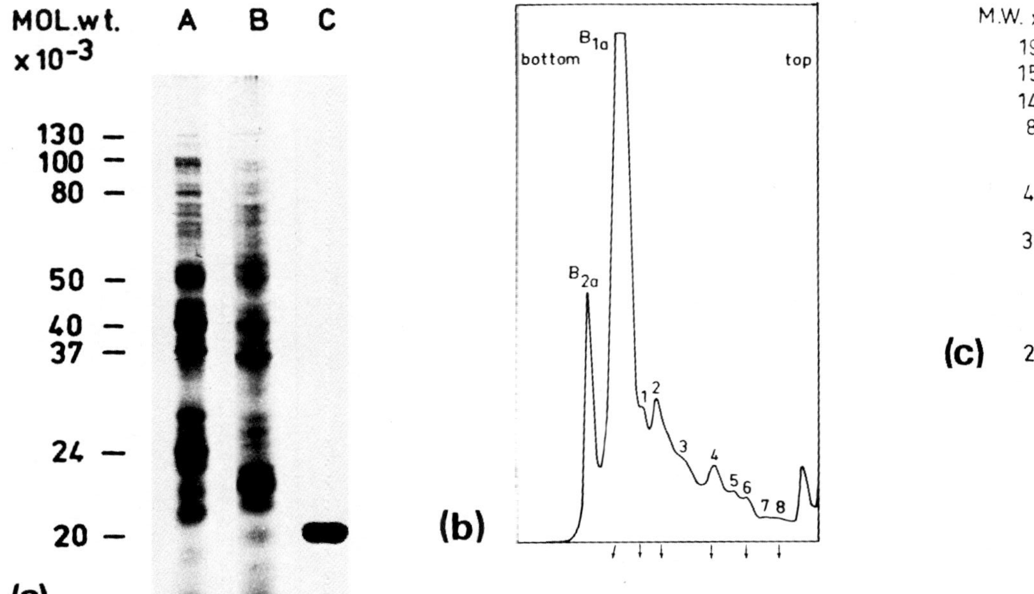


FIGURE 6. (a) In vitro translation products of TYMV and EMV large RNAs. A wheat germ cell-free incubation was used, as described for Figure 1. TYMV RNA from isolated $B_1$ type particles banded in a sucrose gradient. The infectious RNA was a gift from Dr. C. W. A. Pleij. EMV RNA (peaking as 2 to $2.5 \times 10^6$ mol wt on Ultragel filtration) was a gift from Dr. T. C. Hall. Total (virion) TYMV RNA was also tested, also high salt-treated EDTA, SDS, and heat-treated TYMV and EMV RNAs. The results were similar to those shown here. (A) TYMV RNA, 5 $\mu g/100\ \mu\ell$ assay. After [$^{35}S$] methionine incorporation, $25/\mu\ell$ was applied to a discontinuous SDS polyacrylamide gel.[101] (B) EMV RNA, 5 $\mu g/100\ \mu\ell$ assay; 30 $\mu\ell$ applied to the gel. (C) BMV RNA 4 translation product (see Figure 1) as a control and as a polypeptide size marker. (b) A TYMV preparation after a second cycle centrifugation on CsCl (nucleoprotein fractions which centrifuged with a density of 1.38 g/m$\ell$ in the first cycle). The numbered fractions were further purified up to four CsCl cycles of centrifugation. RNA was isolated from the purified fractions for translation. (c) Reticulocyte lysate translation products of some RNA fractions from (c). Even after purification by four CsCl cycles, shoulders 3 and 5 were mixed (aggregates). Other fractions: A, $B_{1a}$ RNA (intact). B, fraction 1 ($B_{ob}$) RNA. C, fraction 2 ($B_{oa}$) RNA. D, fraction 4 ($B_{oob}$) RNA. E, fraction 6 ($B_{eoa}$) RNA. Fraction 7 ($B_{ooob}$) is not shown, but some of its 36,000 dalton product can be seen also in E. F, fraction 8 ($B_{oooa}$) RNA. (The data in (b) and (c) are adapted from results of Mellema, Bénicourt, Haenni, Noort, Pleij, and Bosch,[120] by kind permission of Drs. Mellema, Bosch, and Haenni.

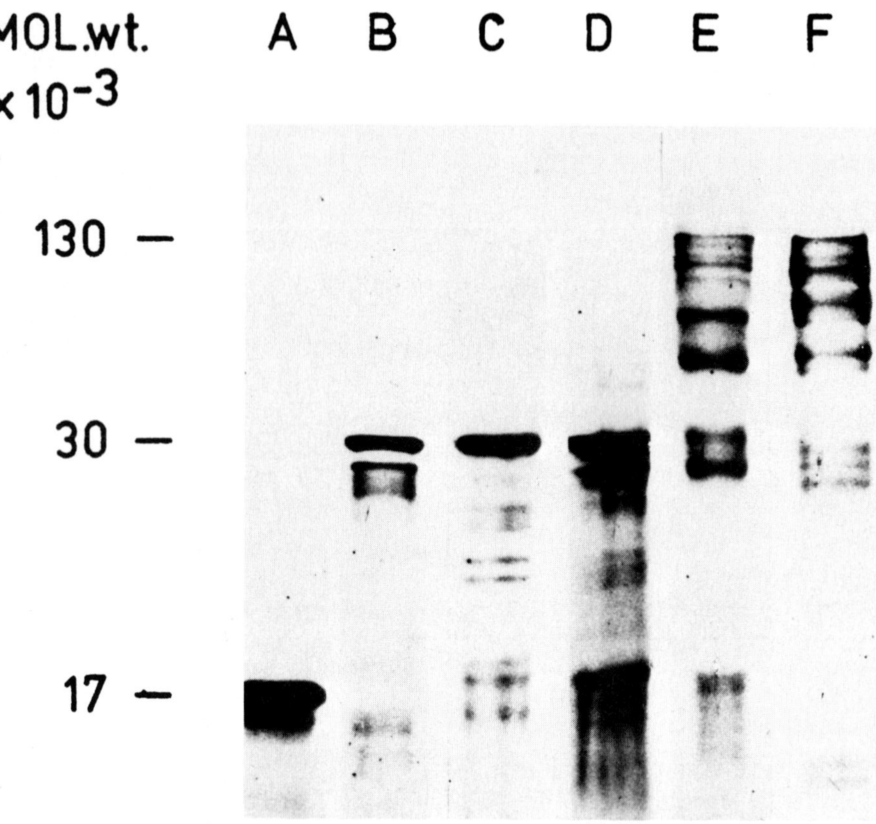


FIGURE 7. In vitro translation products of TMV RNAs. TMV RNA from long rods and rod fragments was translated in wheat germ extracts.[89] These results were provided by Dr. Milton Zaitlin. The polypeptides are electrophoresed on 12% polyacrylamide gels.[101] [$^3$H]-leucine was incorporated, and scintillation-autoradiography (fluorography) was used. (A) Small rod RNA (RNA S) from $C_c$TMV, (B) RNA $I_2$ (intermediate rod) from $C_c$TMV, (C) RNA $I_2$ from strain K TMV (Kansas wheat isolate), (D) RNA $I_2$ from strain U 1 TMV, (E) RNA L (long rod) from U 1 TMV, (F) RNA L from $C_c$TMV. The product designated 130,000 mol wt is probably the same as referred to in the text as 140,000. However, note that there is no larger product (150,000 or 165,000). The position marked 17,000 is the exact position to which TMV coat protein migrates (not shown). (Modified from Beachy, R. N. and Zaitlin, M., *Virology*, 81, 160, 1977. With permission from Academic Press.)

As with the early prokaryote in vitro TMV RNA translation attempts,[1,3,7] translation in wheat embryo or wheat germ cell-free extracts produced only heterogeneous products;[17-19] little or no TMV coat protein could be detected. However, synthesis of a coat protein peptide has been claimed[18] using RNA from TMV mutant Ni 568 which has a methionine substitution for threonine in the coat protein. It has been reported that the products of intact TMV RNA translation in wheat germ extracts range from quite small (~ 10,000 mol wt) to large (~ 140,000 mol wt),[24,28] whereas in frog oocytes[82] a major product of 140,000 mol wt is synthesized and in rabbit reticulocyte lysates a product of 165,000 mol wt is also synthesized.[22,28] Appearance of this product is variable. It is absent from the wheat germ extract products in Figure 7. It is not known if the in vitro 165,000 mol wt product is a real virus product, which occurs also in vivo, nor if it contains TMV coat protein sequences. The latter would seem unlikely, since the molecular weight does not correspond to the translation of the full length of the RNA. Furthermore, a small RNA (low molecular weight component, LMC) has been detected in infected tobacco leaves, which codes for coat protein as shown by its in vitro translation.[20-22] A current model for TMV RNA translation strategy is, therefore, that there could be two "open" initiation sites on the intact RNA, near the 5′-terminus, which give rise to the 140,000 and 165,000 mol wt polypeptides (which thus have overlapping genes). That there are two open initiation sites on the large RNA is not proven, but was supported by analyses of dipeptide synthesis in the presence of sparsomycin, which restricts protein synthesis to the formation of methionine-containing dipeptides.[22,83] However, recent evidence suggests two termination sites. Pelham[82a] has demonstrated that the larger (160,000 mol wt) polypeptide is generated in vitro in reticulocyte lysates, by partial ("leaky") read-through of an amber (UAG) termination codon. Presence of an amber suppressor tRNA increases the yield of this polypeptide. This explains why this product is not made in wheat germ extracts (see Figure 7), since an amber termination is recognized in the wheat extracts[14] that presumably have no such suppressor tRNA. Towards the 3′-terminus, the RNA carries the gene for coat protein, which has a "closed" initiation site.[21,22] Coat protein cannot be translated until the low molecular weight (LMC) RNA, the monocistronic messenger for coat protein, is produced in the infected cell. The synthesis of coat protein is thus delayed (late function), while the larger polypeptides presumably can be translated earlier in infection.

A cowpea strain of TMV ($C_c$TMV) forms not only the typical long (∿ 300 nm) rod-like virions, but also short rods only about one tenth as long.[84,85] This strain is the same as the "bean form" TMV (B-TMV).[86] After "passaging" in *Vigna sinensis*, $C_c$TMV is usually grown in *Phaseolus vulgaris*. Short and long particles are also found in *Nicotiana tabacum* infected with this strain.

It was suggested[84] that the small RNA in the short particles may be analagous to the LMC RNA in TMV U1 infected *N. tabacum*[20-22] (see above). Indeed, it was subsequently demonstrated that this small RNA ($0.3 \times 10^6$ mol wt, about the same as the small RNA component in the tripartite genome viruses) from the short rods can be translated in wheat germ cell-free extracts into a polypeptide very similar to $C_c$TMV coat protein, as identified by tryptic peptide fingerprinting,[88] the ability to reconstitute along with authentic $C_c$TMV coat protein and $C_c$TMV RNA,[85] and electrophoretic mobility[85,88] (molecular weight estimation). One determination[86] of the molecular weight of the authentic coat protein from the short rods was 16,500, whereas that of the long rods was 18,000. Tryptic peptide maps of these two types of coat protein showed them to be essentially the same, but the smaller of the two lacked an amino acid sequence present in the larger.[85] The wheat germ product of the small rod RNA translation was actually slightly larger than the longer (18,000 mol wt) virion coat protein, according to the tryptic peptide analysis[88] of immunoprecipitates of the in vitro product. The "fingerprint" revealed two extra tryptic peptides. One of these could be the N-terminal peptide. It was also reported that the in vitro product contained methionine, which is absent from the natural coat protein(s) of this strain of TMV.

A third, intermediate-size class of $C_c$TMV rod has also been identified, and the intermediate-sized RNA of these rods, $I_2$RNA, translated in wheat germ extracts.[85] This RNA directs the synthesis of at least one polypeptide, of molecular weight about 30,000 (Figure 7). This polypeptide is quite distinct from coat protein, as shown by tryptic peptide maps. Recently, the U1 strain (wild type) has also been shown to possess an intermediate-sized ($0.51 \times 10^6$) RNA which is also translated into a 30,000 mol wt polypeptide in vitro.[89] Again we see what appears to be at least a superficial resemblance to multicomponent viruses such as BMV and AMV, whose intermediate-size RNA directs the synthesis of a 34,000 mol wt polypeptide in vitro (see Section IV).

The large RNA of $C_c$TMV directed the synthesis (at 130 m$M$ $K^+$) of a range of polypeptides up to a molecular weight of approximately 130,000, rather similar to the translation products of wild-type (U1) TMV RNA. Some 150,000 mol wt product was also detected.[85] Thus, some characteristics of $C_c$TMV translation strategy are apparently similar to those of the tobacco strains of TMV, while others bear at least superficial resemblance to multicomponent (multipartite genome) virus translation strategy, perhaps reflecting a common but varied evolutionary trend among plant viruses. Among the large number of plant viruses known, one might expect to find strains which represent different stages of evolution or even which possess a completely different strategy. It has already been seen that RNA 3 of some tripartite viruses may be a type of dicistronic messenger, as may TRV RNA 2 (see Section V.A). The 80S cytoplasmic ribosomes of wheat germ can translate the polycistronic Q$\beta$ RNA,[14] albeit not efficiently. The possibility that some plant viral RNAs and plant mRNAs (see Section II, Volume I by Hall) may be polycistronic should not be overlooked.

### D. Tobacco Necrosis Virus RNA

Tobacco necrosis virus (TNV) is the type member of the tobacco necrosis virus group. Some members have little or no serological relationship and some TNV isolates have such widely differing nucleotide compositions that they may be separate viruses rather than strains. The Urbana, Yarwood, and AC 36 strains have been studied for their relationship with the associated STNV (see Section III.A). The proportion of TNV to STNV is host dependent. Common hosts are *Nicotiana tabacum* and *Vigna sinensis.*

It has recently been reported[90] that TNV (Yarwood) RNA of molecular weight 1.4 $\times$ $10^6$ is translated in vitro (wheat germ) as if it were a polycistronic messenger. The major product appears to be coat protein, judging by its molecular weight (30,000), cross-reaction with TNV strain AC 36 antiserum, and tryptic peptide analysis. There were also minor products of molecular weights 26,000, 43,000, and 63,000.[90] Assuming that there is not an in vitro pretranslational processing of the RNA into functional monocistronic messengers, or hidden breaks in the apparently intact large RNA, it could be concluded that this TNV RNA serves as a polycistronic messenger, at least in vitro. To be certain of this, proof is needed that the RNA is intact, that there is more than one ribosome binding site, and that noncoat protein products are not in vitro artifacts but can be detected in vivo. Even so, that the coat protein is produced in vitro by translation of TNV RNA seems certain, which is an unusual but perhaps not unique feature.

### E. Carnation Mottle Virus RNA

Carnation mottle virus (CarMV) has no serological cross-reaction with other viruses so far tested. It is generally considered as an "ungrouped" virus. In some physical criteria the virus resembles TNV (28 nm diameter isometric particles; particle weight 7 $\times$ $10^6$ daltons; 18% RNA of $1.4 \times 10^6$ mol wt).

Recent data suggest a translation strategy (in vitro) of CarMV not unlike that suggested for TNV above. Salomon et al.[91,91a] claim that three discrete (no serological cross-reactivity) polypeptides are sequentially produced when CarMV RNA is translated in wheat germ extracts. One of these products was identified as CarMV coat protein, by comigration in gel electrophoresis and isoelectric focusing, and by immunoprecipitation with antibody to disrupted CarMV particles. The other two polypeptides have molecular weights of 77,000 and 30,000, but are of unknown function and not yet identified in vivo. Limited phosphorolysis by *E. coli* polynucleotide phosphorylase, to remove nucleotides from the 3′-end of CarMV RNA, resulted in loss of the 77,000 mol wt product. Phosphorolysis for 1 min (removing an average of very ap-

TABLE 1

**Some Examples of in vitro Coat Protein-like Polypeptide Synthesis Directed by Plant Viral RNAs**

| Viral RNA[a] | Molecular weight RNA (× $10^{-6}$) | Cell-free extract[b] | Molecular weight coat protein (approx.) | Identification of coat protein[c] | Ref. |
|---|---|---|---|---|---|
| STNV RNA | 0.4 | WE | 22,800 | s, tp, Nt | 12, 33 |
| BMV RNA 4 | 0.28 | WE | 20,000 | s, tp | 23 |
| CCMV RNA 4 | 0.23 | WG | 19,000 | s | 19 |
| BBMV RNA 4 | 0.36 | WG | 21,000 | s | 19 |
| CMV RNA 4 | 0.33 | WE | 25,000 | s, cbp | 50, 51 |
| | | RR | | s | 51 |
| AMV RNA 4 | 0.28 | WG | 24,280 | s, tp | 58 |
| | 0.34 | RR | | s, tp, i | 27,59 |
| | | KA | | s, tp, i | 27, 59 |
| TYMV | | | | | |
| "Class V" RNA | 0.25 | WG | 20,000 | s | 74 |
| "Light" RNA | 0.3 | WG | | s, tp, i | 75 |
| TMV LMC RNA | 0.25 | WG | 17,500 | s, tp, a | 20, 21 |
| $C_c$TMV small rod RNA | 0.3 | WG | | s, tp, i, a | 85, 88 |
| TRV RNA 2 | 1.0 | MLC | 22,000 | s, tp | 26 |
| | | WG | | s, i, a | 64 |
| | | RR | | s, tp, a | 65 |
| BSMV RNA 2 | 1.35 | WE | 25,000 | s, i | 93 |
| TNV | 1.4 | WG | 30,000 | s, tp, i | 90 |
| CarMV | 1.4 | WG | 38,000 | s, ief, i | 91 |

[a] For abbreviations of virus names, see text.

[b] WE = wheat embryo. WG = commercial wheat germ. RR = rabbit reticulocyte. KA = Krebs II ascites (mouse) cells. MLC = mouse L cells.

[c] s = size (molecular weight by SDS-acrylamide electrophoresis). tp = tryptic peptide analysis. Nt = N-terminal analysis. cbp = cyanogen bromide peptide analysis. i = immunoprecipitation. a = aggregation or assembly into particles. ief = isoelectric focusing.

proximately 3 to 13 nucleotides) caused a threefold reduction in the amount of this large polypeptide relative to the other products. Phosphorolysis for 10 min (120 to 140 nucleotides removed) decreased the amount of this product produced to less than 10% of the untreated RNA translation. This phosphorylase-treated (10 min) RNA was not infectious. These results[91] would seem to indicate that not only is coat protein produced directly from the virion RNA, as may be the case with TRV and TNV, but that the RNA is a polycistronic messenger. A rigorous investigation of translation of CarMV RNA in vitro is awaited.

## VII. DISCUSSION

### A. Translation of Coat Protein Cistrons

Several examples of plant viral coat protein synthesis in cell-free extracts have been given in this chapter, and are summarized in Table 1. Many coat proteins have molecular weights between 17,000 and 25,000 and consequently have small cistrons. These often occur on small RNAs of molecular weights between 0.2 and 0.4 × $10^6$. These cistrons are translated very efficiently in vitro, in terms of both quality and quantity, providing they are in this monocistronic form. Presumably, in infected cells, the coat protein which is the viral protein required in the largest quantity is translated by the most efficient method. Thus, the existence of coat protein cistrons as small monocistronic messengers may reflect an adaptation of the viruses to the host plant protein synthesis mechanism. We may expect, therefore, that many (but perhaps not all) messenger RNAs in the plant cell cytoplasm may also be monocistronic whereas perhaps in plastids, which have a prokaryote-like protein synthesis, we may expect to find more polycistronic messenger RNAs.

Of the plant viral coat protein synthesis mechanisms so far investigated, there appear to be only a few exceptions to the small monocistronic messenger trend (see Table 1). One is PRN strain TRV RNA 2 which has a molecular weight of about $1.0 \times 10^6$ and is probably not monocistronic,[65,92] yet apparently is translated into coat protein in both wheat germ and reticulocyte cell-free extracts.[65] Whether this RNA is translated as such in vivo remains to be seen. Another example is the rather singular case of barley stripe mosaic virus (BSMV). The Russian strain of this virus also has two RNAs of about 1.35 and $1.5 \times 10^6$ mol wt. These were translated into similar products, including a coat protein-sized product, leading the authors[93] to conclude that there may be a doubling of at least part of the genetic information in the genome. While this clearly requires further investigation, that coat protein genes occur on more than one RNA is known among the multicomponent viruses.

Although more evidence is required to support the suggestion that BSMV coat protein is produced by translation of a relatively large RNA molecule, the evidence[90,91] in the cases of TNV and CarMV is stronger. Whether other polypeptides are also products of the intact RNA, serving as a polycistronic messenger, is not easy to ascertain, especially when the size and function of viral coded products in infected plants is not known. Little is known about the composition, quantity, location, and function of noncoat protein polypeptides coded by plant viral RNAs.

### B. Translation of Noncoat Protein Cistrons

While the small coat protein cistrons mentioned above are translated with efficiency in vitro under a variety of conditions and in cell-free extracts from various sources, the larger plant viral RNAs appear to require more precise in vitro conditions. For example, a higher $K^+$ ion concentration,[30,42,69,85] polyamines such as spermidine or spermine,[25,65,69,75] or tRNA from a different source[28,68] may be required. Under suitable conditions, some larger RNAs are translated as monocistronic messengers into polypeptides larger than coat protein. A problem arises in identifying these products, as little is yet known about noncoat protein polypeptides of plant viruses. However, in some instances, polypeptides of a similar size to those synthesized in vitro have been detected in infected cells or protoplasts. Some examples are given in Table 2. The function of these polypeptides is a matter of speculation. The smaller of the noncoat protein polypeptides of BMV detected in infected barley leaves is 34,500 daltons and, owing to its appearance in infection at the same time as replicase activity and its cofractionation with the enzyme, was suggested as a possible replicase or subunit of replicase.[94] Although this protein and the similar sized in vitro product of BMV RNA were reported several years ago, no attempt seems to have been made to positively identify the in vitro and in vivo products as being identical. Similar sized in vitro and in vivo (protoplast) products have been reported for CCMV.[19,95] Also, a 30,000 mol wt product is made in vitro from $C_c$TMV $I_2$-RNA[85] and U1 RNA,[89] and a 37,000 mol wt product has been detected in TMV-infected leaves.[96] In the case of TMV however, a partially purified replicase[99] cosedimented in glycerol gradient centrifugation as if its molecular weight were 160,000. A polypeptide of 165,000 daltons has been detected in tobacco protoplasts infected with TMV[98] and also as a product of in vitro translation in both reticulocyte[28] and wheat germ[69] extracts. It is not proven however that TMV

TABLE 2

**Some Examples of Noncoat Protein Plant Viral Polypeptides Synthesized In Vitro and In Vivo**

| Virus[a] | *In vitro* polypeptide (molecular weight) | Ref.[b] | *In vivo*[c] polypeptide (molecular weight) | Ref. |
|---|---|---|---|---|
| BMV | 34,000 | 23 | 34,500 | 94 |
| | 120,000 | 42 | 107,000 | 108 |
| | 110,000 | 42 | 100,000 | 108 |
| CCMV | 105,000 | Figure 2 | — | — |
| | 90,000 | Figure 2 | 100,000 | 95 |
| | 35,000 | Figure 2 | 34,000—36,000 | 95 |
| $C_c$TMV | 150,000 | 85 | 150,000 | 85 |
| | 130,000 | 85 | 130,000 | 85 |
| | 30,000 | 85 | — | — |
| TMV | — | — | 37,000 | 96 |
| | 130,000 | 28 | 130,000 | 97 |
| | 165,000 | 28 | 165,000 | 98 |
| | 140,000 | 21, 28 | 135,000 | 98 |

[a] For abbreviations of virus names, see text.
[b] Reference to publications cited in this chapter or to original Figure presented in this chapter.
[c] *In vivo* = infected leaves or protoplasts.

replicase is a large polypeptide, nor that the in vitro polypeptides represent real gene products. Their size is also not certain. Estimation of molecular weight of large proteins by gel electrophoresis is not accurate. Some reports on the synthesis of plant viral polypeptides in vitro estimate molecular weights of polypeptides over 100,000 daltons on polyacrylamide gels which have too small a pore size (10 to 12% gels) to give a linear relationship of size to distance migrated. Often, β-galactosidase (*E. coli*) is used as a standard, assuming a molecular weight of 130,000 for the subunit. However, with reference to other markers such as transferin (80,000 mol wt), phosphorylase A (92,000) and *E. coli* RNA polymerase (β and β′ subunits, 155 and 165,000 mol wt), β-galactosidase migrates as if its molecular weight were about 116,000, which is in agreement with that calculated from the amino acid sequence.[106] Figures published assuming β-galactosidase to be 130,000 mol wt may be 20% in error.

None of the large polypeptides synthesized in vitro from TRV,[65] AMV, TSV (Figure 2), or CPMV RNAs[69] have been shown conclusively to occur in vivo. There is some evidence, however, that polypeptides of 180,000 and 130,000 mol wt can be detected in CPMV infected cowpea protoplasts.[107] These could be related to the in vitro products (see Figure 5).

Products of a similar size (around 100,000) to the large in vitro polypeptides of BMV RNA 1 and 2 translation have also been detected in vivo,[108] and a 100,000 mol wt product has been found in protoplasts infected with the closely related CCMV,[95] the large RNAs of which also yield large in vitro polypeptides (see Figure 2).

Often, with the current trend of expecting plant viral RNAs to be monocistronic, it is assumed that products smaller than the largest polypeptides must be in vitro artifacts. In some cases, this may prove to be a regretted assumption. Sometimes "intermediate"-sized products are the major products in vitro but are ignored, whereas the largest polypeptides, perhaps not the major products in vitro, are given all the attention. Products of about 50,000 and 80,000 mol wt can be seen among the translation products of BMV, CCMV, AMV, TSV, TMV, and other RNAs (see Figures 1 to 7). Are such products incomplete polypeptides, or are they the products of overlapping genes or of posttranslational modification of larger polypeptides? Can they be detected in infected protoplasts or leaves?

Clearly, matching the identity of in vitro and in vivo translation products of plant viral RNAs and identification of their function is an area of research which has much potential.

### C. Translation and the Genetic Code of Plant Viruses

The early partial success of translation of plant viral RNAs in bacterial cell-free systems (see Introduction) indicated that the genetic code utilized by plant viruses must be at least similar to that of prokaryotes and, since the virus has evolved to function in a plant cell, supported the general idea that the genetic code of plants is also similar. There are many examples of faithful translation of plant viral RNAs in animal cell extracts and animal viral RNAs in plant cell extracts, and universality of the genetic code is generally accepted.

The codon AUG is almost certainly the major universal initiation codon. The BMV RNA 4 initiation site was the first plant viral RNA ribosome binding site to be sequenced.[43] This work utilized the wheat germ cell-free protein synthesis extracts to select the BMV RNA 4 fragment containing the AUG codon and neighboring bases. Indeed, the sequence of bases coding for the first 14 amino acids of BMV coat protein was also revealed. Using techniques involving in vitro enzymatic labeling of the 5′ end of AMV RNA 4, the sequence of 74 nucleotides from the 5′ terminus has been sequenced.[105] This includes the codons for the first 12 amino acids of AMV coat protein. The exact codons for some amino acids in TMV coat protein are also known.[21,104] These known specific codons are shown in Table 3. Degeneracy is indicated by the use of more than one codon for one amino acid. In TMV coat protein for example, at least three codons are used for threonine, leucine, and proline and four for valine. However, four aspartic acids in TMV coat protein have the same codon, as do three threonines in BMV coat protein. Whether some codons are used more frequently than others, in different viruses or hosts, remains to be seen. As yet, there are insufficient data to enable a comparison with the many codons known from the sequence of *E. coli* bacteriophage nucleic acids. It will also be interesting to see if in the examples of possible overlapping genes (TMV, TRV, CPMV) the sequence of bases is read "in phase" or "out-of-frame". Examples of out-of-frame reading of the sequence is known in bacteriophages, for example, ϕX174 DNA.[100] The in vitro read-through of AMV RNA 3 is presumably in phase since the read-through product is precipitated by anti-AMV serum,[58a] and the two large TYMV polypeptides made in reticulocyte lysates have common peptide pattern.[76d]

Relatively little is known about the termination process of plant viral translation. The wheat germ cell-free system appears to recognize both UAG and UGA as termination codons[14] and it is likely that these or UAA are used in plant viral messengers.

Further codon assignments and understanding of the involvement of RNA structure in plant viral RNA translation awaits the sequence data. It is likely that small (coat protein cistron) RNAs, such as STNV RNA, and those of TYMV, BMV, AMV, and TMV, will be completely sequenced in the not too distant future.

### D. Translation Strategies and Evolution

The isolation of separate plant viral RNAs and their subsequent translation in cell-free extracts has revealed the existence of translation strategies not only different from the polycistronic bacteriophage RNAs, but also different among various plant viruses.

TABLE 3

**Some Examples of Codons for Plant Viral Coat Protein Amino Acids**

| | The codon and position[a] of its amino acid in the coat protein | | |
|---|---|---|---|
| Amino acid | BMV[b] | TMV[c] | AMV[d] |
| Methionine | AUG (initiation) | AUG (initiation) | AUG (initiation) |
| Serine | UCG (1), UCA (3) | UCA (55), AGU (65) | AGU (1), UCU (2), UCA (3) |
| Threonine | ACU (2, 5, 9) | ACG (103), ACU (59, 89), ACA (81) | — |
| Glycine | GGA (4), GGU (6) | GGU (85) | GGU (8), GGG (9) |
| Lysine | AAG (7) | AAA (53), AAG (68) | AAG (5), AAA (6, 10) |
| Arginine | CGC (10), CGU (13) | AGA (90, 92), AGG (61) | — |
| Alanine | GCG (11) | GCA (86), GCG (74) | GCU (7, 11) |
| Glutamine | CAG (12) | CAA (57) | CAA (4) |
| Isoleucine | — | AUA (93, 94, 125), UAG (129) | — |
| Asparagine | — | AAU (126, 127), AAC (101) | — |
| Leucine | — | UUA (83, 128), CUA (79), CUA (84) | — |
| Proline | — | CCU (54), CCA (56), CCC (102) | — |
| Valine | — | GUC (80), CUA (58), GUU (60), GUG (69, 75) | — |
| Tyrosine | — | UAC (70, 72) | — |
| Aspartic acid | — | GAC (64, 65, 77, 88) | — |
| Phenylalanine | — | UUU (87, 67), UUC (62) | — |
| Tryptophan | — | UGG (52) | — |

[a] Position of amino acid from the N-terminal amino acid (number 1) of the coat protein.
[b] Data from Dasgupta et al.[43]
[c] Data from Hunter et al.[21] and Guilley et al.[104]
[d] Data from Koper-Zwarthoff et al.[105]

Models depicting some of the mechanisms so far in evidence are presented in Figure 8.

A common feature in Models 1 to 3 is the existence of a small encapsidated RNA coding for the coat protein. This includes the bromoviruses, CMV, AMV, TSV, and the cowpea strain of TMV. In TYMV (Model 2), the small RNA may occur in some virions but not all (probably in minor components, but not all main particles[74a]). What mechanism separates this RNA in the cell before translation is not known. This situation appears to be between the deliberately encapsidated RNAs of Model 1 and the nonencapsidated type of Models 4 and 5. In this latter type, the small (LMC) RNA is produced in infected cells before translation, probably by differential transcription of that portion of the large RNA carrying the same sequence. This LMC RNA is not encapsidated (see Section III, Volume II by Zaitlin). Several of these systems (Models 1 to 5, Figure 8) produce two large polypeptides of molecular weights between 100 and 165,000. In Models 2, 3, 4, and 5, these two polypeptides probably are both coded by the same the large RNA. Thus, assuming both polypeptides have a real function, the genes overlap. Two initiation sites, rather than one initiation site and two termination sites, are assumed but not proven (see Reference 82a). The types with no small coat protein message (Models 6, 7, and 8) also all produce polypeptides of around 120,000 to 170,000 mol wt. Models 6 and 8 also have a larger polypeptide of about 200,000 mol wt. The mechanism of coat protein production in these types is not known.

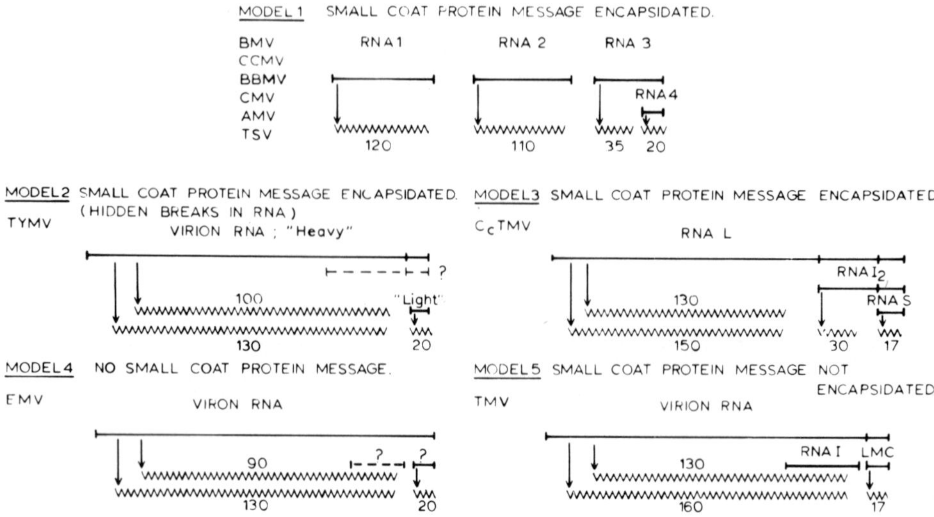


FIGURE 8. Models of gene locations and translation strategies of several plant viruses. Models are presented of the possible distribution or location of genes in some plant viral RNA genomes and the means by which functional translation units of the RNA are derived or polypeptides processed to yield the virus specific proteins. (—) RNA; ( ΛΛΛΛ ) polypeptide. The numbers given for the polypeptides are the approximate molecular weight × $10^{-3}$. Estimations depend on the gel system and markers used;[102,103] most values have ± 10% error. **Model 1** — This is based mainly on the data for BMV, but also represents other bromoviruses, and CMV, AMV, and probably TSV. The sizes given for the products of RNA 1 and 2 (120 and 110,000) are as published for BMV. In Figure 1, these products were estimated as 110,000 and 100,000. RNA 4, the coat protein message, is somehow derived from RNA 3, and is encapsidated in a virion. **Model 2** — TYMV; based on the data for RNA heated in EDTA or SDS[75] the "light" RNA coding for coat protein is assumed to be analogous to LMC RNA of TMV (see Model 5), but is encapsidated (but perhaps not in all virions). Distinction between cleavage of RNA and selective (differential) transcription has not been made. An intermediate-size class of RNA (class IV)[74] was suggested by TYMV translation after density gradient centrifugation of heated RNA. This may be equivalent to RNA 3 and RNA $I_2$ of Models 1 and 2. Polypeptide sizes are based on wheat germ translation. In rabbit reticulocytes, the largest products are 150,000 and 195,000 daltons. **Model 3** — Cowpea strain of TMV. RNA L represents the largest form of the RNA, found in the long rods. Location of the sequence of $I_2$ RNA (dicistronic) in the 3′ region of RNA L and the small RNA "S" in the 3′ region of RNA $I_2$ is assumed.[85] By analogy to wild-type TMV (see Model 5), it is highly probable that the coat protein cistron is near the 3′ end. The two larger polypeptides probably have overlapping genes. RNAs, the coat protein gene RNA, is encapsidated. **Model 4** — EMV is closely related to TYMV (both classified as tymoviruses), but existence of a small ("light") RNA has not been shown. Such an RNA may be producedn vivo; thus this virus could be more TMV-like than TYMV. However, a small amount of labeled product (about 20,000 mol wt) can be detected (see Figure 6). It is not known if this is coat protein. **Model 5** — TMV. The model is based on the translation data for wild-type TMV.[21,22] The exact sizes of the large polypeptides are not certain (see Table 2). Existence of an intermediate "RNA 3", equivalent to RNA $I_2$ of Model 3, has been demonstrated.[89] It codes for a 30,000 mol wt product (not shown here).

Model 7 is based on TRV. There are two large RNAs, which in the case of strain PRN are a similar size to CPMV RNAs (Model 8). However, the translation strategy seems to be very different. The smaller RNA, RNA 2, appears to be dicistronic.[65,92] In vitro, this RNA is translated into coat protein and a 31,000 mol wt polypeptide. The order of the two cistrons is not certain. The CAM strain has an RNA 2 of only 0.68 × $10^6$ daltons, which would be just about the necessary coding capacity for coat protein and a 31,000 mol wt product, which, however, is not produced in vitro. If this RNA is dicistronic, a similarity is seen between it and RNA 3 of Model 1 and the polycistronic strategy of Model 9. Like the RNAs of Models 1 to 4, this RNA is capped with $m^7G^{5'}ppp$. However, the other RNA is not capped.[63] In these respects, Model 7 would seem to be intermediate between Models 5 (TMV) and 8 (CPMV).

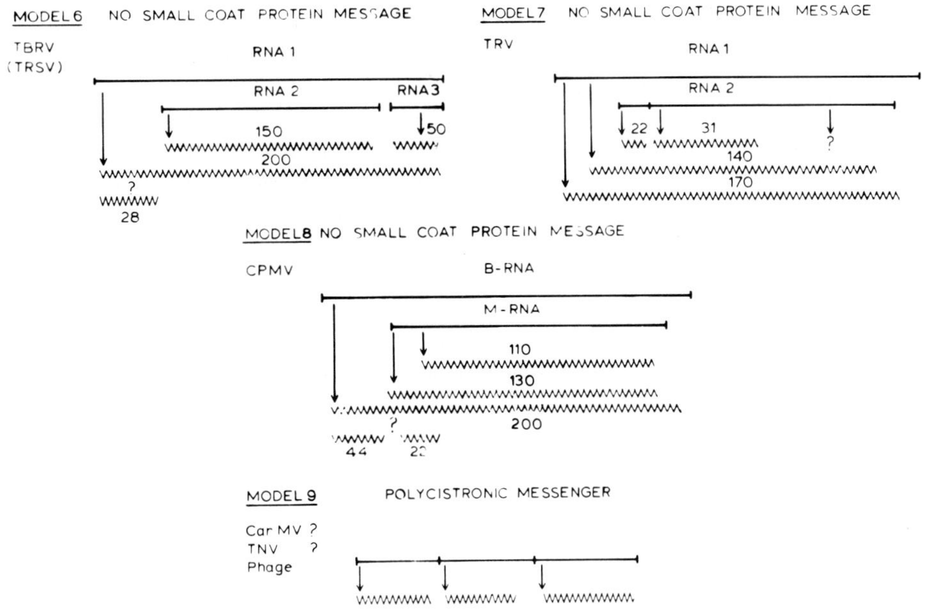


FIGURE 8. (continued). **Model 6** — TBRV. The evidence available so far[112] suggests that the large RNAs are translated into large polypeptides, analogous to those of CPMV (Model 8). No small coat protein message has yet been detected, but its existence in vivo cannot be ruled out. The alternative strategy would be posttranslational cleavage of large precursor proteins. An "RNA 3" is indicated, but this may represent a satellite-like entity occurring in some strains only. **Model 7** — TRV. Like Models 6 and 8 there does not seem to be a small monocistronic RNA coding for coat protein. In this, and the size of the two virion RNAs, TRV bears some similarity to CPMV (Model 8), which however has no $m^7G$ cap, whereas at least the smaller of the two TRV RNAs is capped.[63] It could be regarded therefore as an intermediate form of translation strategy, between Models 5 or 6 and 8 or 9. The gene order of RNA 2 is entirely arbitrary. **Model 8** — CPMV. The possibility of posttranslational processing of large precursor polypeptides is not suggested for any other plant virus, with the possible exception of TBRV (see Model 6). In vivo cleavage or differential transcription of the RNA cannot be completely ruled out. Location of the coat protein-coding sequences are not known, and are arbitrarily placed at the 5′ region of the B-RNA. Recent evidence[110] suggests that the size of the largest product is more like 185,000 daltons. **Model 9** — Polycistronic messenger. RNA bacteriophages, such as R17, MS2, and Qβ of *E. coli*, have a tricistronic genome which serves as a polycistronic messenger. Recently, evidence from in vitro translation experiments suggests that some plant viruses may have an analogous translation strategy. These include TNV and CarMV.

CPMV (Model 8) RNAs lack the $m^7G^{5'}ppp$ cap.[71] The large RNA (B-RNA) from the bottom component of the virus is translated in vitro into a roughly 200,000 mol wt polypeptide of unknown function. The other CPMV RNA (M-RNA) is translated into two polypeptides whose encoded sequences appear to be overlapping. So far, no separate smaller RNAs have been detected for CPMV, and since the coat protein sequences are most likely translated as part of the large polypeptides (for which almost the entire sequence of the RNAs is used), it is currently proposed that the large polypeptides are precursors for the viral proteins which are produced by posttranslational cleavage of the precursors. If this is true, then Model 8 is somewhat removed from Models 1 to 7.

The extent to which the Models represented in Figure 8 reflect parallel evolution or direct relationships is difficult to argue at the present time, but it is tempting to speculate in which direction the evolution may be progressing. One might assume that bacteria and hence their bacteriophages existed before higher plants and that therefore the polycistronic messenger (Model 9), like the tricistronic $1.0 \times 10^6$ mol wt RNA of MS 2 and R17 RNA or its ancestors, may be considered as a primitive type. Furthermore, if eukaryotic plant cells are equipped mainly to translate monocistronic messengers, particularly small ones, then viruses with permanently separated monocistronic RNAs may be considered more advanced than those which have no functional cleavage or transcription step to produce smaller RNAs in infected cells before translation can efficiently occur. By such argument, translation strategies represented in Figure 8 could be regarded as a pattern of evolution proceeding from Model 9 type viruses to Model 1 type viruses. This can be nothing more than speculation until more information is available. It should not be overlooked that some plant viruses bear resemblance, in both structure and molecular functions including translation, to some animal viruses. Evolution pathways may have crossed host kingdom barriers. Such comparisons and speculations are beyond the scope of this chapter.

The reader may wish to consult recent publications concerning CMV-associated RNA 5,[113] AMV RNA translation,[114] CPMV RNA structure,[115] TYMV coat protein RNA,[116] EMV RNA translation,[117] TMV noncoat protein polypeptides,[118] and CarMV RNA translation.[119]

## ABBREVIATIONS

| | |
|---|---|
| TMV | Tobacco mosaic virus |
| STNV | Satellite of tobacco necrosis virus |
| AMV | Alfalfa mosaic virus |
| BMV | Brome mosaic virus |
| –SH | Sulfhydryl group |
| HEPES | *N*-2-Hydroxyethylpiperazine-*N*'-2-ethane sulfonate |
| tris | Tris(hydroxymethyl)aminomethane |
| DNase | Deoxyribonuclease |
| TCA | Trichloroacetic acid |
| EDTA | Ethylene diaminetetraacetic acid (disodium) |
| SDS | Sodium dodecyl sulfate |
| EGTA | Ethylene glycol-bis(2-aminoethylether)-*N,N*'-tetraacetic acid |
| RNase | Ribonuclease |
| CCMV | Cowpea chlorotic mottle virus |
| BBMV | Broad bean mottle virus |
| CMV | Cucumber mosaic virus |
| TSV | Tobacco streak virus |
| TRV | Tobacco rattle virus |
| TBRV | Tomato black ring virus |
| CPMV | Cowpea mosaic virus |
| TYMV | Turnip yellow mosaic virus |
| EMV | Eggplant mosaic virus |
| TMV | Tobacco mosaic virus |
| CcTMV | Cowpea strain of TMV |
| TNV | Tobacco necrosis virus |
| CarMV | Carnation mottle virus |
| BSMV | Barley stripe mosaic virus |

## ACKNOWLEDGMENTS

I wish to thank Dr. A. van Kammen, Dr. E. J. Stuik, Dr. L. van Vloten-Doting, Dr. E. M. J. Jaspers, and Dr. L. Bosch for the help and comments, and all colleagues who provided information about unpublished results. I also thank Mr. A. Hoogeveen for photographs and drawings, and Ms. Wil Landeweerd for typing.

# REFERENCES


1. **Nirenberg, M. W. and Matthaei, J. H.,** The dependence of cell-free protein synthesis in *E. coli* upon naturally occurring or synthetic polyribonucleotides, *Proc. Natl. Acad. Sci. U.S.A.*, 47, 1588, 1961.
2. **Nirenberg, M. W.,** Cell-free protein synthesis directed by messenger RNA, in *Methods in Enzymology*, Vol. 6, Colowick, S. P. and Kaplan, N. O., Eds., Academic Press, New York, 1963, 17.
3. **Tsugita, A., Fraenkel-Conrat, H., Nirenberg, M. W., and Mattaei, J. H.,** Demonstration of the messenger role of viral RNA, *Proc. Natl. Acad. Sci. U.S.A.*, 48, 846, 1962.
4. **Nathans, D., Notani, G., Schwartz, J. H., and Zinder, N. D.,** Biosynthesis of the coat protein of coliphage f2 by *E. coli* extracts, *Proc. Natl. Acad. Sci. U.S.A.*, 48, 1424, 1962.
5. **Nathans, D.,** Cell-free protein synthesis directed by coliphage MS 2 RNA: synthesis of intact viral coat protein and other products, *J. Mol. Biol.*, 13, 521, 1965.
6. **Ohtaka, Y. and Spiegelman, S.,** Translational control of protein synthesis in a cell-free system directed by a polycistronic viral RNA, *Science*, 142, 493, 1963.
7. **Aach, H. G., Funatsu, G., Nirenberg, M. W., and Fraenkel-Conrat, H.,** Further attempts to characterize products of TMV RNA-directed protein synthesis, *Biochemistry*, 3, 1362, 1964.
8. **Clark, J. M., Chang, A. Y., Spiegelman, S., and Reichmann, M. E.,** The *in vitro* translation of a monocistronic message, *Proc. Natl. Acad. Sci. U.S.A.*, 54, 1193, 1965.
9. **Van Ravenswaay Claassen, J. C., van Leeuwen, A. B. J., Duijts, G. A. H., and Bosch, L.,** *In vitro* translation of alfalfa mosaic virus RNA, *J. Mol. Biol.*, 23, 535, 1967.
10. **Stubbs, J. D. and Kaesberg, P.,** Amino acid incorporation in an *Escherichia coli* cell-free system directed by bromegrass mosaic virus ribonucleic acid, *Virology*, 33, 385, 1967.
11. **Rice, R. and Fraenkel-Conrat, H.,** Fidelity of translation of satellite tobacco necrosis virus ribonucleic acid in a cell-free *Escherichia coli* system, *Biochemistry*, 12, 181, 1973.
12. **Klein, W. H., Nolan, C., Lazar, J. M., and Clark, J. M.,** Translation of satellite tobacco necrosis virus ribonucleic acid. I. Characterization of *in vitro* procaryotic and eukaryotic translation products, *Biochemistry*, 11, 2009, 1972.
13. **Davies, J. W.,** The multipartite genome of brome mosaic virus: aspects of *in vitro* translation and RNA structure, *Ann. Microbiol. (Paris)*, 127A, 131, 1976.
14. **Davies, J. W. and Kaesberg, P.,** Translation of virus mRNA: synthesis of bacteriophage Q$\beta$ proteins in a cell-free extract from wheat embryo, *J. Virol.*, 12, 1434, 1973.
15. **Marcus, A., Luginbill, B., and Feeley, J.,** Polysome formation with tobacco mosaic virus RNA, *Proc. Natl. Acad. Sci. U.S.A.*, 59, 1243, 1968.
16. **Marcus, A.,** Tobacco mosaic virus ribonucleic acid dependent amino acid incorporation a wheat embryo system *in vitro*: analysis of the rate-limiting reaction, *J. Biol. Chem.*, 245, 955, 1969.
17. **Efron, D. and Marcus, A.,** Translation of TMV RNA in a cell-free wheat embryo system, *Virology*, 53, 343, 1973.
18. **Roberts, B. E., Matthews, M. B., and Bruton, C. J.,** Tobacco mosaic virus RNA directs the synthesis of a coat protein peptide in a cell-free system from wheat, *J. Mol. Biol.*, 80, 733, 1973.
19. **Davies, J. W. and Kaesberg, P.,** Translation of virus mRNA: protein synthesis directed by several virus RNAs in a cell-free extract from wheat germ, *J. Gen. Virol.*, 25, 11, 1974.
20. **Siegel, A., Hari, V., Montgomery, I., and Kolacz, K.,** A messenger RNA for capsid protein isolated from tobacco mosaic virus-infected tissue, *Virology*, 73, 363, 1976.
21. **Hunter, T. R., Hunt, T., Knowland, J., and Zimmern, D.,** Messenger RNA for the coat protein of tobacco mosaic virus, *Nature (London)*, 260, 759, 1976.
22. **Knowland, J., Hunter, T., Hunt, T., and Zimmern, D.,** Translation of tobacco mosaic virus RNA and isolation of the messenger for TMV coat protein, in *In Vitro Transcription and Translation of Viral Genomes*, Haenni, A. and Beaud, G., Eds., Inserm, Paris, 1975, 211.
23. **Shih, D. S. and Kaesberg, P.,** Translation of brome mosaic virus ribonucleic acid in a cell-free system derived from wheat embryo, *Proc. Natl. Acad. Sci. U.S.A.*, 70, 1799, 1973.
24. **Roberts, B. E. and Paterson, B. M.,** Efficient translation of tobacco mosaic virus RNA and rabbit globin 9S RNA in a cell-free system from commercial wheat germ, *Proc. Natl. Acad. Sci. U.S.A.*, 70, 2330, 1973.
25. **Marcu, K. and Dudock, B.,** Characterization of a highly efficient protein synthesizing system derived from commercial wheat germ, *Nucleic Acids Res.*, 1, 1385, 1974.
26. **Ball, L. A., Minson, A. C., and Shih, D. S.,** Synthesis of plant virus coat proteins in an animal cell-free system, *Nature (London) New Biol.*, 246, 206, 1973.
27. **Mohier, E., Hirth, L., Le Meur, M., and Gerlinger, P.,** Translation of alfalfa mosaic virus RNAs in mammalian cell-free system, *Virology*, 68, 349, 1975.
28. **Pelham, H. R. B. and Jackson, R. J.,** An efficient mRNA-dependent translation system from reticulocyte lysates, *Eur. J. Biochem.*, 67, 247, 1976.

29. **Marcus, A., Efron, D., and Weeks, D. P.,** The wheat embryo cell-free system, in *Methods in Enzymology,* Vol. 30F, Modave, K. and Grossman, L., Eds., Academic Press, New York, 1974, 749.
30. **Efstratiadis, A. and Kafatos, F. C.,** The chorion of insects. VI. Translation of mRNA in the wheat germ cell-free system, in *Eukaryotes at the Subcellular Level,* Vol. 8, Last, J. A., Ed., Marcel Dekker, New York, 1976, 68.
31. **Hunt, T. and Jackson, R. J.,** The rabbit reticulocyte lysate as a system for studying mRNA, in *Modern Trends in Human Leukaemia,* Neth, R., Gallo, R. C., Spiegelman, S., and Stohlman, F., Eds., Lehmanns Verlag, Munich, 1974, 300.
32. **Klein, W. H. and Clark, J. M.,** N-terminal sequence of the eucaryotic *in vitro* products made upon translation of satellite tobacco necrosis virus ribonucleic acid, *Biochemistry,* 12, 1528, 1973.
33. **Leung, D. W., Gilbert, C. W., Smith, R. E., Sasavage, N. L., and Clark, J. M.,** Translation of satellite tobacco necrosis virus ribonucleic acid by an *in vitro* system from wheat germ, *Biochemistry,* 15, 4943, 1976.
34. **Horst, J., Fraenkel-Conrat, H., and Mandeles, S.,** Terminal heterogeneity at both ends of the satellite tobacco necrosis virus ribonucleic acid, *Biochemistry,* 10, 4748, 1971.
35. **Lesnaw, J. and Reichmann, M.,** Identity of the 5′-terminal RNA nucleotide sequence of the satellite tobacco necrosis virus and its helper virus: possible role of the 5′ terminus in the recognition by virus-specific RNA replicase, *Proc. Natl. Acad. Sci. U.S.A.,* 66, 140, 1970.
36. **Hickey, E. D., Weber, L. A., and Baglioni, C.,** Inhibition of initiation of protein synthesis by 7-methyl guanosine-5′-monophosphate, *Proc. Natl. Acad. Sci. U.S.A.,* 73, 19, 1976.
37. **Roman, R., Brooker, J. D., Seal, S. N., and Marcus, A.,** Inhibition of the transition of a 40 S ribosome-met-tRNA$_i^{met}$ complex to an 80 S ribosome-met-tRNA$_i^{met}$ complex by 7-methylguanosine-5′-phosphate, *Nature (London),* 260, 359, 1976.
38. **Marcus, A.,** personal communication.
39. **Owens, R. A. and Schneider, I. R.,** Satellite of tobacco ringspot virus RNA lacks detectable mRNA activity, *Virology,* 79, 1977.

40a. **Lane, L. C.,** The bromoviruses, in *Advances in Virus Reserch,* Vol. 19, Smith, K. M. and Lauffer, M. A., Eds., Academic Press, New York, 1974, 151.

40b. **Lane, L. C. and Kaesberg, P.,** Multiple genetic components in bromegrass mosaic virus, *Nature (London) New Biol.,* 232, 40, 1971.

41. **Shih, D. S., Davies, J. W., and Kaesberg, P.,** The *in vitro* synthesis of viral protein and its regulation, in *Proc. 1st. Int. Congr. IAMS,* Vol. 3, Hasegawa, T., Ed., Science Council of Japan, Tokyo, Japan, 1975, 83.
42. **Shih, D. S. and Kaesberg, P.,** Translation of the RNAs of brome mosaic virus; the monocistronic nature of RNA 1 and RNA 2, *J. Mol. Biol.,* 103, 77, 1976.
43. **Dasgupta, R., Shih, D. S., Saris, C., and Kaesberg, P.,** Nucleotide sequence of a viral RNA fragment that binds to eukaryotic ribosomes, *Nature (London),* 256, 624, 1975.
44. **Dasgupta, R., Harada, F., and Kaesberg, P.,** Blocked 5′ termini in brome mosaic virus RNA, *J. Virol.,* 18, 260, 1976.
45. **Both, G. W., Banerjee, A. K., and Shatkin, A. J.,** Methulation-dependent translation of viral messenger RNAs *in vitro, Proc. Natl. Acad. Sci. U.S.A.,* 72, 1189, 1975.
46. **Muthukrishnan, S., Both, G. W., Furuichi, Y., and Shatkin, A. J.,** 5′-Terminal 7-methyl-guanosine in eukaryotic mRNA is required for translation, *Nature (London),* 255, 33, 1975.
47. **Rose, J. K. and Lodish, H. F.,** Translation *in vitro* vesicular stomatitis virus messenger RNA lacking 5′-terminal 7-methyl guanosine, *Nature (London),* 262, 32, 1976.
48. **Shih, D. S., Dasgupta, R., and Kaesberg, P.,** 7-Methyl-guanosine and efficiency of RNA translation, *J. Virol.,* 19, 637, 1976.
49. **Shine, J. and Dalgarno, L.,** Identical 3′-terminal octanucleotide sequences in 18 S ribosomal ribonucleic acid from different eukaryotes, *Biochem. J.,* 141, 609, 1974.
50. **Schwinghamer, M. W. and Symons, R. H.,** Fractionation of cucumber mosaic virus RNA and its translation in a wheat embryo cell-free system, *Virology,* 63, 252, 1975.
51. **Schwinghamer, M. W. and Symons, R. H.,** Translation of the four major RNA species of cucumber mosaic virus in plant and animal cell-free systems and in toad oocytes, *Virology,* 79, 88, 1977.
52. **Kaper, J. M., Tousignant, M. E., and Lot, H.,** A low molecular weight replicating RNA associated with a divided genome plant virus: defective or satellite RNA?, *Biochem. Biophys. Res. Commun.,* 72, 1237, 1976.
53. **Owens, R. A. and Kaper, J. M.,** Cucumber mosaic virus-associated RNA 5. II. *In vitro* translation in a wheat germ protein synthesis system, *Virology,* 80, 196, 1977.
54. **Van Vloten-Doting, L.,** Coat protein is required for infectivity of tobacco streak virus: biological equivalence of the coat proteins of tobacco streak and alfalfa mosaic viruses, *Virology,* 65, 215, 1975.
55. **Lister, R. M., Ghabrial, S. A., and Saksena, K. N.,** Evidence that particle size heterogeneity is the cause of centrifugal heterogeneity in tobacco streak virus, *Virology,* 49, 290, 1972.

56. **Fulton, R. W.,** The role of particle heterogeneity in infection by tobacco streak virus, *Virology,* 41, 288, 1970.
57. **Bol, J. F., van Vloten-Doting, L., and Jaspers, E. M. J.,** A functional equivalence of top component-a RNA and coat protein in the initiation of infection by alfalfa mosaic virus, *Virology,* 46, 73, 1971.
57a. **Heijtink, R. A., Houwing, C. J., and Jaspers, E. M. J.,** Molecular weights of particles and RNAs of alfalfa mosaic virus. Number of subunits in protein capsids, *Biochemistry,* 16, 4684, 1977.
58. **Van Vloten-Doting, L., Rutgers, T., Neeleman, L., and Bosch, L.,** *In vitro* translation of the RNAs of alfalfa mosaic virus, in *In Vitro Transcription and Translation of Viral Genomes,* Haenni, A. and Beaud, G., Eds., Inserm, Paris, 1975, 225.
58a. **Van Vloten-Doting, L.,** Similarities and differences between viruses with a tripartite genome, *Ann. Microbiol. (Paris),* 127A, 119, 1976.
58b. **Neeleman, L., Rutgers, A. S., and van Vloten-Doting, L.,** Internal initiation of protein synthesis on RNA of a eukaryotic virus? in Translation of Natural and Synthetic Polynucleotides, Legocke, A. B., Ed., University of Agriculture in Poznan, Poznan, Poland, 1977, 292.
59. **Thang, M. N., Dondon, L., Thang, D. C., Mohier, E., Hirth, L., Le Meur, M. A., and Gerlinger, P.,** Translation of alfalfa mosaic virus RNAs in plant cell and mammalian cell extracts, in *In Vitro Transcription and Translation of Viral Genomes,* Haenni, A. and Beaud, G., Eds., Inserm, Paris, 1975, 225.
60. **Mohier, E., Hirth, L., Le Meur, M., and Gerlinger, P.,** Analysis of alfalfa mosaic virus 17 S RNA translational products, *Virology,* 71, 615, 1976.
61. **Gerlinger, P., Mohier, E., Le Meur, M. A., and Hirth, L.,** Monocistronic translation of alfalfa mosaic virus RNAs, *Nucleic Acids Res.,* 4, 813, 1977.
62. **Sänger, H. M.,** Characteristics of tobacco rattle virus. I. Evidence that its two particles are functionally defective and mutually complementing, *Mol. Gen. Genet.,* 101, 346, 1968.
63. **Haidar, M. A. and Hirth, L.,** 5′-terminal structure of tobacco rattle virus RNA: Evidence for polarity of reconstitution, *Virology,* 76, 173, 1977.
64. **Mayo, M. A., Fritsch, C., and Hirth, L.,** Translation of tobacco rattle virus RNA *in vitro* using wheat germ extracts, *Virology,* 77, 408, 1976.
65. **Fritsch, C., Mayo, M. A., and Hirth, L.,** Further studies on the translation products of tobacco rattle virus RNA *in vitro, Virology,* 77, 722, 1977.
65a. **Fritsch, C., Mayo, M., and Murant, A. F.,** Translation of the satellite RNA of tomato black ring virus in vitro and in tobacco protoplasts, *J. Gen. Virol.,* 40, 587, 1978.
66. **Mayo, M. A. and Robinson, D. J.,** Revision of estimates of the molecular weights of tobravirus coat proteins, *Intervirology,* 5, 313, 1975.
67. **Wu, G. and Bruening, G.,** Two proteins from cowpea mosaic virus, *Virology,* 46, 596, 1971.
68. **Pelham, H. R. B. and Stuik, E. J.,** Translation of cowpea mosaic virus RNA in a messenger-dependent cell-free system from rabbit reticulocytes, in *Proc. Coll. Nucleic Acids Protein Synthesis Plants,* Centre National de la Recherche Scientifique, Paris, 1977, 691.
69. **Davies, J. W., Aalbers, A. M. J., Stuik, E. J., and van Kammen, A.,** Translation of cowpea mosaic virus RNA in a cell-free extract from wheat germ, *FEBS Lett.,* 77, 265, 1977.
69a. **Stuik, E. J. and Davies, J. W.,** unpublished results.
70. **El Manna, M. M. and Bruening, G.,** Polyadenylate sequences in the ribonucleic acids of cowpea mosaic virus, *Virology,* 56, 198, 1973.
71. **Klootwijk, J., Klein, I., Zabel, P., and van Kammen, A.,** Cowpea mosaic virus RNAs have neither $m^7$GpppN . . . , nor mono-, di, or triphosphates at their 5′-ends, *Cell,* 11, 73, 1977.
72. **Lee, Y. F., Nomoto, A., Detjen, B. M., and Wimmer, E.,** A protein covalently linked to poliovirus genome RNA, *Proc. Natl. Acad. Sci. U.S.A.,* 74, 59, 1977.
73. **Benicourt, C. and Haenni, A. L.,** Translation of TYMV RNA in a wheat germ cell-free system, in *In Vitro Transcription and Translation of Viral Genomes,* Haenni, A. and Beaud, G., Eds., Inserm, Paris, 1975, 189.
74. **Pleij, C. W. A., Neeleman, A., van Vloten-Doting, L., and Bosch, L.,** Translation of turnip yellow mosaic virus RNA *in vitro*: a closed and open coat protein cistron, *Proc. Natl. Acad. Sci. U.S.A.,* 73, 4437, 1976.
74a. **Pleij, C. W. A., Mellema, J. R., Noort, A., and Bosch, L.,** The occurrence of the coat protein messenger RNA in the minor components of turnip yellow mosaic virus, *FEBS Lett.,* 80, 19, 1977.
75. **Klein, C., Fritsch, C., Briand, J. P., Richards, K. E., Jonard, G., and Hirth, L.,** Physical and functional heterogeneity in TYMV RNA: evidence for the existence of an independent messenger coding for coat protein, *Nucleic Acids Res.,* 3, 3043, 1976.
75a. **Klein, C., Fritsch, C., and Hirth, L.,** unpublished results.
76. **Matthews, R. E. F.,** Properties of nucleoprotein fractions isolated from turnip yellow mosaic virus preparations, *Virology,* 12, 521, 1960.

76a. **Matthews, R. E. F.,** Some properties of TYMV nucleoproteins isolated in cesium chloride density gradients, *Virology,* 60, 54, 1974.

76b. **Higgins, T. J. V., Whitfeld, P. R., and Matthews, R. E. F.,** Size distribution and *in vitro* translation of the RNAs isolated from turnip yellow mosaic virus nucleoproteins, *Virology,* 84, 153, 1978.

76c. **Ricard, B., Barreau, C., Renaudin, H., Mouches, C., and Bové, J.-M.,** Messenger properties of TYMV-RNA, *Virology,* 79, 231, 1977.

76d. **Bénicourt, C., Péré, J.-P., and Haenni, A.-L.,** Translation of TYMV RNA into high molecular weight proteins, *FEBS Lett.,* 86, 268, 1978.

77. **Fraenkel-Conrat, H.,** Plant viruses, in *Comprehensive Virology,* Vol. 1, Fraenkel-Conrat, H. and Wagner, R. R., Eds., Plenum Press, New York, 1974, 97.

78. **Leis, J. P. and Keller, E. B.,** Protein chain initiation by methionyl-tRNA, *Biochem. Biophys. Res. Commun.,* 40, 416, 1970.

79. **Marcus, A.,** Tobacco mosaic virus ribonucleic acid-dependent amino acid incorporation in a wheat embryo system *in vitro.* Formation of a ribosome-messenger "initiation" complex, *J. Biol. Chem.,* 245, 962, 1970.

80. **Seal, S. N., Bewley, J. D., and Marcus, A.,** Protein chain initiation in wheat embryo. Resolution and function of the soluble factors, *J. Biol. Chem.,* 247, 2592, 1972.

81. **Weeks, D. P., Verma, D. P. S., Seal, S. N., and Marcus, A.,** Role of ribosomal subunits in eukaryotic protein chain initiation, *Nature (London),* 236, 167, 1972.

82. **Knowland, J.,** Protein synthesis directed by the RNA from a plant virus in a normal animal cell, *Genetics,* 78, 383, 1974.

82a. **Pelham, H. R. B.,** Leaky UAG termination codon in tobacco mosaic virus RNA, *Nature (London),* 212, 469, 1978.

83. **Jackson, R. J. and Hunter, A. R.,** Role of methionine in the initiation of haemoglobin synthesis, *Nature (London),* 227, 672, 1970.

84. **Morris, T. J.,** Two ribonucleoprotein components associated with the cowpea strain of TMV, *Am. Phytopathol. Soc. Proc.,* 1, 83, 1974.

85. **Bruening, G., Beachy, R. N., Scalla, R., and Zaitlin, M.,** *In vitro* and *in vivo* translation of the ribonucleic acids of a cowpea strain of tobacco mosaic virus, *Virology,* 71, 498, 1976.

86. **Whitfeld, P. R. and Higgins, T. J. V.,** Occurrence of short particles in beans infected with the cowpea strain of TMV. I. Purification and characterization of short particles, *Virology,* 71, 471, 1976.

87. **Jackson, A. O., Zaitlin, M., Siegel, A., and Francki, R. I. B.,** Replication of tobacco mosaic virus. III. Viral RNA metabolism in separated leaf cells, *Virology,* 48, 655, 1972.

88. **Higgins, T. J. V., Goodwin, P. B., and Whitfeld, P. R.,** Occurrence of short particles in beans infected with the cowpea strain of TMV. II. Evidence that short particles contain the cistron for coat protein, *Virology,* 72, 486, 1976.

89. **Beachy, R. N. and Zaitlin, M.,** Characterization and *in vitro* translation of the RNAs from less-than-full-length, virus-related, nucleoprotein rods present in tobacco mosaic virus preparations, *Virology,* 81, 160, 1977.

90. **Salvato, M. S. and Fraenkel-Conrat, H.,** Translation of tobacco necrosis virus and its satellite in a cell-free wheat germ system, *Proc. Natl. Acad. Sci. U.S.A.,* 74, 2288, 1977.

91. **Salomon, R., Soreq, H., Gozes, I., Bar-joseph, M., and Littauer, U. Z.,** *In vitro* translation of carnation mottle virus RNA. Israel Biochem. Soc. and Israel Soc. Path. Chem., Abstract, Annual Meeting, Tel-Aviv, Israel, 1977.

91a. **Salmon, R.,** personal communication.

92. **Robinson, D. J.,** A variant of tobacco rattle virus: evidence for a second gene in RNA-2, *J. Gen. Virol.,* 35, 37, 1977.

93. **Dolja, V. V., Negruk, V. I., and Atabekov, J. G.,** Doubling of genes in the RNA fragments of barley stripe mosaic virus, *FEBS Lett.,* 65, 47, 1976.

94. **Hariharasubramanian, V., Hadidi, A., Singer, B., and Fraenkel-Conrat, H.,** Possible identification of a protein in a brome mosaic virus infected barley as a component of viral RNA polymerase, *Virology,* 54, 190, 1973.

95. **Sakai, F., Watts, J. W., Dawson, J. R. O., and Bancroft, J. B.,** Synthesis of proteins in tobacco protoplasts infected with cowpea chlorotic mottle virus, *J. Gen. Virol.,* 34, 285, 1977.

96. **Zaitlin, M. and Hariharasubramanian, B.,** A gel electrophoretic analysis of proteins from plants infected with tobacco mosaic and potato spindle tuber viruses, *Virology,* 47, 296, 1972.

97. **Scalla, R., Boudon, E., and Rigaud, J.,** Sodium dodecyl sulphate polyacrylamide gel electrophoretic detection of two high molecular weight proteins associated with tobacco mosaic virus infection in tobacco, *Virology,* 69, 339, 1976.

98. **Paterson, R. and Knight, C. A.,** Protein synthesis in tobacco protoplasts infected with tobacco mosaic virus, *Virology,* 64, 10, 1975.

98a. **Sakai, F. and Takebe, I.,** Protein synthesis in tobacco mesophyll protoplasts induced by tobacco mosaic virus infection, *Virology*, 62, 426, 1974.
99. **Zaitlin, M., Duda, C. T, and Petti, M. A.,** Replication of tobacco mosaic virus. V. Properties of the bound and solubilized replicase, *Virology*, 53, 300, 1973.
100. **Sanger, F., Air, G. M., Barrell, B. G., Brown, N. L., Coulson, A. R., Fiddes, J. C., Hutchison, III, C. A., Slocombe, P. M., and Smith, M.,** Nucleotide sequence of bacteriophage × 174 DNA, *Nature (London)*, 265, 687, 1977.
101. **Laemmli, U. K.,** Cleavage of structural proteins during the assembly of the head of bacteriophage T4, *Nature (London)*, 227, 680, 1970.
102. **Weber, K. and Osborn, M.,** The reliability of molecular weight determinations by dodecylsulfate-polyacrylamide gel electrophoresis, *J. Biol. Chem.*, 244, 4406, 1969.
103. **Swaney, J. B., van de Woude, G. F., and Bachrach, H. L.,** Sodium dodecyl-sulfate-dependent anomalies in gel electrophoresis: alterations in the banding patterns of foot-and-mouth disease virus polypeptides, *Anal. Biochem.*, 58, 337, 1974.
104. **Guilley, H., Jonard, G., Richards, K. E., and Hirth, L.,** Observations concerning the sequence of two additional specifically encapsidated RNA fragments originating from the tobacco mosaic virus coat protein cistron, *Eur. J. Biochem.*, 54, 145, 1975.
105. **Koper-Zwarthoff, E. C., Lockard, R. E., Rajbhandary, U. L., de Weerd, B., and Bol, J. F.,** Nucleotide sequence of the 5′-terminus of alfalfa mosaic virus RNA 4, leading into the coat protein cistron, *Proc. Natl. Acad. Sci. U.S.A.*, 74, 5504, 1977.
106. **Fowler, A. V. and Zabin, I.,** The amino acid sequence of β-galactosidase of *Escherichia coli*, *Proc. Natl. Acad. Sci. U.S.A.*, 74, 1507, 1977.
107. **Rottier, P., Rezelman, G., and van Kammen, A.,** unpublished results.
108. **Sakai, F., Dawson, J. R. O., and Watts, J. W.,** Synthesis of proteins in tobacco protoplasts infected with brome mosaic virus, *J. Gen. Virol.*, 42, 323, 1979.
109. **Verduin, B. J. M. and Davies, J. W.,** unpublished results.
110. **Stuick, E. J. and Davies, J. W.,** unpublished results.
111. **Rezelman, G. and Rottier, P.,** unpublished results.
112. **Fritsch, C., Mayo, M., and Hirth, L.,** unpublished results.
113. **Richards, K. E., Jonard, G., Jacquemond, M., and Lot, H.,** Nucleotide sequence of cucumber mosaic virus-associated RNA 5 *Virology*, 89, 395, 1978.
114. **van Tol, R. G. L. and van Vloten-Doting, L.,** Translation of alfalfa mosaic virus RNA 1 in the mRNA-dependent translation system from rabbit reticulocyte lysates, *Eur. J. Biochem.*, 93, 461, 1979.
115. **Stanley, J., Rottier, P., Davies, J. W., Zabel, P., and van Kammen, A.,** A protein linked to the 5′ termini of both RNA components of the cowpea mosaic virus genome, *Nucleic Acids Res.*, 5, 4505, 1978.
116. **Guilley, H., and Briand, J. P.,** Nucleotide sequence of turnip yellow mosaic virus coat protein mRNA, *Cell*, 15, 113, 1978.
117. **Ricard, B., Renaudin, H., and Bové, J.-M.,** Translation of eggplant mosaic virus RNA in wheat germ extracts and reticulocyte lysates, *Virology*, 91, 305, 1978.
118. **Scalla, R., Romaine, P., Asselin, A., Rigaud, J., and Zaitlin, M.,** An in vivo study of a non-structural polypeptide synthesized upon TMV infection and its identification with a polypeptide synthesized in vitro from TMV RNA, *Virology*, 91, 182, 1978.
119. **Salomon, R., Bar-Joseph, M., Soreq, H., Gozes, I., and Littauer, U. Z.,** Translation in vitro of carnation mottle virus RNA. Regulatory function of the 3′-region, *Virology*, 90, 288, 1978.
120. **Mellema, J. R., Bénicourt, C., Haenni, A.-L., Noort, A., Pleij, C. W. A., and Bosch, L.,** Translational studies with turnip yellow mosaic virus RNAs isolated from major and minor virus particles, unpublished.

# Section IV
# *Viroids, Plasmids, and Genetic Engineering*

# VIROIDS: INFECTIOUS RNA IN PLANTS

**Elizabeth Dickson**

## TABLE OF CONTENTS

## I. INTRODUCTION

Viroids are the smallest microbial agents currently known. Made entirely of RNA, they have a genome size less than one tenth the size of the smallest bacteriophage of *Escherichia coli.*[1] Despite this small size and the apparent lack of any kind of protein- or lipid-containing protective coat, viroids replicate in a variety of plants and cause severe disease symptoms in a few. Because viroids can cause significant crop loss, an increased understanding of their mode of action could be of great importance to the agricultural industry. At the same time, elucidation of the molecular basis of replication and pathogenicity could reveal novel roles for RNA in the control of cellular processes. Previous reviews[2-9] include a recent review of methods for studying viroids[8] and a comprehensive general review.[9]

### A. Potato Spindle Tuber Disease

The spindle tuber disease of potato was first described in 1922,[10,11] and later shown to be the same disease as that referred to as tomato bunchy top disease.[12] In its most severe form, this disease causes general stunting of plant growth, deformity of the upper foliage, and production of disfigured potatoes (tubers are elongated, contain prominent eyes, and are frequently cracked).[10] Productivity studies showed that these symptoms could lead to crop losses of up to 64% in the affected plants.[13-15] Transmission was shown to occur horizontally by means of harvesting equipment[16] or contact between seed pieces of seed potatoes,[16] and vertically by high frequency transmission through the true seed[12,17-21] (see section III-A). Because symptoms exhibited in the foliage of diseased potato plants were sometimes hard to detect,[15] it was important to identify another plant susceptible to the spindle tuber disease which would display unambiguous symptoms. Raymer and O'Brien[22] were the first to demonstrate that the potato spindle tuber disease could be transmitted mechanically to a host other than potato. Certain cultivars of tomato were found which developed easily recognized symptoms within about 3 weeks of inoculation (severe stunting of upper foliage; see Figure 1), thus providing the required indicator system for the infectious agent.

Initial attempts to isolate the infectious substance from diseased plant tissue proceeded on the assumption that the disease was caused by a virus. However, it soon became clear that the pathogenic agent was not a conventional virus.[23-25] In 1967, Diener and Raymer[26] showed that the infectious entity isolated from diseased tomato plants (*Lycopersicon esculentum* Mill. cv. Rutgers) survives phenol extraction, chloroform extraction, ethanol precipitation, and DNase treatment, but is sensitive to RNase treatment in buffers of low ionic strength. Combining these observations with the fact that most of the infectious material sedimented in sucrose gradients at approx-

FIGURE 1. Comparison of healthy and PSTV-infected tomato plants. Top Panel: Seedlings of *Lycopersicon esculentum* cv. Rutgers were inoculated with PSTV at the four-leaf stage and grown in the greenhouse for 3 weeks. The plant on the left hand side is the PSTV-infected specimen. The healthy plant on the right is the same age. Bottom Panel: Comparison of leaves removed from the PSTV-infected plant shown above with those removed from the uninfected plant of the same age.

imately 10S,[26] it was concluded that the infectious agent had the properties of a free nucleic acid, probably in the form of a small double-stranded RNA molecule.

Over the course of the next 4 years it was shown that (1) PSTV (potato spindle tuber virus, as it was known at that time) could be concentrated by ethanol precipitation to yield infectivity at dilutions up to $10^{-7}$ in tomato plants,[27,28] (2) PSTV is predominantly localized in nuclei and associated with chromatin,[28,29] (3) either virus particles do not exist in PSTV-infected plant tissue, or else they are extremely labile,[23-25,29] (4) no helper virus can be implicated in the replication of PSTV,[23,24] (5) antisera prepared to react specifically with double-stranded RNA or with RNA:DNA hybrids do not reduce infectivity,[30] and (6) upon electrophoresis, the infectious entity enters 20% polyacrylamide gels and moves through the gels as a single, well-defined band.[31]

### B. Viroid: A Definition

The dramatic differences observed between the low molecular weight, naked RNA molecule that causes the potato spindle tuber disease and conventional virus systems led Diener[23,32] in 1971 to propose that the term "viroid" be adopted to designate potato spindle tuber "virus" RNA and all other RNAs with similar properties.

In studies which paralleled some of those outlined above, Singh and colleagues[33,34] provided evidence that infectious low molecular weight RNA could be isolated from both potato (*Solanum tuberosum* L. cv. Saco) and tomato (*Lycopersicon esculentum* Mill. cv. Sheyenne) plants after infection by the potato spindle tuber agent. In 1973, it was proposed that the term "meta-virus" (meaning after or beyond viruses) be applied to such pathogens.[4] As a result, a number of publications appeared in which the spindle tuber agent was referred to as PSTM instead of PSTV.

A third name was proposed by Semancik and co-workers[35,36] during investigations of the infectious agent of citrus exocortis disease. Because they felt that the term "viroid" might imply that these infectious agents had virus-like modes of interaction with their hosts, they suggested that the terms "low-molecular-weight pathogenic RNA" or "pathogene" be used instead, in order to remain open-minded about the possibility that these agents might interact directly with the host genetic machinery and might in fact represent host components that had become autonomous.[36]

Despite these early problems in nomenclature, by the year 1975, all authors publishing in this field were following the lead of Diener in his use of the term "viroid".

### C. Other Diseases Caused by Viroids

In order to show that a particular disease is caused by a viroid, it is necessary to demonstrate that highly purified low molecular weight RNA is the causative agent. In addition to potato spindle tuber viroid (PSTV), four other plant diseases have been linked directly with the presence of a viroid: (1) citrus exocortis viroid (CEV),[35,37] (2) chrysanthemum stunt viroid (CSV),[38,39] (3) chrysanthemum chlorotic mottle viroid (ChCMV),[40,41] and (4) cucumber pale fruit viroid (CPFV).[42,43] Two other diseases, cadang-cadang,[37] a disease of coconut palms, and hop stunt disease (HSD),[43a] may also be incited by viroids. In these cases, however, the disease-specific RNA has not yet been shown to be infectious in highly purified form.

#### *1. Citrus Exocortis Viroid (CEV)*

Citrus exocortis disease of Etrog citron *(Citrus medica)* was first described in 1950 by Benton and co-workers.[47] Characteristic symptoms included necrosis of midveins, epinasty (curling under) of young, developing leaves, and severe stunting of growth in general. In 1968, Weathers and Greer[48,49] demonstrated mechanical transmission of CEV to a number of herbaceous plants. Two of these, petunia (*Petunia hybrida* Vilm. var. Burpee Blue) and purple velvet (*Gynura aurantiaca* DC) began to develop disease symptoms similar to those of Etrog citron within about 30 days of inoculation. Using Gynura as an indicator plant, investigators soon demonstrated that the agent of citrus exocortis disease is a viroid.[35,37,50-52] The finding that CEV is similar to PSTV both in host range and symptom production led to suggestions that these two viroids might be independent isolates of the same pathogen.[53-55]

#### *2. Chrysanthemum Stunt Viroid (CSV)*

A disastrous epidemic of a disease causing inferior blooms which open 7 to 10 days earlier than normal and general dwarfing appeared in cultivated chrysanthemums *(Chrysanthemum morifolium)* in the U.S. in 1945 and soon thereafter in England.[56,57] The infectious agent, transmitted mainly by foliage contact, was shown to be a viroid by Diener and Lawson[38] and by Hollings and Stone[39] in 1973. Although initial studies indicated that CSV could not infect herbaceous species outside the family Compositae,[39] recent studies have shown that this is not the case (see Sections III. D and III. E). Florists' cineraria (*Senecia cruentus* cv. Hansa) has proven to be an excellent indicator plant for CSV since this plant responds to inoculation by the production on its leaves of starch lesions which increase in number when infectivity of the inoculum is increased.[38]

*3. Chrysanthemum Chlorotic Mottle Viroid (ChCMV)*

This second, distinct viroid disease of chrysanthemums was first described in 1971 by Dimock and colleagues,[55] although the etiology of the disease was not known at that time. Romaine and Horst[40] reported the recovery of a majority of the infectious material 3 years later as a low molecular weight fraction (6-14S in sucrose gradients), which was extremely sensitive to ribonuclease treatment, resistant to deoxyribonuclease treatment, caused mottling of leaves of *Chrysanthemum morifolium* cv. Deep Ridge 10 to 16 days after inoculation, and led to extensive mottling or complete chlorosis 3 to 5 weeks postinoculation. These authors concluded that ChCMV was not related to CSV since (1) these two viroids elicit drastically different disease syndromes in cultivars which can be infected by both, (2) ChCMV has a much narrower host range than does CSV, and (3) ChCMV fails to show any cross-protection against subsequent inoculation of the same plant with CSV (a standard test for relatedness of viroid RNAs; see Section III. D below).[34]

*4. Cucumber Pale Fruit Viroid (CPFV)*

The first cases of pale fruit disease of cucumbers were recognized in 1963 in two greenhouses in the Westland of The Netherlands.[42] By 1974, this disease, which causes general stunting of growth, crumpled flowers and leaves, and light green fruit, was observed in many greenhouses throughout the country. Initial host range studies suggested that this viroid may undergo some sort of modification upon passage through some of its hosts.[42] However, in the light of recent molecular studies,[59] it seems unlikely that such host modification of the viroid genome occurs (see Section II. C. 3).

*5. Cadang-Cadang*

Cadang-cadang, also known as yellow mottle decline, first described in 1931 in the Philippines,[60] is an economically important disease of coconut palms (*Cocos nucifera* L.). Early during infection yellow leaf spots appear, plant growth is stunted, and more than the usual number of coconuts develop. As the disease progresses, flower and nut production decline, fronds decline in size and number, leaf spots increase in size and frequency, and the palm eventually dies.[46] During attempts to isolate presumptive viral components, Randles discovered two low molecular weight species of RNA that did not appear to be present in extracts of healthy tissue.[44] Furthermore, recent studies[46] have shown that extracts of diseased tissue contain some agent capable of producing the early symptoms of cadang-cadang in 18 to 22 months, and at the same time cause the appearance of the new species of RNA. If it can be shown that the disease can be incited by inoculating plants with nucleic acid preparations containing only the low molecular weight RNA species, then this would be the first case of viroid disease in a monocotyledonous plant. Further studies on this disease would be greatly facilitated by the identification of a host capable of displaying disease symptoms soon after inoculation instead of the approximately 7 years required to develop the full symptoms of cadang-cadang in coconut palms.

### *6. Hop Stunt Disease (HSD)*

Hop stunt disease was first described in Japanese commercial hop plants (*Humulus lupulus* L. cv. Kirin II) in 1970.[60a] Sap inoculation studies showed that disease symptoms (severe stunting and greatly reduced cone yield) develop only after incubation of 2 to 3 years.[60a] A second host, *Humulus japonicus* Sieb. et Zucc., requires an incubation period of more than 3 months.[60a] Host range studies led to the identification of several cucurbitaceous plants (*Cucumis sativus* L., *C. melo* L., and *Lugenaria siceraria* Standl.) as rapid indicator plants for the disease agent of HSD.[60a] The infectious agent, as isolated from cucumber plants, has the properties of low molecular weight RNA. Since most Japanese commercial hops originate from Europe, and since the symptoms of HSD in cucumber plants closely resemble those of CPFV-infected cucumbers,[42] it is possible that the agents of these two diseases are identical. However, as in the case of PSTV and CEV, an answer to this question will probably not be obtained until RNA fingerprint analysis is carried out (see Section II. C. 2).

## D. Evidence for the Existence of Viroids

There is no question as to the existence of the plant diseases mentioned above, nor would anyone question the fact that infectious entities capable of transmitting these diseases also exist. However, the prospect that such pathogens might be composed entirely of RNA of low molecular weight (the estimates range from 250 to 350 nucleotides), seems hard to believe at the outset.

The most powerful evidence available for the existence of a pathogen with viroid properties comes from the experiments carried out by Diener[61] which correlate infectivity with a band of UV-absorbing material present only in RNA preparations from diseased tissue. These results will be presented here in some detail because they have been of central importance in moving the study of viroids from a study of the phenomenology of plant disease into the arena of molecular biology. RNA was purified from both healthy and PSTV-infected tomato plants (Rutgers) by a procedure including, among other steps, organic extraction, DNase treatment, removal of material which precipitates with 2*M* LiCl treatment, and fractionation of the resulting RNA on cylindrical 20% polyacrylamide gels. After electrophoresis sufficient to move 5S RNA to the bottom of an 8 cm gel, the optical density profile at 260 nm was recorded. Low molecular weight RNA from healthy plants was found to contain three well-resolved RNA species with electrophoretic mobilities slower than that of 5S (see Figure 2A). PSTV-infected plants contained these same three RNA species and an additional one migrating between the two faster components. When the gel was sliced and assayed for the presence of PSTV, a peak of infectivity was observed whose position corresponded exactly with that of the UV-absorbing component present only in PSTV-infected plant RNA (see Figure 2B).

While the formal possibility remains that the major component of the new gel band is a host-specified product that is overproduced in the presence of PSTV and, coincidentally, has precisely the same electrophoretic mobility as the infectious agent, it is much more likely that the new UV-absorbing species is the pathogenic agent itself. Additional support for this latter view can be gained by careful infectivity studies. Using an inoculation procedure identical to that used for viroid inoculation (the inoculum is rubbed onto the upper surface of a leaf in the presence of carborundum as an abrasive agent), $10^6$ particles of tobacco mosaic virus (TMV) must be applied to one

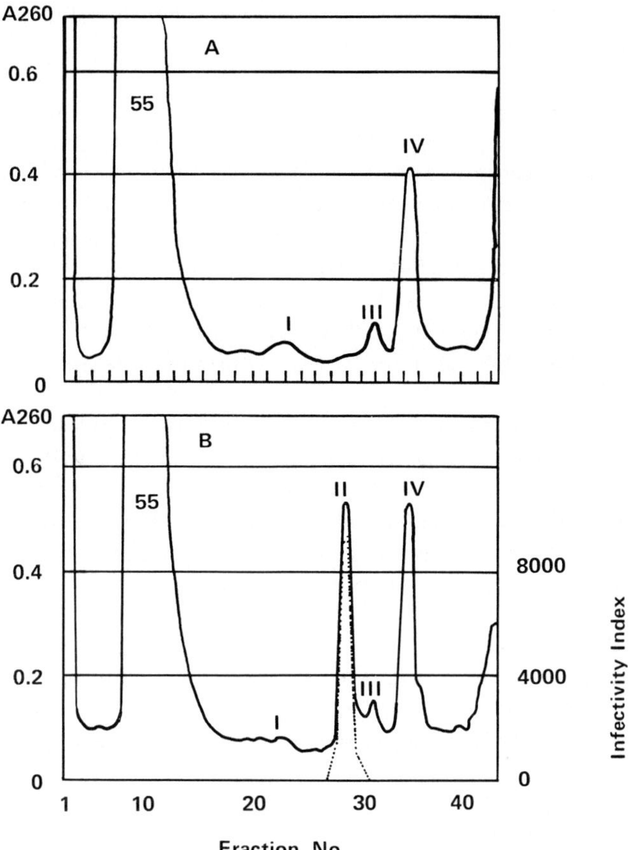


FIGURE 2. A. UV-absorption profile of RNA preparation from healthy tomato leaves after electrophoresis in a 20% polyacrylamide gel, for 7.5 hr at 4°C (5 mA per tube, constant current). B. UV-absorption (——) and infectivity distribution (-----) profiles of RNA preparation from PSTV-infected tomato leaves after electrophoresis in a 20% polyacrylamide gel (same conditions as in A). 5S = 5S ribosomal RNA; I, III, IV = unidentified minor components of cellular RNA; II = PSTV; $A_{260}$ = Absorbance at 260 nm. Electrophoretic movement from right to left. (From Diener, T. O., Virology, 50, 607, 1972. With permission from Academic Press.)

plant to obtain infection.[62] On the other hand, inoculation with only 50 to 100 molecules of PSTV can lead to infection of 10% of the tomato plants tested.[63] Thus, one out of every 50 to 100 molecules of this free RNA species, without the protection of a coat of any kind, can penetrate and infect plant cells under conditions where only one out of every 1 million TMV particles, fully encapsulated in a protective protein coat, is successful. Because of this extremely high infectivity of viroid RNA, it will be assumed for the remainder of this discussion that the major component of the UV-absorbing gel band present at the position of the infectious agent in nucleic acid extracts of viroid-infected plant tissue is made up of viroid RNA.

Investigations of infectious RNAs associated with all (but one) of the other viroid diseases described above (Section I. C) soon led to a demonstration, in each case, that a new RNA species could be correlated with the infectious agent (CEV,[52,64] CSV,[41] ChCMV,[41] CPFV,[43] and cadang-cadang.[44,46] No analyses of RNA from plants inoculated with the HSD agent have been reported.

## II. THE PROBLEM OF VIROID IDENTIFICATION

Conventional methods for recognizing particular viruses include the identification of a number of morphological and antigenic characteristics unique to each virus. This approach does not work for the identification of viroids, since no virions are present.[7,23,44,50,65] Furthermore, attempts to identify antigenic proteins which might be present only in the diseased state have failed.[25] Since the active agent of viroid diseases consists entirely of RNA, an identification procedure must be found which discriminates readily between different RNA species. In addition, since yields of highly purified viroids tend to be low (see Section II. A and Table 1), it would be most convenient if the identification method could be applied to amounts of RNA in the range 0.1 to 1.0 μg to permit characterization of material from individual plants or small quantities of cultured plant tissue.

The development of reproducible conditions for the in vitro labeling of submicrogram quanities of highly purified RNA with $^{125}I$[66,67] together with the demonstration that RNA modified by iodination can be subjected to two-dimensional fingerprint analysis,[68] has provided a straight-forward solution to the identification problem. A description of the three steps involved in this assay system, (1) viroid isolation, (2) iodination, and (3) fingerprinting, are presented below.

### A. Isolation of Viroid RNA

A number of methods for the isolation of viroids have been reported.[23,34,35,37,38,30,43-45,50,51,61,63,69-72] Abbreviated forms of the most recent versions of these methods are presented in Table 1. Several generalizations can be drawn from a survey of the published methods. (1) The youngest leaves frequently give the highest yield of viroid RNA.[69] (2) Fresh,[40,69] frozen (−70°C)[23,64] or freeze-dried[73] material may be used. (3) Tissue may be prepared for extraction by freezing in liquid nitrogen followed by pulverization with a pestle and mortar.[40] (4) Homogenization should take place within the pH range 7.0 to 9.5[40] in the presence of a nuclease inhibitor (Bentonite at a final concentration of 1 mg/mℓ),[23,71] an organic phase (a phenol-chloroform mixture appears best), and a detergent (for example, 1% sodium dodecylsulphate). (5) Polyacrylamide gel electrophoresis should constitute the final fractionation step to separate viroids from other RNA species (5% acrylamide gels are preferred by most).

The first methods developed[23,35,61,70] are also the most time-consuming and produce the lowest yields (see Table 1). However, these methods have also been shown to produce viroid RNA suitable for iodination and fingerprinting (see Section II. C). Minor modifications of a procedure published by Morris and Smith[72] have led to a method for obtaining highly purified preparations of PSTV or CEV from tomato plants in amounts ranging from 0.1 to 0.5 μg/g of fresh plant tissue used.[59] The details of this method are presented here. From a fresh tomato plant (mature leaves are discarded), 6 g of upper leaves and main stem tissue are quickly cut. This chopped tissue is immediately homogenized for 1 min at top speed in a Virtis® homogenizer in the presence of 3 mℓ glycine buffer (0.2 *M* glycine, 0.1 *M* $Na_2HPO_4$, 0.6 *M* NaCl, 1% sodium dodecylsulphate, adjusted to pH 9.5 with 5 *M* NaOH), 6 drops of betamercaptoethanol, 3 mg Bentonite, 12 mℓ phenol (saturated with 0.01 *M* Tris-0.001 *M* ethylenediaminetetraacetic acid (EDTA) pH 7.6), and 12 mℓ chloroform:entanol (25 to 1, v/v). The homogenate is centrifuged for 5 min at 4°C at 10,000 rpm in a Sorvall® SS-34 rotor, and the aqueous supernatant reextracted by vortexing for 1 min with 2.5 mℓ phenol and 1.25 mℓ chloroform-pentanol. Aqueous and organic layers are separated by centrifugation for 3 min at top speed in a clinical centrifuge (swinging bucket rotor). The upper phase (aqueous) is removed, its volume measured, and ¼ volume of 10 *M*

TABLE 1

**Methods for Viroid Extraction**

| Viroid | Plant | Fresh weight (g) | Main steps of method[a] | Approximate yield (fresh tissue) (μg/g) | Ref. |
|---|---|---|---|---|---|
| PSTV | Tomato | 5000 | a (pH 7.5), c, b, b, f, f, g, h, i, f, m, n, f, j, t, f, u(20%), v and w | 0.016 | 23, 61 |
| PSTV | *Scopolia sinensis*, tomato | 50 | a (ph 7.2), b, e, h, h, j,f, l or r or j(2—8M) or s or u (10 and 20%) | Not reported | 34,71 |
| PSTV | Potato, tomato | 1 | a + b + c (pH 9.5), j, t, f, u (5%), v, w, x | 0.2—1.3 | 69, 72 |
| CEV | *Gynura aurantiaca* | 5000 | a + b (pH 7.0), b, f, t, f, t, g, t, j, f, o, t, f, t, u (5%), w, a, o + n, t | 0.02—0.04 | 35, 70 |
| CSV | Cineraria, chrysanthemum | 100 | a (pH 7.5), c,b, b, f, f, n, i, f, u (20%), v | Not reported | 23, 38 |
| CSV | Chrysanthemum | 1 | a (pH 9.0), b + c, h, u, (5%) | Visible band | 41 |
| CPFV | Tomato | 2000 | a + b (pH 8.0), k, h, u (5%), y, a + b, q, f, u to f cycle repeated twice more | 0.1 | 43, 63 |
| ChCMV | Chrysanthemum | 20—50 | a (pH 9.0), b + c, h, j, m, u (5%) | Visible band | 40, 41 |
| ChCMV-NS | Chrysanthemum | 100 | a (pH 9.0), b + c, h, j, m, i, l, u (5%) | Visible band | 40, 41 |
| Cadang-Cadang | Coconut palm | 100 | a, z, b, f, t + protease, b, f, u (2.2%), w or x, a + b, f, h | 0.5 | 44, 45 |

[a] The procedures have been assigned letters as follows and are listed in the order in which they are carried out: a. homogenization in aqueous buffer (pH); b. phenol extraction; c. chloroform extraction; d. phenol plus chloroform extraction; e, ether extraction; f. ethanol precipitation; g. methoxyethanol extraction; h. cetyltrimethyl ammonium bromide precipitation; i. Deoxyribonuclease treatment; j. lithium chloride fractionation; k. NaCl fractionation; l. isopropyl alcohol fraction; m. Sephadex gel filtration; n. hydroxy-apatite chromatography; o. cellulose CF11 chromatography; p. methylated albumin-coated Kieselguhr column chromatography; q. DEAE cellulose chromatography; r. reverse phase chromatography; s. high-pressure liquid chromatography; t. dialysis; u. polyacrylamide gel electrophoresis (% acrylamide); v. assay infectivity of gel slices; w. detection of gel band by UV absorption; x. stain gel in 0.1% toluidine blue 0 in water; y. stain gel in methylene blue; z. polyethylene glycol precipitation.

LiCl added. After 2 hr on ice, the LiCl precipitate (containing high molecular weight RNA) is pelleted by centrifugation (5 min, 10,000 rpm, SS-34 rotor) and the supernatant is dialysed overnight against water. RNA is precipitated by the addition of one tenth volume of 2 *M* sodium acetate (pH 5.2) and 2½ volumes of absolute ethanol and storage for 1 hr at −70°C or overnight at −20°C. Centrifugation for 1 min at 14,000 rpm in siliconized centrifuge tubes ensures essentially 100% recovery of the nucleic acid in the form of a pellet. The liquid is poured off and the pellet dried in a vacuum dessicator. The dried pellet is then resuspended in 0.1 mℓ water and stored at −70°C until electrophoresis is carred out. Each sample is made 10% in sucrose by the addition of 0.025 mℓ 50% sucrose (containing bromophenol blue) and is then loaded into one slot of a 5% 5-slot slab gel made up in 0.04 *M* Tris, 0.02 *M* sodium acetate, 0.001 *M* EDTA (pH 7.2) as described by Adesnik.[74] The gel is run in the same buffer at 4°C, 75 mA (about 140 V for these 20 × 20 × 0.3 cm slab gels) until the bromophenol blue marker has moved about 10 cm from the origin. The gel is stained for 15 min in 20 μg/mℓ ethidium bromide-0.001 *M* EDTA and destained for 15 min in 0.001 *M* EDTA. RNA preparations from uninfected and viroid-infected plants may be run side by side in these slab gels so that the viroid band can be found readily by comparison.

### B. Iodination of Viroids

The chemical reaction developed by Commerford[66] and Prensky[67] for in vitro labeling of cytidylate residues of nucleic acids with $^{125}$I has been shown to be suitable for the labeling of submicrogram quantities of viroids.[59,75,76] Since the reaction is extremely sensitive to minor contamination by acrylamide, the procedure used for extraction of the viroid from polyacrylamide gel bands must ensure removal of this component. Before iodination, all RNA samples are subjected to spectrophotometric analysis between 230 and 300 nm. When there is evidence of acrylamide contamination (acrylamide increases the absorbance in the 230 to 260 nm range of the spectrum), the RNA sample is repurified by cellulose CF11 chromatography[77] or hydrochloric acid precipitation.[78] A relatively straight-forward way to obtain RNA of sufficient purity is as follows:[155] (1) homogenize the gel band in a mixture of phenol and TSE buffer (0.05 *M* Tris, 0.1 *M* sodium chloride, 0.001 *M* EDTA, pH 7.0), (2) separate the RNA from residual acrylamide by chromatography on hydroxylapatite,[79] and (3) concentrate the RNA by cellulose CF11 chromatography[77] and ethanol precipitation. This procedure produces intact RNA suitable for subsequent iodination even when the amount of RNA in the gel band is as low as 0.05 μg.[59] The published methods for extraction of viroids from gel bands[44,45,61,63,70,71,80] do not appear suitable for amounts of RNA below 20μg.[156]

Standard iodination reaction mixtures have a volume of 8 μℓ and final concentrations of $3.6 \times 10^{-4}$ *M* RNA, $4.75 \times 10^{-4}$ thallic nitrate, $0.7 \times 10^{-4}$ *M* carrier-free $^{125}$I-sodium iodide, $4.2 \times 10^{-3}$ nitric acid, and $6.25 \times 10^{-2}$ *M* ammonium acetate (pH 4.7). Reactions are carried out for 3 min at 64°C in a glovebox with a built-in activated charcoal filter system. These conditions of iodination yield intact RNA with specific activities in the range 40 to $160 \times 10^{6}$ dpm/μg.[59]

At the end of the reaction, the sample is diluted to 0.5 mℓ with 0.05 *M* Tris-0.1 *M* NaCl-0.001 *M* EDTA (pH 7.0), heated for 20 min at 64°C to rid the preparation of unstable iodination products,[66,67] and is then purified by cellulose CF11 chromatography[77] to separate free iodide from the iodinated RNA. After ethanol precipitation in the presence of 10 μg of carrier RNA, the sample is resuspended in water and stored at −20°C until use. $^{125}$I-labeled RNA species up to the size of viroids have been observed to remain intact for several months under these conditions.

### C. Fingerprinting of $^{125}$I-Labeled Viroids

When an RNA species is digested to completion by a base-specific ribonuclease such as RNase T1 or pancreatic RNase, a set of oligonucleotides is generated which is unique for that particular RNA species. When these oligonucleotides are subjected to high voltage electrophoresis in the first dimension and then to ascending homochromatography in the second dimension, the oligonucleotides assume unique positions within a two-dimensional array of spots (whose positions are revealed by autoradiography) creating a pattern called a fingerprint.[81,82] Most alterations in the RNA sequence are reflected by alterations in the fingerprint. Thus, RNA species can be positively identified by their fingerprints.

In the case of $^{125}$I-labeled RNA, only the cytidylate (C) residues are labeled.[68] Therefore, any oligonucleotide that lacks C will not appear in the fingerprint. However, this does not present a serious problem since the oligonucleotides most characteristic of each RNA species are also the longest oligonucleotides which, therefore, have the highest probability of containing at least one C residue. This has been borne out by studies showing that fingerprints of $^{125}$I-labeled RNA are nearly identical to those of the same species labeled in all four residues by $^{32}$P.[68,75]

As discussed in detail below, analysis of the two-dimensional fingerprints of $^{125}$I-labeled viroids has led to demonstrations that (1) the complexity of the viroid fingerprints is compatible with molecular weight estimates,[76] (2) PSTV isolated from tomato plants is not identical to CEV isolated from Gynura plants,[76] (3) PSTV and CEV undergo no major sequence changes during replication in two different hosts,[59] and (4) CSV and the RNA species associated with cadang-cadang both have distinctive fingerprints with complexities similar to those of PSTV and CEV.[75] No fingerprints of ChCMV, CPFV or the agent of HSD have been published.

### *1. Fingerprint Complexity*

With the exception of RNA molecules containing repeated sequences, the complexity of a fingerprint increases as the molecular weight increases. Thus, it is possible to estimate the complexity of a particular RNA species by comparing its fingerprint with fingerprints of RNAs of known complexity. Comparison of the two-dimensional fingerprint of an RNase T1 digest of PSTV with those of HeLa cell 5S ribosomal RNA (120 bases) and duck globin alpha and beta mRNA (total complexity of about 1200 bases) shows that, while the viroid RNA is more complex than the HeLa 5S RNA, it is substantially less complex than the globin mRNA (see Figure 3).[75] Thus, the sequence complexity of PSTV is consistent with the size estimate of about 380 bases.[63] This rules out the possibility that this viroid genome is composed of a mixture of several entirely different RNA species.[75,76]

### *2. Are PSTV and CEV Identical?*

The independent recovery of two viroids, PSTV and CEV, which appeared to infect a common host range and to cause similar disease symptoms, led to the suggestion that these two viroids were merely independent isolates of the same plant pathogen.[53-55] Visual inspection of the RNase T1 and pancreatic RNase fingerprints shown in Figures 4 and 5, respectively, reveals that PSTV isolated from tomato plants is not identical to CEV isolated from Gynura.[76] Further analysis of these fingerprints (involving elution, specific enzymatic digestion, and electrophoretic analysis of the oligonucleotides) showed that the difference between the nucleotide sequences of PSTV and CEV is 13.3% or greater.[76] This means that these two viroids must have nucleotide sequences which differ from each other in at least 50 out of the approximately 380 bases. Furthermore, these differences must be distributed throughout the molecules so that all of the major RNase T1-resistant oligonucleotides differ.[75,76]

### *3. Are Viroids Modified by Their Hosts?*

While the mechanism of viroid replication is still unknown, it seems likely that viroids depend upon machinery present in their hosts for their replication.[23] Depending upon the exact mode of replication, it is conceivable that the nucleotide sequence of a

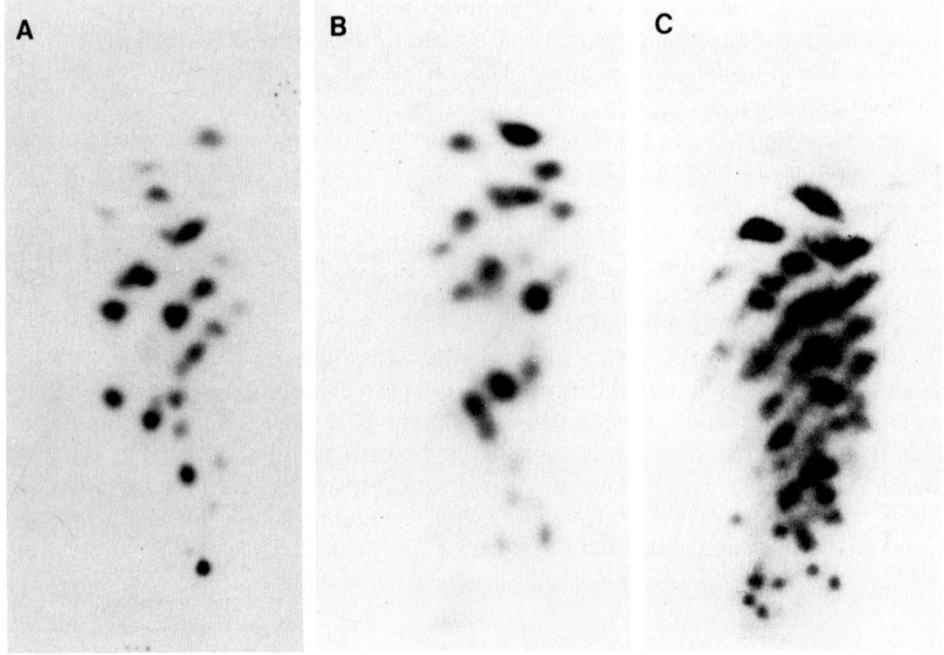


FIGURE 3. Comparison of the complexity of PSTV with that of HeLa 5S RNA and duck globin mRNA. About 1 × $10^6$ dpm of $^{125}$I-labeled RNA were mixed with 10 μg of bacteriophage f2 RNA and digested with 2 μg of RNase T1 in 2 μℓ of 0.01 *M* Tris-HCl (pH 7.5)-1 m*M* EDTA for 40 min at 37°. Reaction mixtures were applied to cellulose acetate strips (Schleicher and Schuell, Keene, N. H.; 2.5 × 57 cm) soaked with 5% glacial acetic acid, containing 5 m*M* EDTA in 7 *M* urea, and subjected to high voltage electrophoresis (25 min, 6 kV). The oligonucleotides were transferred to Machery-Nagel 20 × 40-cm DEAE-cellulose thin-layer chromatography plates by standard procedures.[81] Second dimensions consisted of thin-layer homochromatography, a procedure carried out at 60° in which a partially hydrolyzed solution of yeast RNA (BDH) in 7 *M* urea moves upward over the DEAE-cellulose thin-layer plate causing the oligonucleotides to separate according to size, with the smallest ones moving the longest way up to the plate. In the configuration shown here, the first dimension was from right to left and the second was from bottom to top as shown by the arrows. (A) HeLa 5S T1 fingerprint. (B) PSTV T1 fingerprint. (C) Duck globin mRNA T1 fingerprint. The upper regions of all T1 fingerprints contain the same oligonucleotides (e.g., CG, CUG, ACG, etc.). The lower portions contain the longer oligonucleotides which uniquely characterize each RNA species. The size of the longest RNase T1 products increase, in general, with an increase in the complexity of the NA in question.

viroid could be significantly altered by replication in a second host. For example, the observations (1) that DNA sequences complementary to PSTV appear to be present in comparable amounts in the DNA from both uninfected and PSTV-infected hosts,[83] and (2) that PSTV replication in host nuclei can be inhibited by actinomycin D[84,85] led Diener[9] to suggest that PSTV might act as a specific derepressor of PSTV-specifying host DNA sequences. Furthermore, it was proposed that a viroid infecting a species different from its normal host species might derepress a DNA sequence which specifies an RNA molecule whose nucleotide sequence is not completely identical to that of the infecting viroid.[9]

Thus, it is necessary to consider the possibility that CEV isolated from Gynura only differs from PSTV isolated from tomato plants because it has replicated in a different host. This possibility has been tested by Dickson and co-workers.[59] PSTV and CEV, each purified from tomato and from Gynura, were labeled in vitro with $^{125}$I, digested

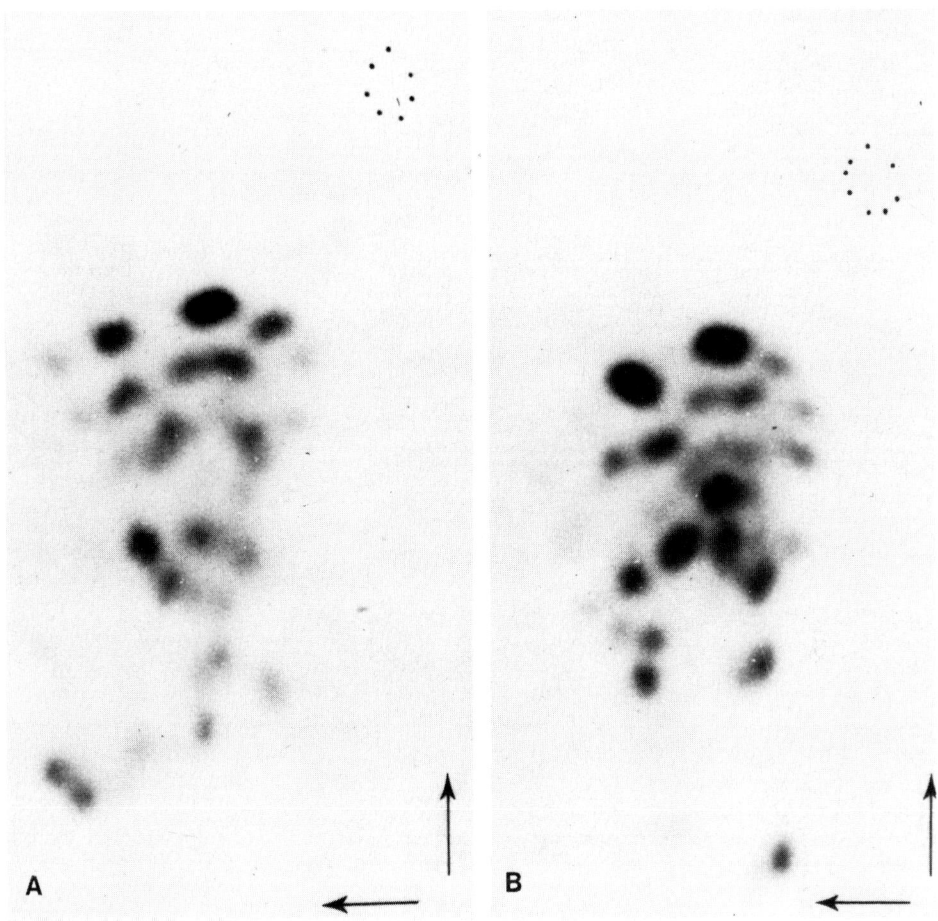


FIGURE 4. Ribonuclease T1 fingerprints of $^{125}$I-labeled PSTV and CEV. About $1 \times 10^6$ dpm of $^{125}$I-labeled RNA were mixed with 10 $\mu$g of bacteriophage f2 RNA and digested with 2 $\mu$g of RNase T1 in 2 $\mu \ell$ of 0.01 *M* Tris. HCl (pH 7.5) — 1 m*M* EDTA for 40 min at 37°C. Reaction mixtures were fingerprinted as described in the legend to Figure 3. In the configuration shown here, the first dimension was from right to left and the second was from bottom to top, as shown by the arrows. (A) PSTV iodinated in vitro to a specific acitvity of $12 \times 10^6$ dpm/$\mu$g. (B) CEV iodinated in vitro to a specific activity of $18 \times 10^6$ dmp/$\mu$g.

with RNase T1, and subjected to two-dimensional fingerprinting analysis. Each viroid retained its own distinctive fingerprint pattern irrespective of the host species from which it was isolated. Thus, as shown above, not only are these two viroids different from each other when isolated from different hosts, but they also fail to turn into each other when they replicate in the same host. Thus, PSTV and CEV are not independent isolates of the same pathogen as originally proposed,[53-55] nor do they undergo extensive variation after propagation in a second host. Recently, these results have been extended by a demonstration that PSTV, CEV, and CSV each retains its characteristic RNA fingerprint pattern following replication in tomato plants and chrysanthemum plants.[85a]

### 4. *Fingerprints of CSV and Cadang-Cadang-Associated RNA*

Figure 6 shows RNase T1 and pancreatic RNase fingerprints of CSV.[75] These finger-

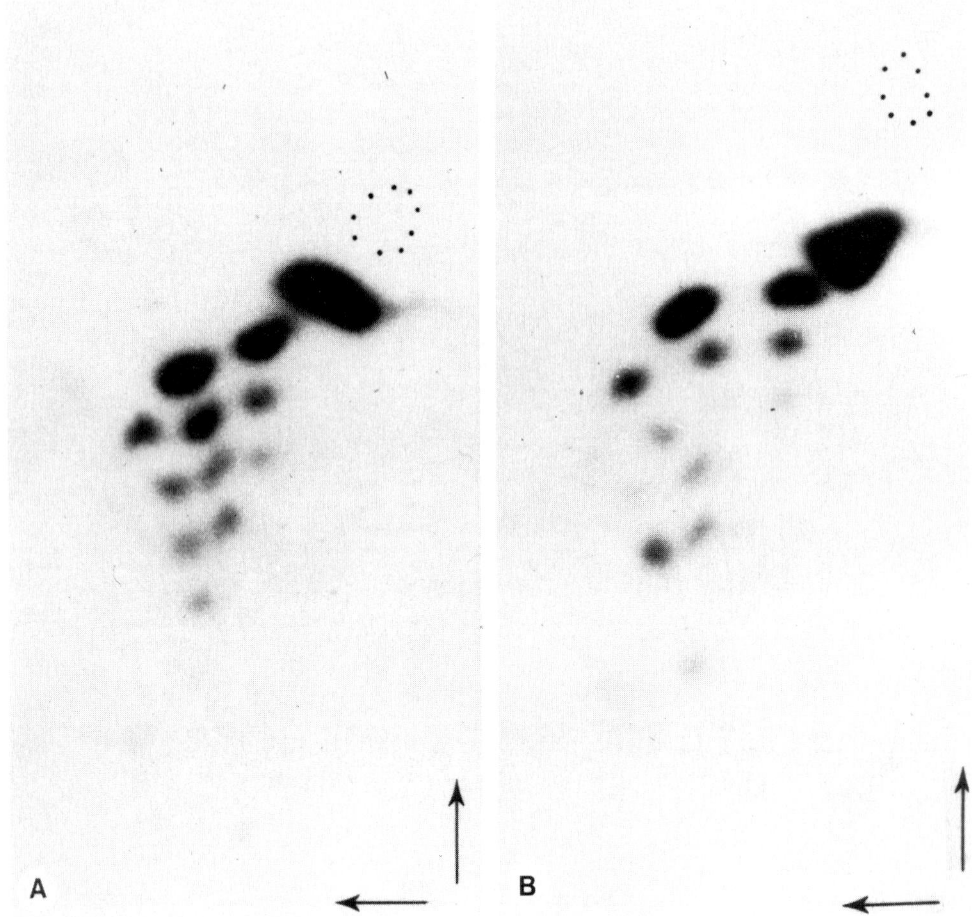


FIGURE 5. Pancreatic ribonuclease fingerprints of $^{125}$I-labeled PSTV and CEV. About $1 \times 10^6$ dpm of $^{125}$I-labeled RNA were mixed with 10 $\mu$g of bacteriophage f2 RNA and digested with 2 $\mu$g of pancreatic ribonuclease in 2 $\mu\ell$ of 0.01 *M* Tris. HCl (pH 7.5) — 1 mM EDTA for 30 min at 37°C. Two-dimensional fingerprinting analyses were carried out as detailed in the legend to Figure 3. The origin is at the lower right; the electrophoretic first dimension was from right to left, and the second dimension (homochromatography) was from the bottom of the picture to the top. (A) PSTV iodinated in vitro to a specific activity of $12 \times 10^6$ dpm/$\mu$g. (B) CEV iodinated in vitro to a specific activity of $18 \times 10^6$ cpm/$\mu$g.

prints have complexities comparable to those of PSTV and CEV, but have patterns which are entirely different in the characteristic (lower) portion of the fingerprint.

In the case of cadang-cadang, as discussed above (Section I. C. 5), neither of the two low molecular weight RNA species associated with this disease (ccRNA-1 and ccRNA-2) has yet been shown to be the infectious agent. While ccRNA-2 is somewhat larger (11 S), ccRNA-1 has a size similar to PSTV, CEV, and CSV (7.5 S).[44,45] Nucleic acid extracts of healthy plants sometimes reveal the presence of a minor band at the position of ccRNA-1.[45] However, when the RNA present in healthy plants was compared by fingerprinting analysis with ccRNA-1, it was found that a relatively complex mixture of host-specific RNA species occurs at the position of ccRNA-1 in low amounts and that ccRNA-1 is a new species of RNA that appears in diseased palms with a fingerprint entirely different from the one obtained from healthy plants (Figures 7 and 8).[75] As shown, ccRNA-1 produces a simple fingerprint after digestion by either

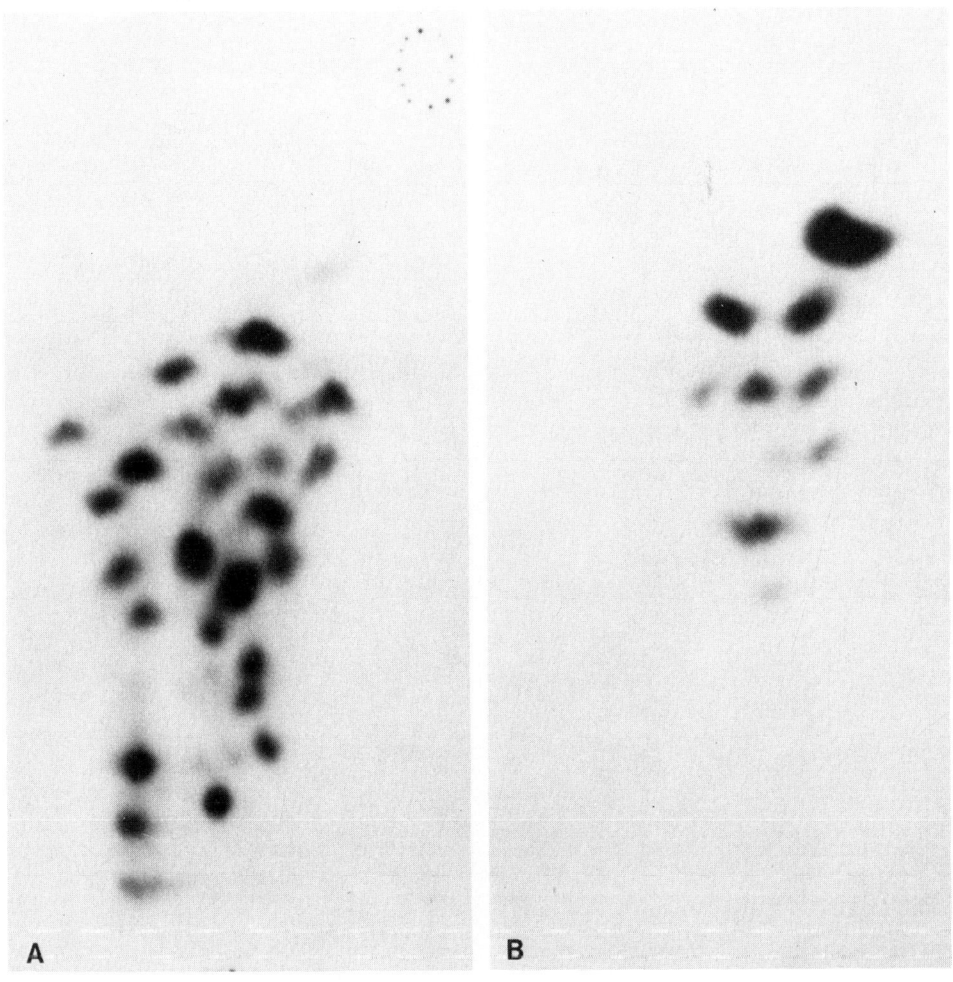


FIGURE 6. T1 and pancreatic RNase fingerprints of $^{125}$I-labeled chrysanthemum stunt viroid. Ribonuclease digestion and fingerprinting were carried out as described in the legends to Figures 3 and 5. (A) T1 fingerprint of CSV. (B) Pancreatic RNase fingerprint of CSV.

RNase T1 (Figure 7A) or pancreatic RNase (Figure 8A) with a complexity comparable to the viroids discussed above. On the other hand, the low level of RNA of the same electrophoretic mobility purified from healthy plants yields a much more complex fingerprint, both after RNase T1 digestion (Figure 7B) and after pancreatic RNase digestion (Figure 8B), indicating the presence of a number of distinct RNA species, each present in very low amounts, in these extracts. These same RNA species are present as contaminants of the ccRNA-1 preparations, as can be seen by overexposure of the ccRNA-1 fingerprints. Under these conditions, distinctive background patterns homologous to those in Figure 7B and 8B can be seen.[75]

### D. Fingerprinting of $^{32}$P-Labeled Viroids

Unlabeled oligonucleotides of viroids can be labeled in vitro with $^{32}$P by the action of polynucleotide kinase in the presence of gamma-$^{32}$P-ATP following digestion with RNase T1 or pancreatic RNase producing oligonucleotides suitable for fingerprinting. This approach has been used by two different laboratories for the study of

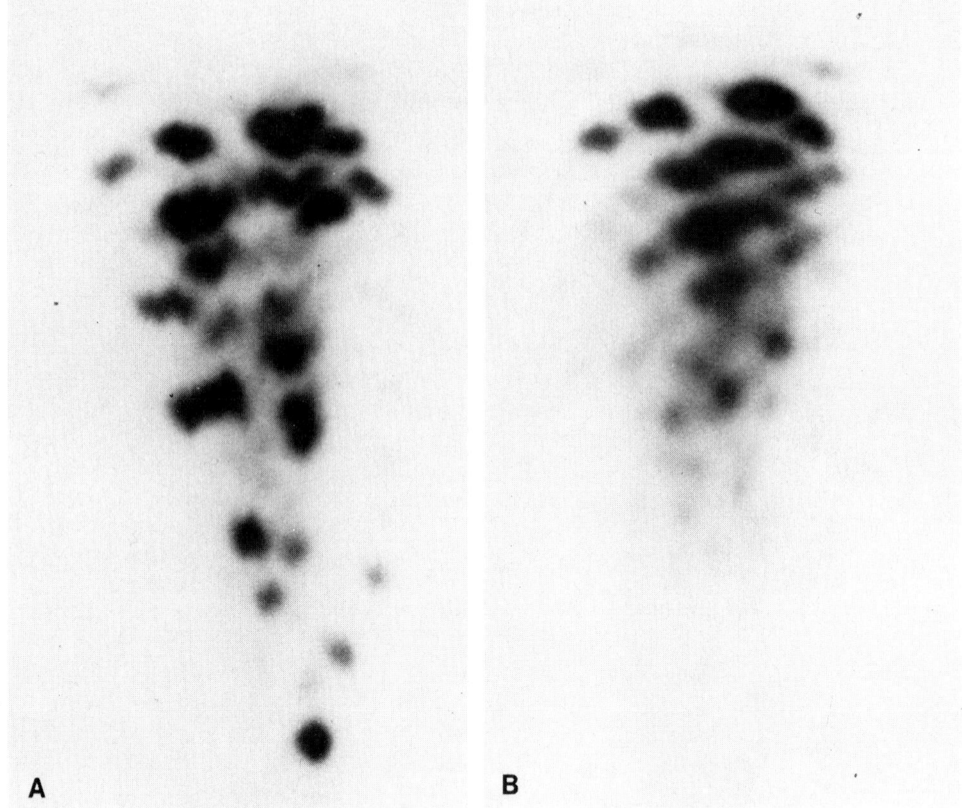


FIGURE 7. T1 fingerprints of $^{125}$I-labeled coconut palm RNA. Ribonuclease T1 digestion and fingerprinting were carried out as described in the legend to Figure 3. (A) T1 fingerprint of ccRNA-1. (B) T1 fingerprint of RNA from healthy coconut palms with a mobility similar to that of ccRNA-1.

viroids.[71,86,87] When Singh and co-workers[88] discovered that three different sizes of infectious RNA could be isolated from PSTV-infected *Scopolia sinensis* (the only known local-lesion host for PSTV[89]), they isolated the two major forms and carried out fingerprinting analysis to see if they were related to each other. They reported that the fingerprint of fraction II RNA (infectious RNA with a size similar to the PSTV isolated from tomato and potato plants by others) was unlike that of fraction I RNA (a larger infectious species).[86] While both RNase T1 and pancreatic RNase fingerprints of fraction II RNA have recently appeared,[71] those of fraction I RNA have not yet been published. The third infectious species (fraction III RNA) has not yet been obtained in highly purified form since it co-purifies with host tRNA and is present in extremely low amounts.[71,88]

Enzymatic digests of PSTV, CEV, and CSV have been labeled in vitro with $^{32}$P using T4 polynucleotide kinase, providing material suitable for RNA fingerprint analysis[87] and sequence analysis in the case of PSTV[87a,87b] (see Section IV. E). The fingerprints obtained had the same patterns as those of these same viroids labeled in vitro with $^{125}$I and fingerprinted in previous studies.[75,76] Thus, these two methods of in vitro labeling are interchangeable for the identification of viroids.

The potential of obtaining isotopically labeled viroids by in vivo labeling methods for RNA fingerprinting studies has not yet been explored. A number of methods have, however, been developed for obtaining viroids labeled in vivo.[64,84,85,90,91] Sänger and

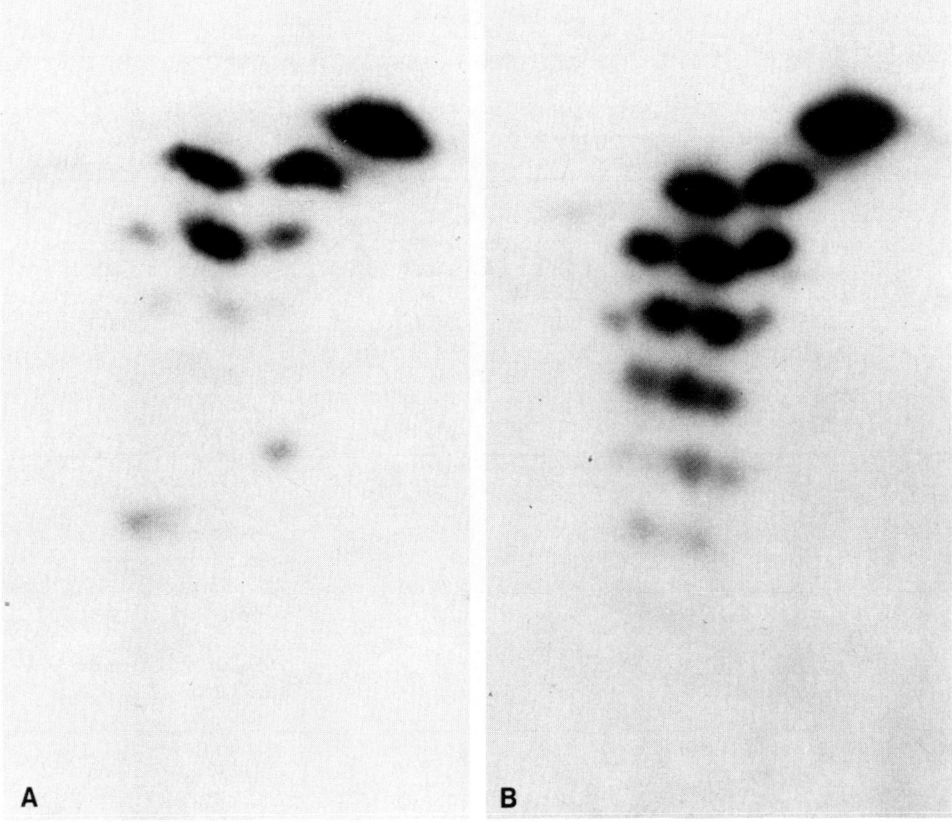


FIGURE 8. Pancreatic RNase fingerprints of $^{125}$I-labeled coconut palm RNA. Pancreatic RNase digestion and fingerprinting were carried out as described in the legends to Figures 3 and 5. (A) Pancreatic RNase fingerprint of ccRNA-1. (B) Pancreatic RNase fingerprint of RNA from healthy coconut palms with a mobility similar to that of ccRNA-1.

Ramm[64] and Hadidi and Diener[91] have incorporated $^{32}$P into viroids by feeding $^{32}$P-labeled orthophosphate to viroid-infected plants through their roots. By incubation with leaf strips[84] and with isolated nuclei[85] of PSTV-infected tomato plants, $^{3}$H-labeled uridine has been incorporated into PSTV. CPFV has been labeled by incubating protoplasts from CPFV-infected tomato plants (cv. Hilda 72) with $^{3}$H-uridine.[90] In this study, Mühlbach and Sänger observed even better incorporation into the CPFV when they prepared protoplasts from uninfected tomato plants and inoculated the protoplasts with CPFV just prior to labeling.[90] Unlike the two in vitro methods of labeling discussed above, in vivo methods utilizing $^{32}$P-labeled orthphosphate should yield viroids labeled uniformly in all four nucleotides, which may prove useful in future molecular studies.

The catalogue of fingerprints represented by Figures 4 to 8 will, no doubt, soon be expanded by the addition of ChCMV and CPFV fingerprints. If viroids breed true in general, as was found to be the case with PSTV and CEV in their two indicator hosts,[59] then any new viroid that is discovered, regardless of host, can be tested rapidly to see if it is the same as one of the known viroids. In addition to being useful for initial identification, RNA fingerprinting provides a direct way to test for cross-contamination of one viroid by another and for host modification or mutation of the viroid genome.

TABLE 2

**Viroid Transmission**

| Viroid | Natural Transmission: Horizontal | Natural Transmission: Vertical (through the seed) | Artificial Transmission |
|---|---|---|---|
| PSTV | Potato<br>Plant lice[11]<br>Grasshoppers[92]<br>Harvesting equipment[16]<br>Grafting and cutting tools[16] | Through pollen and/or ovules<br>Potato: 96—100%[18]<br>0—100%[19]<br>6—12%[20]<br>Tomato: 7.9—11.1%[12]<br>2—23%[20]<br>Through ovules only<br>Scopolia<br>71%[21] | Potato<br>Leaf rubbing[11]<br>Tuber or vine graft[11,16]<br>Tissue plug in seed piece[92]<br>Seed piece contact[16]<br>Contaminated knife[16]<br>Tomato<br>Leaf rubbing[22] |
| CEV | Citron<br>Graft[48]<br>Dodder[48]<br>Contaminated grafting and pruning tools[93] | Not reported | Citron, Petunia and Gynura[48-50,93]<br>Leaf rubbing<br>Grafting<br>Needle puncture (stem)<br>Razor slashing (stem) |
| CSV | Chrysanthemum<br>Foliage contact[39]<br>Aphids[57]<br>Grafting[57] | No seed transmission observed[39] | Chrysanthemum<br>Leaf rubbing[57]<br>Grafting[57]<br>Hypodermic syringe[94] |
| ChCMV | Chrysanthemum<br>Grafting[58] | Not reported | Chrysanthemum<br>Leaf rubbing[58]<br>Grafting[58]<br>Tissue implantation[58,95] |
| CPFV | Cucumber<br>Possible insect vector[42] | No seed transmission observed[42] | Cucumber<br>Grafting[42]<br>Sap inoculation[42]<br>Razor slashing (stem)[42]<br>Tomato<br>Leaf rubbing[90] |
| Cadang-Cadang | Coconut Palm<br>Not known[96] | Not reported | Coconut Palm<br>High pressure injection into petioles[46] |

# III. BIOLOGICAL PROPERTIES OF VIROIDS

## A. Transmission and Bioassay

In nature, viroids may be transmitted vertically through the seed and horizontally by insects, dodder (a parasitic vine), foliage contact, and cultivation (pruning, grafting, and harvesting). Artificial transmission may be accomplished by rubbing sap or purified RNA from infected plants onto the leaves of test plants in the presence of carborundum as an abrasive, by grafting, by tissue implantation, by puncturing the stem with a contaminated needle or slashing it with a contaminated razor blade, and by injecting infectious material with a hypodermic syringe. Table 2 presents a summary of the viroids and their modes of transmission.

Bioassay of viroids should be carried out using indicator plants with the shortest incubation periods and the most pronounced disease symptoms. In addition, symptom expression should not depend on narrowly defined environmental conditions, since most greenhouses vary considerably from one season to another. Times required for symptom production in the commonly used indicator plants for each viroid are listed in Table 3. It has been observed that, as the viroid concentration in the inoculum decreases, the time taken to develop symptoms tends to increase.[8] Several systems have been developed for measuring the relative infectivities of different viroid preparations: Raymer and Diener[27] designed an infectivity index calculated from the dilution end point (the highest dilution of an inoculum capable of infecting test plants), the percentage of plants infected at each dilution, and the time required for symptom expression; Semancik[35] and Weathers defined the "relative infectivity" as the total number of infected-plant-days within the experimental time period; Hörst[97] expressed infectivity as the fraction of plants with visible symptoms after 30 days; and Sänger[37] defined "$ID_{50}$" as the dilution capable of infecting 50% of the plants inoculated. Local lesion hosts have been found both for PSTV *(Scopolia sinensis)*[89] and for CSV (*Senecio cruentus*, cv. Hansa).[98] In general, the number of lesions increases with viroid concentration permitting estimates of relative infectivity to be made.

In the case of PSTV, good symptoms are difficult to obtain during winter months when temperatures and light intensity tend to be suboptimal (see Section III. B). However, assay of such plants can be improved by decapitation and defoliation two weeks after inoculation. Tomato plants treated in this way produce severe symptoms in newly developed axillary shoots 2 weeks later.[12]

Morris and Smith[72] have shown that it is possible to screen 250 potato plants at a time for the presence of PSTV, in an assay which depends upon the appearance of a viroid band in polyacrylamide gels. One tomato plant is inoculated with low molecular weight nucleic acids prepared from pooled tissue samples of 250 potato plants. Two weeks after inoculation, RNA is extracted from the inoculated tomato plant and analyzed by polyacrylamide gel electrophoresis. If even one of the 250 potato plants was PSTV-infected, a viroid band appears in the gel.[72]

### B. Factors That Affect Viroid Propagation

While only 10 days are required for symptom development of PSTV in greenhouse-grown tomato plants in early fall, 42 days or more are required from November to January.[22] This is probably a result of the combined effects of reduced temperature, light intensity, and hours of daylight, as suggested by the following studies. Conditions must be held within sharply defined limits of both temperature (22 to 24°C) and light intensity (400 to 600 footcandles) for development of the local lesion response of *Scopolia sinensis* to PSTV infection.[99] The recovery of PSTV from Rutgers tomato plants increases approximately 300-fold relative to the amount of tRNA recovered when greenhouse temperatures are increased from 20 to 32°C,[64] and the time required for symptom formation drops dramatically as the light intensity is increased.[64] The incubation time for CPFV in cucumber plants decreases from 76 to 21 days as the temperature of the greenhouse increases from 20 to 30°C.[42] Optimum conditions for the replication of ChCMV in chrysanthemum plants include at least 1000 footcandles light intensity, a 12-hr photoperiod, and a temperature of 24 to 27°C.[97] On the other hand, 21°C is the optimum temperature for the action of the latent agent, ChCMV-NS, which protects against the severe strain (see Section III-D).[97]

The only nutrient that has been linked with viroid propagation is manganese. The presence of manganese ions at a concentration of 1500 μg/g or more in tomato tissue enhances symptom formation if the manganese to iron ratio is 12 or more.[100] A similar effect is observed with *Scopolia sinensis* where the most severe symptoms are observed when manganese is at a concentration of 1900 μg/g in the plant and the manganese to iron ratio is 27.[101] This effect cannot be achieved when magnesium is substituted for manganese.[101]

TABLE 3

**Some Hosts of Viroids**

| Viroid | Host | Inoculation procedure | Time for symptom development | Symptoms | Ref. |
|---|---|---|---|---|---|
| PSTV | Potato[a,c] *(Solanum tuberosum)* cv. Irish Cobbler | Leaf rubbing | 50 days | Erect, spindly stalks; small leaves; dark green foliage; misshapen tubers with more numerous and prominent eyes | 10,11,22 |
| | Tomato[b] *(Lycopersicon esculentum)* cv. Rutgers | Grafting | 20 days | Epinasty and mottling of foliage; severe stunting; small fruit | 12, 22 |
| | | Leaf rubbing | 30 days | | |
| | *Scopolia sinensis*[b] | Leaf rubbing | 7—10 days (23°; 400 foot-candles, 18 hr/day; 70% humidity) | Local lesion host; small round necrotic spots on inoculated leaves | 89,99 |
| CEV | Eureka lemon scions[a,c] *(Citrus limon)* | Grafting onto *Poncirus trifoliata* | — | Scaling of bark (scaly butt) | 47 |
| | Etrog citron[c] *(Citrus medica)* | Grafting; dodder *(Cuscuta subinclusa)*; razor slashing; needle puncture | 90 days (75°F) | Stem cracking; leaf twisting; browning of leaf midribs; severe epinasty | 48, 109 |
| CEV | *Gynura aurantiaca*[b] | Leaf rubbing; grafting; razor slashing; needle puncture | 14—21 days | Epinasty; curling of leaves; severe stunting; systemic mosaic | 49, 53 |

| | | | | | |
|---|---|---|---|---|---|
| CSV | *Chrysanthemum morifolium*[a,c] cv. Mistletoe[b] | Leaf rubbing; grafting | Highly variable with temperature and light conditions (2—12 months) | Varies with cultivar; stunting; inferior blooms open 7—10 days earlier than normal; cv. Mistletoe develops yellow spots and flecks | 57, 102 |
| | Cineraria[b] *(Senecio cruentus)* | Leaf rubbing | 12—18 days | Local lesion host; starch lesions appear on inoculated leaves; systemic stunting | 39, 98 |
| ChCMV | *Chrysanthemum morifolium*[b,c] cv. Yellow Delaware[a] | Leaf rubbing; grafting; tissue implantation | 10—16 days (24—27°C) | Variable mottling and chlorosis; stunting | 58, 97, 116 |
| CPFV | Cucumber[a,c] *(Cucumis sativus)* | Leaf rubbing; razor slashing | 20—30 days | Pale fruit; crumpled flowers; twisted leaves; stunting | 42 |
| CPFV | *Benincasa hispida*[b] | Leaf rubbing<br>razor slashing<br>grafting | 21 days (31°C)<br>12 days (31°C)<br>less than 12 days (76 days at 20°C)<br>42—60 days | Veins of new leaves become yellow; plant tip becomes necrotic; plant death | 42 |
| Cadang-Cadang | Coconut palm *(Cocos nucifera)*[a,c] | High pressure injection | 18—22 months for earliest symptoms; 7 years for full symptom expression | ccRNA-1 and ccRNA-2 appear; yellow spots on fronds; slender pinnae; stunted leaves and trunk; no fruit produced after yellow spots appear; roots become discolored and rot | 46, 60 |

[a] Host in which viroid was originally discovered.
[b] Indicator host.
[c] Economically important host.

## C. Localization of Viroids

### *1. Distribution Within the Plant*

In discussing CEV inoculation, Garnsey and Jones[93] write: "The difference between the inconsistent results with mechanical inoculation and the rather high and consistent rate of infection with inoculation via contaminated knife, and with cut stems in the case of petunia, suggests that exocortis virus is largely phloem-limited." Because attempts to cure plants of viroids by combining heat treatment with meristem-tip culture[102] (CSV-infected Mistletoe chrysanthemum) or with axillary bud culture[103] (PSTV-infected potato) had success rates of only 2 to 3%, it seems unlikely that viroids are phloem-limited, in general. However, the observation that inoculation methods such as razor slashing of stems lead to more rapid symptom development than does inoculation of leaves[42] suggests that viroids may be transported throughout the plant via the phloem. When leaves are inoculated, there may be a lag time during which the viroid replicates and spreads slowly from cell to cell in the leaf prior to reaching the phloem for transport to other parts of the plant. This is supported by the observation that, when all the inoculated leaves of Mistletoe chrysanthemums were removed (with petiole) at intervals after inoculation, no transmission was obtained if the leaves were removed after 35 days or less.[39]

The amount of mild PSTV which can be recovered from potato plant tissue ranges from about 0.1 to 1.4 μg per gram tissue. [69] Viroid concentration is low in tuber and stem (0.1 to 0.3 μg/g), medium in mature leaves and petioles (0.3 to 0.6 μg/g), and high in leaf midveins and terminal shoots (0.9 to 1.4 μg/g).[69] Morris and Smith[72] showed, however, that the lower concentration of viroid in the older tissue is essentially proportional to the lower overall nucleic acid content of such tissue (on a gram fresh weight basis) compared to young actively growing tissue. Infectivity of severe PSTV in potato plant parts was shown to be highest 4 to 8 weeks after inoculation, with the earliest infectivity appearing about 1 week after inoculation in the midleaves and roots.[104] At the height of the infectious period, the highest titer was found in mature leaves in the middle of the plants and the lowest was found in the roots and tubers.[104]

### *2. Distribution Within Cells*

Cell fractionation studies have shown that both PSTV and CEV are associated with nuclei and isolated chromatin of infected plants.[28,29,37] Furthermore, a component with electrophoretic mobility identical to that of PSTV has been shown to be synthesized in isolated nuclei from PSTV-infected tomato leaves.[85] Because PSTV is sensitive to ribonuclease treatment *in situ* (leaves are vacuum infiltrated with ribonuclease), Diener[28] suggests that it may be present in cells as a free RNA molecule, rather than as some kind of complex. On the other hand, the disease-specific RNA of cadang-cadang, ccRNA-1, appears to be present in the cell as a complex, since it is no longer precipitated with polyethyleneglycol following treatments designed to remove proteins.[45]

In addition to being present in host nuclei, viroids may be associated with cell plasma membranes. Electron microscopic and cell fractionation studies carried out by Semancik and co-workers[105,106] have demonstrated a correlation between an aberration of the plasma membrane and CEV infection. The frequency of paramural bodies of plasmalemmasome-like structures was correlated with the initiation of symptoms, as well as the recovery of pathogenic RNA.[106] It was suggested that alterations in cell surface properties could constitute a significant phase in viroid replication or pathogenesis.

## D. Mild and Symptomless Viroids

In 1967, Fernow[107] showed that both mild and severe varieties of PSTV (M-PSTV and S-PSTV) could be recovered from field-grown potatoes. It was demonstrated that the disease associated with M-PSTV is caused by a distinct entity; it is not simply caused by an unusually low concentration of severe PSTV. The discovery that inoculation of plants with mild strains could prevent them from developing severe disease symptoms following subsequent inoculation with severe strains led to the development of the challenge-inoculation technique. Also called the tomato "cross-protection test," this assay was used in the following way to detect mild strains of viroids.[108] A seed piece from a potato suspected of having a mild strain of PSTV was rubbed onto the leaves of a tomato seedling. The same tomato plant was challenged with a severe strain of PSTV 14 days later. If the original inoculum contained M-PSTV, no severe symptoms would appear on the tomato test plant. If no M-PSTV was present, severe symptoms would develop about 2 weeks after the challenge inoculation.

Using the tomato cross-protection test, Singh et al.[14] screened 355 potato plants collected from 23 fields in eastern Canada in 1970. Of the potatoes tested, 92% displayed evidence of M-PSTV, while 8% harbored S-PSTV. Close observation of the mild strains demonstrated that different isolates differed from each other in severity, with some so mild they were referred to as latent.[14] A survey of potato fields carried out in 1971[15] showed that 2.5 to 4.6% of the potato plants were infected by PSTV, and among these, M-PSTV was 11 times as prevalent as S-PSTV.

The cross-protection phenomenon has also been observed with ChCMV, where a latent strain, ChMCV-NS, was first detected in certain clones of Deep Ridge chrysanthemums.[97] Total protection from the symptoms of ChCMV could be achieved by inoculating the plant with ChCMV-NS 8 days before inoculating with severe ChCMV.

Sänger has described a Mediterranean isolate of CEV which produces severe symptoms in citrus and petunia, but fails to produce visible symptoms in Gynura.[37] Mild isolates of CEV from other sources have been found to remain mild through many serial transmissions in citron.[109] While a good indicator plant has been found for these mild forms (clone 861-S-1 of Arizona 861 citron),[109] no experiments to see whether these strains can protect against severe strains of CEV have been reported.

The mild forms of PSTV and ChCMV have both been correlated with gel bands whose mobilities are like those of the severe forms.[41,69] RNA fingerprinting studies have demonstrated that mild and severe strains of PSTV have only minor differences in their nucleotide sequences.[110] To determine whether cross protection is confined to viroids that exhibit a high degree of sequence homology in their RNAs, Niblett et al.[85a] tested the abilities of various viroids to protect against each other in a variety of combinations. On tomato, M-PSTV protected against the expression of symptoms by S-PSTV and CEV; CSV delayed the onset of S PSTV and CEV symptons by about 16 days. On chrysanthemum, CSV, M-PSTV AND S-PSTV protected against CEV, but ChCMV failed to protect against S-PSTV, CEV, or CSV; and S-PSTV did not protect against ChCMV. Thus, cross protection can occur among viroids that differ significantly in nucleotide sequence. The molecular basis of cross protection is yet to be determined.

### E. Host Range of Viroids

An excellent survey of the literature concerning the host range of viroids has recently been complied by Diener and Hadidi.[9] Table 3 presents a brief summary of original, indicator, and economically important hosts, together with the method of inoculation, typical symptoms, and the time required for symptom development of each viroid.

The earliest recorded attempt at interspecific transfer of a viroid-incited disease occurred in 1931.[92] Although Goss[92] succeeded in transmitting PSTV among many potato cultivars, he failed to transmit it to tomato plants. The demonstration by Raymer and O'Brien[22] in 1962 that PSTV could be transmitted by mechanical inoculation from potato to tomato (cv. Rutgers) was soon followed by a wide variety of other PSTV host range studies.[12,13,17,24,55,89,99,111-115] A great number of potato[13,22,111] and tomato[12,17,113,115] cultivars were tested, many other solanaceous species were tested,[24,89,99,111,114] a local lesion host was found *(Scopolia sinensis)*,[89,99] symptomless hosts were described,[24,99,112,115] and species outside the solanaceous family were found.[55,99] Altogether, 117 species within the family Solanaceae and 11 species in ten other families were shown to support the replication of PSTV.[9] Among these are three of the common hosts of CEV: *Gynura aurantiaca, Petunia hybrida,* and *Citrus medica.*[54,55] Niblett et al.[115a] have recently reported that both mild and severe strains of PSTV replicate and cause disease symptoms in *Chrysanthemum morifolium* (Ramat.) Hemsl.

The host range of CEV has not been as extensively explored. Nine species within the family Rutacea in addition to *Citrus medica,* one in addition to *Gynura aurantiaca* in the family Compositae, and 20 in addition to *Petunia hybrida* in the family Solanacea and *Chrysanthemum morifolium*[115a] are susceptible to CEV infection.[9,47-49,52-55,109] Many of the solanaceous plants were first shown to be hosts of PSTV.[53-55]

The host range of CSV at first appeared to be restricted to the family Compositae.[39,57,98,102] Hollings and Stone[39] found that 8 out of 29 Compositae species tested were susceptible to CSV while none of the 116 species, chosen from 47 other families, was susceptible. However, the recent host range studies of Niblett et al.[85a,115a] demonstrate that CSV replicates in both tomato and Gynura plants.

The only hosts reported for ChCMV are *Chrysanthemum morifolium*[58,116] and *Chrysanthemum Zawadskii* Herbich.[115a] Out of 44 cultivars tested for susceptibility to ChCMV by grafting to stocks of infected Yellow Delaware, 19 developed disease symptoms within 9 weeks and 25 did not. Those with no visible symptoms were subsequently shown to be symptomless carriers of ChCMV by their ability to transmit the disease to Yellow Delaware and Blue Ridge when scions of these cultivars were grafted onto them.[58] The latent infectious agent, ChCMV-NS, originally recovered from certain clones of cultivar Deep Ridge,[97] was later shown to be present in three more cultivars of *Chrysanthemum morifolium* (Albatross, #2 Yellow Fred Shoesmith, and #2 Good News).[95]

CPFV infects 30 of the 37 species of the family Cucurbitaceae which have been tested.[42] Twelve of these are symptomless carriers.[42] Recently Sänger and co-workers[43,63,115b] have propagated CPFV in a number of tomato cultivars to obtain large quantities of the viroid for use in physical studies.

While ccRNA-1 appears to be a diagnostic marker for cadang-cadang[44,45] and has a number of properties in common with established viroids,[45] it has not yet been shown to be the causative agent of the disease. However, the disease can be transmitted to coconut palm seedlings by high pressure injection of nucleic acids isolated from polyethyleneglycol-precipitated material from homogenates of infected palms (see Section II. A and Table 1).[46] ccRNA-1 was detected in a majority of the inoculated seedlings 19 months after inoculation. All of these plants displayed early symptoms of cadang-cadang (bright yellow spots on the young fronds) by 22 months after inoculation. None of the uninoculated control plants contained ccRNA-1 in detectable quantities. In these studies, purified ccRNA-1 did not transmit the disease. Thus, while it appears that the agent of cadang-cadang is infectious as a free nucleic acid, it is still not known whether ccRNA-1 is the agent. No other hosts of cadang-cadang have been reported.

The experimentally determined host ranges summarized above may be influenced by the following factors: (1) when the inoculum is composed of sap instead of purified viroid, substances that inhibit infection of some other plant species might be present, (2) the method of inoculation might not be appropriate for some of the species tested,

(3) suboptimal environmental conditions (light, temperature and nutrition; see Section III. B) could render a response undetectable, and (4) plants may contain latent infectious agents (like ChCMV-NS) that protect them from the viroid disease whose host range is being studied. As a result of these uncertainties, we can draw no general conclusions about the potential host ranges of viroids. It is possible, for example, that all viroids could replicate in all plants if suitable conditions were found. It is unlikely that this question will be resolved before the mechanisms of cellular uptake and replication of viroids are understood at the molecular level.

As pointed out previously (Section III. A), the best hosts to use as viroid indicators are those which respond rapidly by the production of highly visible disease symptoms. On the other hand, since many of these disease syndromes include necrosis, these same hosts are probably not the best choice for molecular studies. A correlation has been observed between the presence of necrotic regions and the incidence of RNA breakdown products in nucleic acid preparations.[156] Furthermore, since it is clear that viroid replication can occur either with or without the production of disease, it would appear logical to investigate the mechanisms of viroid replication in symptomless hosts. In this way, it is hoped that alterations in cellular metabolism resulting from various pathogenic processes could be avoided.

### F. Agents Which Interfere With Viroid Propagation

Piperonyl butoxide, an ingredient of Raid (a house and garden bug killer manufactured by S. C. Johnson & Sons Ltd.), acts as a potent inhibitor of potato and *Scopolia sinensis* infection by PSTV.[117,118] An emulsion containing 1% piperonyl butoxide, 0.1% TritonX100, and water, applied to the foliage of potato and Scopolia plants, prevents infection of the plants by PSTV if mechanical inoculation of the sprayed leaves is carried out within about 4 days. Similar treatment of tomato plants results in death of the plant.[118]

Administration of the antibiotics tetracycline and penicillin to coconut palms under conditions known to be effective against procaryotic pathogens failed to cure the coconut palms of cadang-cadang or to affect the progress of the disease significantly.[16]

Since it is known that symptomless viroids can provide effective protection against severe forms,[97] one rationale for eliminating viroid disease in a particular crop would be to inoculate the plants with such a latent strain. If the latent strain is transmitted with high frequency through the seed (see Table 1) or if the crop is propagated vegetatively, the latent strain should persist and reinoculation should not be necessary. Thus, a resistant plant would have been produced. One drawback to this plan concerns the potential biological hazard. As discussed previously (Section III. D and III. E), viroids that are latent in one host may produce severe disease symptoms in another. Thus, by propagation of a viroid in a symptomless host for the purposes of crop protection, a vast reservoir of the infectious agent would be created and could provide the inoculum, in nature, for an epidemic in another cultivar or species.

A more conventional means of eliminating viroid disease is to screen plants prior to propagation. For example, potatoes can be assayed using Fernow's[107,108] challenge inoculation technique (see Section III. D), or using the procedure developed by Morris and Smith[72] for detection of viroid gel bands.

A combination of heat treatment and meristem tip culture[102] or axillary bud culture[103] has led to the recovery of viroid-free plants in the case of CSV-infected Mistletoe chrysanthemums and PSTV-infected potato plants, respectively. Only 2 to 4% of the regenerated plants were free of viroid. Developed for eliminating conventional viruses, this technique is based on the assumption that virus multiplication is inhibited at elevated temperatures while meristematic tissue continues to develop.[119] It is perhaps

TABLE 4

**Viroid Base Composition**

| Viroid | Plant | GMP | CMP | AMP | UMP | G:C | A:U | Ref. |
|---|---|---|---|---|---|---|---|---|
| CEV | Gynura | 28.8[a] | 29.4 | 21.5 | 19.9 | 0.98 | 1.10 | 70 |
| | | 27.3[b] | 28.3 | 21.5 | 23.0 | 0.96 | 0.93 | 70 |
| PSTV | Tomato | 28.9[c] | 28.3 | 21.7 | 20.9 | 1.02 | 1.04 | 121 |
| PSTV | Tomato | 27.8[d] | 30.4 | 20.4 | 21.4 | 0.91 | 0.95 | 87b |

[a] Determined by $OD_{260}$ measurement following fractionation of the nucleoside monophosphates by the method of Morris and Semancik.[120]

[b] Determined by $^{32}P$ scintillation counting after treatment described in footnote a.

[c] Determined by the procedure of Randerath et al.[122]

[d] Determined following complete nucleotide sequence analysis.[87b]

surprising that any success was observed since viroid replication appears to be preferentially enhanced at elevated temperatures (see Section III. B).[64]

## IV. PHYSICAL AND CHEMICAL PROPERTIES OF VIROIDS

The recent success of several groups in isolating large quantities of highly purified viroids has facilitated studies of the physical and chemical properties of viroids. As detailed below, viroids appear to (1) be composed entirely of ribonucleotides, (2) have sizes in the range 330 to 380 nucleotides, (3) have structures with the properties of both single-stranded and double-stranded RNA, and (4) contain linear and covalently closed circular forms. Many of these properties have been confirmed in the complete nucleotide sequence of PSTV reported recently by Gross et al.[87b] as described in Section E below.

### A. Chemical Composition

The base composition of CEV has been determined using two methods of quantitative analysis.[70] Unlabeled and $^{32}P$-labeled CEV were subjected to hydrolysis in 0.3 *M* potassium hydroxide for 16 hr at 37°C. The nucleoside monophosphates were then fractionated by polyacrylamide gel electrophoresis.[120] The quantity of each was determined either spectrophotometrically ($OD_{260}$) or by scintillation counting. CEV was found to have a G + C content of 55.6 to 58.2% and ratios of G to C and A to U that approach unity (see Table 4). The results obtained for PSTV are essentially identical (see Table 4).[121]

There has been no report of modified nucleotides in any of the viroids. However, the observation that both linear and circular forms of PSTV,[124,125] and PSTV infectivity itself,[21,23] are resistant to exonuclease treatment (see Section IV. D), suggests that the linear form may have chemically blocked ends. The inability of pretreatment with alkaline phosphatase to render the linear form susceptible to snake venom phosphodiesterase digestion indicates that the chemical block at the 3′-terminal is not a phosphate group.[124,125] Viroids have not yet been tested for the presence of either a Cap structure[126] or a protein linker[127] at their 5′ termini. The possibility that viroids have an amino acid at their 3′ ends as was found to be the case with a number of RNA viruses of plants[128] has, however, been explored. Hall et al.[129] demonstrated that CEV cannot be charged with any of the 20 common amino acids by any of the in vitro amino acylation systems tested (*Escherichia coli, Sacharomyces cerevisiae,* and *Phaseolus vulgaris*) under conditions appropriate for the amino acylation of tRNA, brome mosaic virus (tyrosine), and tobacco mosaic virus (histidine).[127] Furthermore, CEV did not compete with these test reactions. Thus, the linear forms of viroids do not appear to be charged at their 3′ ends by an amino acid and the nature of the chemical block at both ends of the RNA remains a mystery.

In an attempt to determine whether CEV has any properties in common with cellular messenger RNA, Semancik[130] examined CEV for the presence of poly (A). Tritium-labeled poly (U) failed to hybridize to CEV under conditions where its hybridization to bean pod mottle virus RNA and cowpea mosaic virus RNA were readily detected. Poly (C) also failed to hybridize to CEV. Thus, it was concluded that CEV does not contain regions of poly (A) or poly (G). PSTV also appears to lack poly (A). The assay system used to investigate PSTV for the presence of poly (A) is as follows. *E. coli* DNA polymerase I is capable of synthesizing DNA complementary to an RNA template if a small DNA primer, complementary to part of the template, is supplied.[131] Because Hadidi et al.[132] observed no increase in synthesis of DNA complementary to PSTV when they added either oligo $(dT)_{10}$ or oligo $(dG)_{12-18}$ to the reaction, it was concluded that PSTV lacks both poly (A) and poly (C). However, extensive secondary and tertiary structure of the viroid could prevent effective hybridization to the added homopolymer hybridization probes[130] and template primers.[132]

## B. Molecular Weight of Viroids

Properties including polyacrylamide gel electrophoretic mobility,[23,31,36,37,44,45,133] contour length upon examination in the electron microscope,[124,125,134] target size as estimated from ionizing radiation inactivation response,[36] and sedimentation coefficient[63] have been used to estimate the molecular weights of viroids. A summary of these estimates is presented in Table 5. Although no molecular weight determinations have been reported for CSV or ChCMV, Diener and Lawson[38] have observed that CSV migrates slightly faster than PSTV in 20% polyacrylamide gels.

As shown in Table 5, molecular weight estimates for viroids range from 0.25 to 1.30 $\times 10^5$ daltons. These values correspond to genome sizes ranging from 75 to 390 nucleotides. The maximum estimate is still 7 to 8 times smaller than the smallest known bacteriophage genome.[1] Thus, it is important to ask wether viroids consist of a number of different RNA molecules of similar size, enabling them, in this way, to have a sequence complexity like that of the RNA bacteriophages. Target sizes estimated from the ultraviolet light sensitivity of PSTV[135] and the inactivation of CEV by ionizing radiation[36] indicate that the size of the biologically active unit is approximately equal to that determined by polyacrylamide gel electrophoretic mobility. Furthermore, two-dimensional RNA fingerprints of the viroids (see Section II. C) are not sufficiently complex to permit the presence of even two completely different 300-nucleotide molecules as equal representatives of purified viroid. This assay would not necessarily detect the presence of components representing less than 5% of the RNA. However, if such minor components are obligate parts of viroid genomes, then they must have an infectivity approaching 100% since Sänger et al.[63] showed that 10% of the tomato plants inoculated by rubbing 50 to 100 molecules of CPFV over the surface of a leaf become infected. Thus, it seems likely that viroids have sequence complexities consistent with their molecular weights.

When total low molecular weight RNA from viroid-infected plants is analyzed on polyacrylamide gels, a species of RNA which cannot be detected in uninfected plants is present and coincides with the major peak of infectivity.[52,61,64] This is the material which has been extracted and purified for detailed chemical and physical studies. Various proportions of this RNA (20 to 99%; see Section IV. D) may be isolated as covalently closed circles. Perhaps the most homogeneous RNA populations obtained for

TABLE 5

**Molecular Weights of Viroids**

| Viroid | Method | Molecular weight (daltons × $10^{-5}$) | Ref. |
|---|---|---|---|
| PSTV (tomato) | 3, 5, 7.5, and 10% polyacrylamide gel | 0.25—1.10 (many peaks of infectivity observed) | 23 |
| | 20% polyacrylamide gel | 0.5 | 31 |
| | Mobility of formylated RNA in 5% gel | 0.75—0.85 | 133 |
| | Electron microscopy | 0.8—0.9 | 134 |
| | Electron microscopy of formylated and unformylated RNA | Linear: 1.1; circular: 1.3—1.4 | 124, 125 |
| | Sedimentation coefficient | Circular: 1.27 ± 0.04 | 63 |
| CEV (Gynura) | 2.5, 5, 10, 15, and 20% polyacrylamide gels | 0.5—0.6 | 37 |
| | Polyacrylamide gels ranging from 4 to 7% acrylamide | 1.0 | 36 |
| | Polyacrylamide gels ranging from 10 to 17% acrylamide | 0.5 | 36 |
| | Target size calculation from ionizing radiation inactivation | 1.1 | |
| | 5% polyacrylamide gels | 1.3 | 52 |
| CEV (Gynura) | Electron microscopy | Linear: 1.1—1.2; circular: 1.3—1.4 | 124 |
| CEV (Gynura and tomato) | Sedimentation coefficient | Circular: 1.19 ± 0.04 | 63 |
| CPFV (tomato) | Sedimentation coefficient | Circular: 1.10 ± 0.05 | 63 |
| Cadang-Cadang (coconut palm) | 2.5 and 3.3% polyacrylamide gels | ccRNA-1: 0.84; ccRNA-2: 2.04 | 44 |
| | 3 and 20% polyacrylamide gels | ccRNA-1: 0.63—0.73 | 45 |

study are those isolated by Sänger et al.[63] PSTV, CEV, and CPFV were subjected to three consecutive fractionations in 5% polyacrylamide gels yielding preparations at least 99% composed of circular molecules.[63] Highly reproducible values for the molecular weights of these species (accurate to ±12 nucleotides) were calculated from their sedimentation coefficients (381 (PSTV), 357 (CEV), and 330 (CPFV) nucleotides). If, as shown by McClements[124] and McClements and Kaesberg,[125] the circular forms of PSTV and CEV are 27 and 17.5% longer than the linear forms (determined by contour length measurements in the electron microscope), then the size of the linear forms of these two viroids should be 300 to 304 ± 12 nucleotides, respectively. Since viroids prepared using the methods of Diener or Semancik are composed of at least 80% linear molecules,[124,136] which Hadidi and Diener propose[136a] arise by in vivo cleavage of circular molecules, then results of physical studies carried out using such preparations would be dominated by the properties of the linear form. In contrast, studies carried out using viroids prepared using the method of Sänger et al.[63] would reflect the properties of circular forms only. The exact relationship between circular and linear forms of viroids has yet to be determined.

Infectivity studies have shown that the majority of infectivity resides in the disease-specific RNA gel band, and that the linear and circular forms present in this gel band may both be infectious.[136] However, it is reasonable to suspect that there might be other size classes of infectious RNA representing replication intermediates or metabolic products of the major viroid species. Thus, it is important not to overlook studies in which total low molecular weight RNA was fractionated by gel electrophoresis and assayed for infectivity throughout the gel.[23,71,88] Although CEV isolated from Gynura appears to produce only one infectious species,[36,37] PSTV isolated from either Rutgers tomato plants[23] or *Scopolia sinensis*[71,88] appears to contain two major peaks of infectivity and one or more minor peaks. The smallest infectious species detected are about the size of tRNA.[23,71,88]

### C. Thermal Stability

In 1931, Goss reported[92] that the agent of potato spindle tuber disease could be inactivated by heating infected potato plugs to 65°C prior to using them for tissue implantation inoculations of uninfected potato seed pieces. The infectivity of nondeproteinized tissue extracts was shown by Singh and Bagnall[33] to disappear after a 10 min incubation at 75°C. In contrast, these investigators showed that the infectivity of isolated nucleic acids could resist 10 min at 90°C and could only be inactivated by treatments such as boiling directly over a flame for 5 min or autoclaving at 120°C for 20 min.[33] Semancik and Weathers[35] demonstrated that CEV infectivity was not affected by 10 min treatments at temperatures up to 110°C, but that it decreased rapidly to a thermal death point at about 140°C. The infectivity of sap from CSV-infected chrysanthemum plants is not significantly altered by 10 min incubation at 98°C.[39] Singh and Bagnall[33] have suggested that, invivo, viroids are associated with some component (for example, a protein) which undergoes heat denaturation at 65 to 75°C and traps the viroid. Once the viroid preparation has been deproteinized, this can presumably no longer occur, and the temperature required for heat inactivation rises.[33]

A number of properties of viroids, including their ability to resist the attack of host nucleases during inoculation, have led to suggestions that they are composed mainly of double-stranded regions (see Section IV. D). Spectrophotometric analysis of viroid thermal denaturation suggests that the extent of base-pairing and helicity of viroids is much more like that observed in tRNA than that of double-stranded RNA. Midpoints of the thermal transitions ($T_m$) for PSTV,[9,64a] CEV,[70,137,64a] CPFV,[137,137a,64a] CSV,[64a] ChCMV[64a] and ccRNA-1,[45] measured either in 0.1 × SSC or 0.01 *M* sodium cacodylate-0.001 *M* EDTA (with sodium ion concentrations of 0.0165 and 0.014 molar, respectively) all fall within the range 48 to 54°C. This is 3° higher than the $T_m$ of *Escherichia coli* $tRNA^{Met}$ and 17° higher than yeast $tRNA^{Phe}$ under the same conditions,[70,137] and 31 to 37 C° lower than the $T_m$ of perfect double-stranded RNA.[70] A summary of spectrophotometric estimates of viroid $T_m$'s is presented in Table 6.

### D. Structure of Viroids

Two issues need to be considered regarding the structure of viroids: (1) the extent and distribution of base-pairing, and (2) the existence of both circular and linear forms.

#### *1. The Secondary Structure of Viroids*

PSTV and CEV are like double-stranded RNA (dsRNA) in that their G to C and A to U ratios approach unity (see Section IV-A and Table 4). On the other hand, their melting points are much closer to those typical of tRNA than to that of perfectly base-paired dsRNA (see Section IV. C and Table 6), suggesting that they do not have structures characteristic of perfect double strands. In fact, as will be discussed below, it is likely that viroids have extended, rodlike structures characterized by short regions of Watson-Crick base-pairing interrupted by loops.

TABLE 6

**Spectrophotometric Estimates of Thermal Stability**

| RNA | Buffer | $T_m$[c] (°C) | Hyper-chromicity[d] (%) | Hypochrom-icity[d] (%) | Ref. |
|---|---|---|---|---|---|
| PSTV | 0.01 × SSC[a] | 50 | 24 | 17 | 61 |
| | 0.1 × SSC | 54 | [b] | [b] | 9 |
| | .01 M sodium cacodylate-0.001 M EDTA | 51 | [b] | [b] | 64a |
| CEV | 0.1 × SSC | 52 | 22 | [b] | 70 |
| | 0.1 × SSC-0.005 M magnesium | 79 | 18 | [b] | 70 |
| | 1 × SSC | 79 | 18 | [b] | 70 |
| | 0.01 M sodium cacodylate-0.001 M EDTA | 52 | 21 | [b] | 137 |
| | 0.01 M sodium cacodylate-0.001 M EDTA | 51 | [b] | [b] | 64a |
| CPFV | 0.01 M sodium cacodylate-0.001 M EDTA | 51 | 21 | [b] | 137 |
| | 0.01 M sodium cacodylate-0.001 M EDTA | 51 | [b] | [b] | 64a |
| CSV | 0.01 M sodium cacodylate-0.001 M EDTA | 48.5 | [b] | [b] | 64a |
| ChCMV | 0.01 M sodium cacodylate-0.001 M EDTA | 48.5 | [b] | [b] | 64a |
| ccRNA-1 | 0.01 × SSC | 52 | 12 | 11 | 45 |
| | 0.1 × SSC | 54 | 14 | 12 | 45 |
| | 0.1 × SSC-0.005 M magnesium | 83 | 22 | 13 | 45 |
| $tRNA^{Phe}$ (yeast) | 0.1 × SSC | 37 | 24 | [b] | 70 |
| | 0.1 × SSC-0.005 M magnesium | 76 | 21 | [b] | 70 |
| | 0.01 M sodium cacodylate-0.001 M EDTA | 35 | 20 | [b] | 137 |
| $tRNA^{Met}$ (*E. coli*) | 0.1 × SSC | 45 | 23 | [b] | 70 |
| ϕ6 RNA (dsRNA) | 0.1 × SSC | 85 | 32 | [b] | 70 |

[a] SSC is 0.15 M Sodium chloride-0.015 M sodium citrate.
[b] Not reported.
[c] Mid-point of thermal transition of optical density.
[d] % Hyperchromicity is the total increase in optical density experienced by a nucleic acid sample during melting expressed as a percent of the initial optical density of the sample. % hypochromicity is the decrease in optical density observed when a melted nucleic acid sample is allowed to cool to the original density of the sample (prior to melting).

Numerous assays demonstrate that viroids have properties intermediate between those of dsRNA and ssRNA (single-stranded RNA) and do not contain DNA: (1) when PSTV,[26,28] CEV,[70] CSV,[39] or ccRNA-1[45] are tested prior to separation from the other plant nucleic acids, they are found to be more resistant to pancreatic RNase treatment in buffers of high ionic strength than in buffers of low ionic strength as would be expected for either dsRNA or RNA:DNA hybrids; (2) DNase pretreatment of crude nucleic acid preparations from PSTV-infected plants does not render the infectious RNA sensitive to RNase treatment in buffers of high ionic strength[28] indicating that the viroid is not an RNA:DNA hybrid; (3) no significant quantity of dsRNA or RNA:DNA hybrids was detected in a highly infectious preparation of PSTV using complement fixation assays and rabbit antisera to these two nucleic acid structures;[30] (4) PSTV,[26,28] and CEV[51] elute from cellulose CF11 columns partially with the ssRNAs (15% ethanol fraction) and partially with the dsRNAs (0% ethanol fraction); (5) diethylpyrocarbonate treatment sufficient to reduce the infectivity of cowpea mosaic virus RNA (a single-stranded species) to zero only reduced the infectivity of CEV by 20 to 60%;[35] (6) PSTV and CEV elute from methylated albumin-coated kieselguhr columns at sodium chloride concentrations in the range 0.70 to 0.72, between DNA (0.69 *M*) and dsRNA (0.78 *M*) and some distance from ssRNA (for example, ribosomal RNA elutes at 0.82 to 0.89 *M*);[26,28,35,50,51] and (7) the buoyant densities measured in cesium sulphate density gradients of PSTV (1.62 g/cc),[28] CEV (1.58 to 1.65 g/cc)[36] and ccRNA-1 (1.62 g/cc)[45] are in the range expected for RNA (ssRNA, 1.59 to 1.69; dsRNA, 1.57 to 1.64)[138] and considerably higher than expected for DNA (1.41 to 1.46 g/cc).[138] Thus, viroids have a variety of properties in common with both dsRNA and ssRNA.

Because viroids have relatively high thermal stabilities (see Section IV. C), it is conceivable that, in the native state, they are composed of two or more strands of RNA held together by hydrogen bonds. Semancik et al.[36] have shown that this is not the case for CEV. This viroid migrates identically in 5% polyacrylamide gels whether or not it is heated at 100°C and cooled rapidly prior to running the gel. These denaturation conditions were shown to be appropriate for the irreversible denaturation of dsRNA from bacteriophage $\phi$6.[36] Sänger et al.[63] demonstrated that the sedimentation velocities of native and denatured CPFV are identical, indicating that this viroid is also composed of a single RNA molecule.

Two approaches have permitted estimation of the base composition of base-paired regions of viroids. Low field nuclear magnetic resonance spectroscopy provides a measure of the hydrogen-bonded ring NH protons and indicates the identity of the neighboring base pair. Using this technique, Semancik et al.[70] found that G:C base pairs represent 70 to 80% of the base pairs in CEV. Since the G + C content of CEV is 55.6-58.2% (see Section IV-A), this means that the nonbase-paired portions of CEV must be enriched in A and U. Henco et al.[137] have described a slow relaxation process in CPFV and CEV, thought to correspond to an uninterrupted dsRNA region 52 base pairs long. By analyzing the wavelength dependence of hypochromicity with the aid of a temperature jump apparatus, the proposed dsRNA regions were shown to be composed of 72% G:C and 28% A:U base pairs with no evidence of non-Watson-Crick base pairs.[137] Similar analysis of a fast relaxation process, thought to correspond to 15 to 25 base pairs distributed in short hairpins, indicated that the base pairs in these regions are 50% G:C and 50% A:U.[137] These authors suggest that viroids contain a rigid DNA-like part composed of the 52 or more uninterrupted base pairs and a more flexible tRNA- or mRNA-like part at both ends of the rod. This model gains support from the observation that, in electron micrographs of native CPFV,[63] most of the RNA molecules are shaped like dumb-bells that are asymmetric with respect to the size of their terminal knobs. The finding that *E. coli* RNase III (an enzyme capable of cleaving dsRNA 25 or more base pairs in length)[139,140] fails to cleave PSTV or CEV,[75] indicates that the proposed 52 base-pair region must be interrupted at least a few times by structures that are not Watson-Crick base pairs. In fact, the nucleotide sequence of PSTV[87b] appears to contain no uninterrupted regions of potential perfect Watson-Crick base longer than six base pairs. A recent further refinement of the analysis of thermodynamic data has allowed Sänger and colleagues[64a,137a] to conclude that the melting behavior of viroids indicates the presence of defective double helices and the absence of extended regions of perfect base pairing.

*2. Circular and Linear Forms of Viroids*

In 1970, Diener,[2,123] and Semancik and Weathers[51] demonstrated that the infectious agents of potato spindle tuber and citrus exocortis diseases were resistant to the exonucleolytic actions of bovine spleen phosphodiesterase (which digests 5′-hydroxyl termini of RNA) and snake venom phosphodiesterase (which digests 3′-hydroxyl termini of RNA), both alone and with bacterial alkaline phosphatase pretreatment. It was concluded that PSTV and CEV must either be circular or have blocked termini. Both conjectures eventually turned out to be true. The first electron micrographs (obtained by Sogo et al.[134] in 1973) failed to show any circular forms of PSTV. The major form analyzed was the 500 Å rod-like species that appears to represent the two (or more) stranded native form of PSTV. By spreading formaldehyde-treated PSTV under fully denaturing conditions (45% formamide), McClements discovered in 1975[124] that about 20% of the molecules in PSTV preparations are circular. McClements and Kaesberg[125] showed that the relative amounts of circular and linear molecules of PSTV were not altered by snake venom phosphodiesterase treatment, with or without bacterial alkaline phosphatase pretreatment. Furthermore, Owens et al.[136] assayed the infectivity of both circular and linear forms of PSTV following their electrophoretic separation in 5% polyacrylamide gels containing 6 *M* urea and 50% formamide and concluded that both forms are infectious. Such results must be viewed with caution at this time, however, since infectivity of viroids is so high and the separated forms are mutually cross-contaminated.

More information about the structure and possible relatedness of the circular and linear forms of viroids may be obtained by analysis of partial denaturation products in the electron microscope. Sänger et al.[63] have described such a study of CPFV. At least 99% of the molecules in highly purified preparations of CPFV were reported to be covalently closed circles which were shown to be resistant to proteinase K and pronase treatment.[63] It appears likely that the linear forms seen by others (see below) were systematically removed by Sänger et al.[63] during the three consecutive electrophoretic fractionation steps carried out in 5% polyacrylamide-8 *M* urea gels. These authors found 350 Å rod-like shapes, rods with a knob at each end ("dumb-bells"), circles with a variable length handle at one side ("tennis rackets"), and fully denatured circles ("balloons"). The only linear structures seen in these studies (0.5 to 1.0% of the molecules) were thought to arise by random nicking of the circular molecules. These investigators reported that they had obtained identical results for PSTV and CEV.[63] If the knobs on both ends of the rod-like structures represent tRNA-like regions at both ends of an extended stretch of perfect base-pairing as was suggested,[63] it is somewhat surprising that no rods are seen with either a knob on one end and a balloon on the other, or a balloon on each end.

Studies were carried out by McClements[124] and by McClements and Kaesberg[125] using preparations of PSTV and CEV in which the linear form predominated. The ratio of linear to circular molecules was 4 to 1 for PSTV and 6 to 1 for CEV.[124] In these studies, 500 Å rods, 900 Å linear forms, and 1100 Å circles were observed in preparations of nonformaldehyde-treated PSTV. When the PSTV was treated with formaldehyde, the linears were 1100 Å and the circles were 1400 Å in length. Because the circles were reproducibly and significantly longer than the linears (27% for PSTV; 17.5% for CEV), it was concluded that it is unlikely that the linear molecules arise by random nicking of the circular molecules. Furthermore, tennis racket shapes were observed while no Y shapes were detected. Thus, contrary to the finding that hairpins of poly(dAdT) denature first at the open end,[141] the rod-like native form of linear PSTV always denatures first at the loop end of the rod. This finding provides additional support for the suggestion that the linear molecules have defined ends, since the end containing the free 3′ and 5′ termini must have a specific stable composition. If the linear forms of PSTV and CEV differ from their circular forms by the absence of a particular region of 50 or more bases, then this would be detectable by two-dimensional RNA fingerprinting analysis unless the missing bases are identical to another region of the viroid which is retained in the linear form. Such results have not yet been reported.

Although Haddidi and Diener[136a] have studied the kinetic relationship between the circular and linear forms of PSTV in vivo, and find evidence supporting the notion that the circular form is cleaved in vivo to yield the linear form, the question concerning the exact relationship between these two forms will not be resolved until detailed nucleotide sequence studies are carried out.

### E. Nucleotide Sequence of PSTV

The complete nucleotide sequence of PSTV has recently been published by Domdey et al.[87a] and Gross et al.[87b] In close agreement with the results of earlier chemical and physical studies, PSTV was found to be made up of 359 ribonucleotide residues in the form of a covalently closed ring, with 27.8% G, 30.4% C, 20.4% A, and 21.4% U. The sequence lacks AUG, the protein synthesis initiator codon, lending additional support to conclusions that viroids do not function as mRNA.[129,141a,142] While no modified nucleotides were found, the sequencing method used would only have allowed detection of such nucleotides at the 5′ termini of the PSTV fragments analyzed. Gross et al.[87b] proposed a secondary structure consisting of a serial arrangement of double helical sections (ranging in length from 3 to 8 base pairs including many G:U interactions) and internal loops to form an extended rod.

## V. REPLICATION AND PATHOGENICITY

### A. Possible Modes of Replication

Four possible modes of replication for viroids will be considered below: the viroid specifies its own replicase, the host specifies an RNA-dependent RNA polymerase, the host specifies an RNA-dependent DNA polymerase, or the viroid is transcribed from preexisting regions of the host DNA.

#### *1. Do Viroids Specify Their Own Replicase?*

It is clear that viroids do not carry a replicase because they are composed entirely of RNA. However, their genomes might contain information for a replicase or for a polypeptide capable of converting a host enzyme to the appropriate specificity. For example, the RNA bacteriophage Qβ, directs translation of a 65,000 dalton polypeptide that combines with three host subunits (ribosomal protein S1, and protein synthesis elongation factors EF-Tu and EF-Ts) to create a Qβ-specific replicase.[143] If translated, viroids would have the capacity to code for polypeptides of 10, 20, or 30,000 daltons, depending upon whether one, two, or three reading frames were used. (Cases where two overlapping reading frames are used to code for two entirely different gene products have recently been documented in the DNA bacteriophage ϕX174.[144])

The possibility that viroids have messenger RNA activity has been tested in two laboratories. Davies et al.[142] demonstrated that PSTV does not support in vitro translation in the wheat germ, *E. coli,* or *Bacillus stearothermophilus* systems under conditions where brome mosaic virus mRNAs and Qβ RNA stimulate translation. Hall et al.[129] compared the abilities of CEV and brome mosaic virus coat protein mRNA to stimulate protein synthesis in systems from *Bacillus stearothermophilus, E. coli, Pseudomonas aeruginosa,* and *Triticum aestivum.* In all cases, CEV failed to stimulate protein synthesis.[129] Despite these negative results, it is still possible that viroids contain the information for a replicase. Viroid genomes could be the minus strands of messenger RNA. Alternatively, some specific processing or modification of the viroid RNA might be required before translation can proceed. The absence of AUG from the nucleotide sequence of PSTV (see Section IV. E) supports the notion that it may not function as mRNA.

#### *2. Does the Host Specify An RNA-Dependent RNA Polymerase?*

RNA-dependent RNA polymerase activity has been detected in tobacco plant extracts by Duda et al.[145] and seems to be present in increased quantities in preparations from tobacco mosaic virus-infected plants (see also the chapter by Zaitlin). If activities of this kind are a general property of plant tissue, then viroids may use this route for replication. The in vitro product of this enzyme activity is dsRNA.[145] If a complementary strand mechanism is in operation, then it should be possible to detect minus strand RNA by hybridization.[146] Hadidi et al.[83] failed to detect any RNA complementary to PSTV RNA in the total RNA extracted from either PSTV-infected or healthy tomato plants. On the other hand, RNA complementary to CEV appears to be present in RNA preparations from CEV-infected Gynura plants and absent from RNA isolated from uninfected plants.[146a] Further study along these lines could lead to identification of the replicating form of viroids.

It is conceivable that viroids are extraordinarily efficient templates for replication by one of the host DNA-dependent RNA polymerases. An example of an RNA that multiplies indefinitely has been selected by *E. coli* DNA-dependent RNA polymerase from a mixture of random copolymers.[147] The RNA selected was small (100 to 150 nucleotides), had a sharp thermal transition ($T_m$ = 50°C in 0.2 *M* KCl), and a 20% hyperchromicity upon melting.[147] A recent study of RNA polymerases in Gynura shows that the addition of CEV to various polymerase preparations stimulates as much synthesis as does the addition of an equal amount of native DNA.[148] However, no assay was carried out to see whether the RNA synthesized was template-specific.

### *3. Is There A Host-Specified Reverse Transcriptase?*

The unresolved possibility that RNA-dependent DNA polymerases might be present in normal eukaryotic cells[149] permits speculation that DNA could be synthesized from the viroid RNA template as the first step in viroid replication.[7,37] Either with or without integration into the host chromosome, this new DNA could be transcribed by host RNA polymerase to synthesize new viroid molecules. Semancik and Geelen[150] have reported a great enrichment of DNA complementary to CEV in DNA from CEV-infected Gynura and tomato plants as compared with DNA from uninfected plants. However, subsequent experiments by these workers[146a] demonstrated that RNA present in the DNA preparation was responsible for the hybridization observed. Hadidi et al.[83] reported that DNAs of healthy and PSTV-infected tomato plants do not differ with respect to their content of DNA complementary to PSTV, and that probably one or a few copies of PSTV are present in the tomato genome. The possibility that the apparent hybridization of PSTV to tomato DNA could be an artifact caused by the presence of breakdown products of host ribosomal RNA in the PSTV preparation must be investigated further. One way to resolve this question is to recover the hybridized RNA and subject it to RNA fingerprint analysis. If there is any complementarity between viroids and their hosts' genomes, the precise extent of complementarity must be known in order to evaluate the various proposed mechanisms of replication and pathogenicity. For example, host DNA may contain regions related to only part of the viroid genome. Once again, hybridization followed by RNA fingerprinting analysis of the hybridized RNA would provide a way of determining the identity of the hybridized fraction (see Section II. C).

### *4. Is the Viroid Transcribed From Preexisting Host DNA?*

If, as suggested by Hadidi et al.,[83] DNA complementary to the viroid genome is present in the host DNA prior to infection, then it is possible that viroid replication involves transcription of that region of the DNA by a host RNA polymerase. In this scheme of events, viroids would be their own inducers. Diener and co-workers[84,85] have demonstrated an inhibition of PSTV replication (as measured by incorporation of $^{3}$H-labeled uracil into RNA with the mobility of PSTV in polyacrylamide gels) both in leaf strips and in isolated nuclei of PSTV-infected tomato leaves in the presence of actinomycin D (30 μg/mℓ in leaf strips; 50 μg/mℓ in nuclei). Mitomycin C, a specific inhibitor of DNA polymerase, did not inhibit incorporation of radioactivity into this RNA species.[85] Thus, it is possible that viroid RNA replicates by inducing the transcription of a region of host DNA containing the viroid sequence.

### B. Possible Sources of Pathogenicity

Whatever may be the mode of replication, the occurrence of this replication cycle is not sufficient to produce disease symptoms in the host, as evidenced by the existence of symptomless hosts (see Sections III. D and III. E). Thus, it seems more likely that the pathogenic effect is produced as a result of interaction of the viroid RNA in a sequence-specific manner with some host component rather than as a result of some generalized toxicity of the class of molecules with viroid-like structure. Because nothing is currently known about the mechanism of pathogenicity, a detailed treatment of this subject would not be appropriate at this time. A few proposals will be discussed briefly below.

In 1970, Diener[123] suggested that "PSTV may be an aberrant form of normally occurring exonuclease-resistant RNA." Semancik and Weathers[53] 2 years later proposed that "the term pathogene might be useful to describe the possibility of a host RNA becoming autonomous." In a study of the possible connection between RNA processing and potential regulatory roles for RNA in cellular development, Robertson and Dickson[151] and Dickson and Robertson[152] considered the features of RNA that make it especially suitable for a regulatory role, suggested possible sources for regulatory RNA in the nucleus, and described possible modes of action of the proposed regulatory molecules. It was suggested that viroid systems are prime candidates in which to test these ideas. Reanney[153] has proposed that "elements of the genomes of certain RNA virions constitute regulatory signals which have ceased to carry out their original functions of gene activation during normal ontogeny," and that, in particular, viroids may be "regulatory RNA of cellular origin which has escaped from the controls to which its production is subjected in nonmutated systems." Thus, there appears to be a consensus that viroids are escaped regulatory molecules of some kind.

One way in which viroid RNA could interact in a sequence-specific manner with its host is as a primer of DNA or RNA synthesis.[151,152] Thus, the appearance of a particular class of RNA fragments in the nucleus could control the onset of DNA replication. In an analogous way, an RNA fragment removed from the extra region of a primary transcript of one gene could prime transcription of a second gene, thereby achieving coordinate expression. Numerous RNA processing enzymes capable of reproducible removal of the extra regions from RNA precursors have been known for some time.[151] A second potential pathway for generating RNA fragments from extra regions of primary transcripts has recently been suggested by studies showing that adenovirus 2 mRNAs are composed of RNA from at least four separate locations in the genome.[154] If primary transcripts of these mRNAs have internal regions which are dispensed with during some kind of splicing event, the mechanism of splicing (if analogous to DNA excision) could result in the production of circular forms of the eliminated internal regions. Perhaps circular forms of viroids arise in this way and disrupt their hosts by mimicking the action of host RNA species originating in this way.

If, as discussed above, viroids are aberrant regulatory molecules, then elucidation of the molecular mechanisms involved in their replication and pathogenicity could lead to important new findings concerning a class of, as yet, undiscovered cellular regulatory elements.

## ACKNOWLEDGMENTS

The author wishes to express sincere thanks to Louise Pape and Edgar C. Lawson, III for their help with the organization of bibliographic materials, to Andrea D. Branch for critical reading of the manuscript, and to Dr. Hugh D. Robertson for his untiring participation in discussions of the subject matter.

## REFERENCES

1. **Boedtker, H. and Gesteland, R. F.,** Physical properties of RNA bacteriophages and their RNA, in *RNA Phages*, Zinder, N. D., Ed., Cold Spring Harbor Laboratory, Cold Spring Harbor, New York, 1975, 1.
2. **Diener, T. O.,** A plant virus with properties of a free ribonucleic acid: potato spindle tuber virus, in *Comparative Virology*, Maramorosch, K. and Kurstak, E., Eds., Academic Press, New York, 1971, 433.
3. **Marx, J. L.,** "Viroids": a new kind of pathogen?, *Science*, 178, 734, 1972.
4. **Bagnall, R. H., Singh, R. P., and Clark, M. C.,** The strange case of potato spindle tuber, *Can. Agric.*, Spring, 1973.
5. **Diener, T. O.,** Viroids: the smallest known agents of infectious disease, *Annu. Rev. Microbiol.*, 28, 23, 1974.
6. **Semancik, J. S.,** Citrus Exocortis Disease — 1965 to 1975, in *Proceedings of the Seventh Conferenece of the International Organization of Citrus Virologists*, Calaron, E. L., Ed., IOCV, Riverside, California, 1974, 79.
7. **Diener, T. O.,** Viroids, in *Modifications of the Information Content of Plant Cells*, Markham, R., Davies, D. R., Hopwood, D. A., and Horne, R. W., Eds., North-Holland, Amsterdam, 1975, 215.
8. **Diener, T. O., Hadidi, A., and Owens, R. A.,** Methods for studying viroids, in *Methods in Virology*, Vol. 6, Maramorosch, K. and Koprowski, H., Eds., Academic Press, New York, 1977, 185.
9. **Diener, T. O. and Hadidi, A.,** "Viroids," in *Comprehensive Virology*, Vol. 11, Fraenkel-Conrat, H. and Wagner, R. R., Eds., Plenum Press, New York, 1977, 285.
10. **Martin, W. H.,** "Spindle Tuber", a new potato trouble, Hints to potato growers, New Jersey State Potato Association, New Brunswick, New Jersey, 3, 8, 1922.
11. **Schultz, E. S. and Folsom, D.,** A "spindling-tuber" disease of Irish potatoes, *Science*, 57, 149, 1923.
12. **Benson, A. P., Raymer, W. B., Smith, W., Jones, E., and Munro, J.,** Potato diseases and their control, *Potato Handb.*, 10, 32, 1965.
13. **Hunter, J. E. and Rich, A. E.,** The effect of potato spindle tuber virus on growth and yield of Saco potatoes, *Am. Potato J.*, 41, 113, 1964.
14. **Singh, R. P., Finnie, R. E., and Bagnall, R. H.,** Relative prevalence of mild and severe strains of potato spindle tuber virus in Eastern Canada, *Am. Potato J.*, 47, 289, 1970.
15. **Singh, R. P., Finnie, R. E., and Bagnall, R. H.,** Losses due to the potato spindle tuber virus, *Am. Potato J.*, 48, 262, 1971.
16. **Goss, R. W.,** Transmission of potato spindle-tuber by cutting knives and seed piece contact, *Phytopathology*, 16, 299, 1926.
17. **Singh, R. P.,** Studies on Potato Spindle Tuber Virus, Ph.D. thesis, North Dakota State University, Fargo, North Dakota, 1966.
18. **Hunter, D. E., Darling, H. M., and Beale, W. L.,** Seed transmission of potato spindle tuber virus, *Am. Potato J.*, 46, 247, 1969.
19. **Fernow, K. H., Peterson, L. C., and Plaisted, R. L.,** Spindle-tuber virus in seeds and pollen of infected potato plants, *Am. Potato J.*, 47, 75, 1970.
20. **Singh, R. P.,** Seed transmission of potato spindle tuber virus in tomato and potato, *Am. Potato J.*, 47, 225, 1970.
21. **Singh, R. P. and Finnie, R. E.,** Seed transmission of potato spindle tuber metavirus through the ovule of *Scopolia sinensis*, *Can. Plant Dis. Surv.*, 53, 153, 1973.
22. **Raymer, W. B. and O'Brien, M. J.,** Transmission of potato spindle tuber virus to tomato, *Am. Potato J.*, 39, 401, 1962.
23. **Diener, T. O.,** Potato Spindle Tuber "Virus." IV. A replicating, low molecular-weight RNA, *Virology*, 45, 411, 1971.

24. **Diener, T. O., Smith, D. R., and O'Brien, M. J.,** Potato spindle tuber viroid. VII. Susceptibility of several Solanaceous plant species to infection with low molecular-weight RNA, *Virology*, 48, 844, 1972.
25. **Zaitlin, M. and Hariharasubramanian, V.,** A gel electrophoretic analysis of proteins from plants infected with tobacco mosaic and potato spindle tuber viruses, *Virology*, 47, 296, 1972.
26. **Diener, T. O. and Raymer, W. B.,** Potato spindle tuber virus: a plant virus with properties of a free nucleic acid, *Science*, 158, 378, 1967.
27. **Raymer, W. B. and Diener, T. O.,** Potato spindle tuber virus: a plant virus with properties of a free nucleic acid. I. Assay, extraction, and concentration, *Virology*, 37, 343, 1969.
28. **Diener, T. O. and Raymer, W. B.,** Potato spindle tuber virus: a plant virus with properties of a free nucleic acid. II. Characterization and partial purification, *Virology*, 37, 351, 1969.
29. **Diener, T. O.,** Potato spindle tuber virus: a plant virus with properties of a free nucleic acid. III. Subcellular location of PSTV-RNA and the question of whether virions exist in extracts of *in situ*, *Virology*, 43, 75, 1971.
30. **Stollar, B. and Diener, T. O.,** Potato spindle tuber viroid. V. Failure of immunological tests to disclose double-stranded RNA or RNA-DNA hybrids, *Virology*, 46, 168, 1971.
31. **Diener, T. O. and Smith, D. R.,** Potato spindle tuber viroid. VI. Monodisperse distribution after electrophoresis in 20% polyacrylamide gels, *Virology*, 46, 498, 1971.
32. **Diener, T. O.,** letter to the editor, Virus terminology and the viroid: a rebuttal, *Phytopathology*, 63, 1328, 1973.
33. **Singh, R. P. and Bagnall, R. H.,** Infectious nucleic acid from host tissues infected with the potato spindle tuber virus, *Phytopathology*, 58, 696, 1968.
34. **Singh, R. P. and Clark, M. C.,** Infectious low-molecular weight ribonucleic acid from tomato, *Biochem. Biophys. Res. Commun.*, 44, 1077, 1971.
35. **Semancik, J. S. and Weathers, L. G.,** Exocortis virus: an infectious free-nucleic acid plant virus with unusual properties, *Virology*, 47, 456, 1972.
36. **Semancik, J. S., Morris, T. J., and Weathers, L. G.,** Structure and conformation of low molecular weight pathogenic RNA from exocortis disease, *Virology*, 53, 448, 1973.
37. **Sänger, H. L.,** An infectious and replicating RNA of low molecular weight: the agent of the exocortis disease of citrus, *Adv. Biosci.*, 8, 103, 1972.
38. **Diener, T. O. and Lawson, R. H.,** Chrysanthemum stunt: a viroid disease, *Virology*, 51, 94, 1973.
39. **Hollings, M. and Stone, M.,** Some properties of chrysanthemum stunt, a virus with the characteristics of an uncoated ribonucleic acid, *Ann. Appl. Biol.*, 74, 333, 1973.
40. **Romaine, C. P. and Horst, R. K.,** Suggested viroid etiology for chrysanthemum chlorotic mottle disease, *Virology*, 64, 86, 1975.
41. **Hörst, R. K. and Kawamoto, S. O.,** Detection of viroids affecting *Chrysanthemum morifolium* by polyacrylamide gel electrophoresis, *Annual Proceedings of the American Phytopathological Society*, 1977.
42. **Van Dorst, H. J. M. and Peters, D.,** Some biological observations on pale fruit, a viroid-incited disease of cucumber, *Neth. J. plant Pathol.*, 80, 85, 1974.
43. **Sänger, H. L. and Riesner, D. K.,** Purification and some different properties of five different viroids, *J. Gen. Virol.*, in press.
43a. **Sasaki, M. and Shikata, E.,** On some properties of hop stunt disease agent, a viroid, *Proc. Jpn Acad.*, 53, 109, 1977.
44. **Randles, J. W.,** Association of two ribonucleic acid species with cadang-cadang disease of coconut palm, *Phytopathology*, 65, 163, 1975.
45. **Randles, J. W., Rillo, E. P., and Diener, T. O.,** The viroid-like structure and cellular location of anomalous RNA associated with the cadang-cadang disease, *Virology*, 74, 128, 1976.
46. **Randles, J. W., Boccardo, G., Retuerma, M. L., and Rillo, E. P.,** Transmission of the RNA species associated with cadang-cadang of coconut palm, and the insensitivity of the disease to antibiotics, *Phytopathology*, 67, 1211, 1977.
47. **Benton, R. J., Bowman, F. T., Fraser, L., and Kebby, R. G.,** Stunting and scaly butt of citrus, *Agric. Gaz. N.S.W.*, 61, 20, 1950.
48. **Weathers, L. G., Greer, F. C., Jr., and Harjung, M. K.,** Transmission of exocortis virus of citrus to herbaceous plants, *Plant Dis. Rep.*, 51, 868, 1967.
49. **Weathers, L. G. and Greer, F. C., Jr.,** Additional herbaceous hosts of the exocortis virus of citrus, *Phytopathology*, 58, 1071, 1968.
50. **Semancik, J. S. and Weathers, L. G.,** Exocortis virus of citrus: association of infectivity with nucleic acid preparations, *Virology*, 36, 326, 1968.
51. **Semancik, J. S. and Weathers, L. G.,** Properties of the infectious forms of exocortis virus of citrus, *Phytopathology*, 60, 732, 1970.

52. **Semancik, J. S. and Weathers, L. G.,** Exocortis disease: evidence for a new species of "infectious" low molecular weight RNA in plants, *Nature (London) New Biol.*, 237, 242, 1972.
53. **Semancik, J. S. and Weathers, L. G.,** Pathogenic 10 S RNA from exocortis disease recovered from tomato bunchy-top plants similar to potato spindle tuber virus infection, *Virology*, 49, 622, 1972.
54. **Semancik, J. S., Magnuson, D. S., and Weathers, L. G.,** Potato spindle tuber disease produced by pathogenic RNA from citrus exocortis disease: evidence for the identity of the causal agents, *Virology*, 52, 292, 1973.
55. **Singh, R. P. and Clark, M. C.,** Similarity of host response to both potato spindle tuber and citrus exocortis viruses, *FAO Plant Prot. Bull.*, 21, 121, 1973.
56. **Dimock, A. W.,** Chrysanthemum stunt, *N.Y. State Flower Grow. Bull.*, 26, 2, 1947.
57. **Brierley, P. and Smith, F. F.,** Chrysanthemum stunt, *Phytopathology*, 39, 501, 1949.
58. **Dimock, A. W., Geissinger, C. M., and Horst, R. K.,** Chlorotic mottle: a newly recognized disease of chrysanthemum, *Phytopathology*, 61, 415, 1971.
59. **Dickson, E., Diener, T. O., and Robertson, H. D.,** Potato spindle tuber and citrus exocortis viroids undergo no major sequence changes during replication in two different hosts, *Proc. Natl. Acad. Sci. U.S.A.*, 75, 951, 1978.
60. **Ocfemia, G. O.,** The probable nature of cadang-cadang disease of coconut, *Philipp. Agric.*, 26, 338, 1937.
60a. **Sasaki, M. and Shikata, E.,** Studies on the host range of hop stunt disease in Japan, *Proc. Jpn. Acad.*, 53, 103, 1977.
61. **Diener, T. O.,** Potato spindle tuber viroid, VIII. Correlation of infectivity with the UV-absorbing component and thermal denaturation properties of the RNA, *Virology*, 50, 606, 1972.
62. **Siegel, A. and Zaitlin, M.,** Infection process in plant virus diseases, *Annu. Rev. Phytol.*, 2, 179, 1964.
63. **Sänger, H. L., Klotz, G., Riesner, D., Gross, H. J., and Kleinschmidt, A. K.,** Viroids are single-stranded covalently closed circular RNA molecules existing as highly base-paired rod-like structures, *Proc. Natl. Acad. Sci. U.S.A.*, 73, 3852, 1976.
64. **Sänger, H. L. and Ramm, K.,** Radioactive labelling of viroid-RNA, in *Modifications of the Information Content of Plant Cells*, Markham, R., Davies, D. R., Hopwood, D. A., and Horne, R. W., Eds., Second John Innes Symposium, North-Holland, Amsterdam, 1975, 229.
64a. **Langowski, J., Henco, K., Riesner, D., and Sanger, H. L.,** Common structural features of different viroids: serial arrangement of double helical sections and internal loops, *Nucleic Acids Res.*, 5, 1589, 1978.
65. **Randles, J. W.,** Detection in coconut of rod-shaped particles which are not associated with disease, *Plant Dis. Rep.*, 59, 349, 1975.
66. **Commerford, S. L.,** Iodination of nucleic acids *in vitro*, *Biochemistry*, 10, 1993, 1971.
67. **Prensky, W.,** The radioiodination of RNA and DNA to high specific activities, in *Methods in Cell Biology*, Vol. XIII, Prescott, D., Ed., Academic Press, New York, 1975, 121.
68. **Robertson, H. D., Dickson, E., Model, P., and Prensky, W.,** Application of fingerprinting techniques to iodinated nucleic acid, *Proc. Natl. Acad. Sci. U.S.A.*, 70, 3260, 1973.
69. **Morris, T. J. and Wright, N. S.,** Detection on polyacrylamide gel of a diagnostic nucleic acid from tissue infected with potato spindle tuber viroid, *Am. Potato J.*, 52, 57, 1975.
70. **Semancik, J. S., Morris, T. J., Weathers, L. G., Rodorf, B. F., and Kearns, D. R.,** Physical properties of a minimal infectious RNA (viroid) associated with the exocortis disease, *Virology*, 63, 160, 1975.
71. **Singh, R. P., Michniewicz, J. J., and Narang, S. A.,** Separation of potato spindle tuber viroid ribonucleic acid from *Scopolia sinensis* into three infectious forms and the purification and oligonucleotide pattern of fraction II RNA, Part IX, *Can. J. Biochem.*, 54, 600, 1976.
72. **Morris, T. J. and Smith, E. M.,** Potato spindle tuber disease: procedures for the detection of viroid RNA and certification of disease-free potato tubers, *Phytopathology*, 67, 145, 1977.
73. **Singh, R. P. and Finnie, R. E.,** Stability of potato spindle tuber viroid in freeze-dried leaf powder, *Phytopathology*, 67, 283, 1977.
74. **Adesnik, M.,** Polyacrylamide gel electrophoresis of viral RNA, in *Methods in Virology*, Vol. 5, Maramorosch, K. and Koprowski, H., Eds., Academic Press, New York, 1972, 125.
75. **Dickson, E.,** Studies of Plant Viroid RNA and Other RNA Species of Unusual Function, Ph. D. thesis, Rockefeller University, New York, 1976.
76. **Dickson, E., Prensky, W., and Robertson, H. D.,** Comparative studies of two viroids: analysis of potato spindle tuber and citrus exocortis viroids by RNA fingerprinting and polyacrylamide-gel electrophoresis, *Virology*, 68, 309, 1975.
77. **Franklin, R. M.,** Purification and properties of the replicative intermediate of the RNA bacteriophage R17, *Proc. Natl. Acad. Sci. U.S.A.*, 55, 1504, 1966.

78. **Pieczenik, G., Model, P., and Robertson, H. D.,** Sequence and symmetry in ribosome binding sites of bacteriophage f1 RNA, *J. Mol. Biol.*, 90, 191, 1974.
79. **Bernardi, G.,** Chromatography of nucleic acids on hydroxy-apatite columns, in *Methods in Enzymology XXI-D*, Grossman, L. and Moldave, K., Eds., Academic Press, New York, 1971, 95.
80. **Diener, T. O.,** A method for the purification and reconcentration of nucleic acids eluted or extracted from polyacrylamide gels, *Anal. Biochem.*, 55, 317, 1973.
81. **Brownlee, G. G. and Sanger, F.,** Chromatography of $^{32}P$-labelled oligonucleotides on thin layers of DEAE-cellulose, *Eur. J. Biochem.*, 11, 395, 1969.
82. **Barrell, B. G.,** Fractionation and sequence analysis of radioactive nucleotides, in *Procedures in Nucleic Acid Research*, Vol. 2, Cantoni, G. A. and Davies, D. R., Eds., Harper & Row, New York, 1971, 751.
83. **Hadidi, A., Jones, D. M., Gillespie, D. H., Wong-Staal, F., and Diener, T. O.,** Hybridization of potato spindle tuber viroid to cellular DNA of normal plants, *Proc. Natl. Acad. Sci. U.S.A.*, 73, 2453, 1976.
84. **Diener, T. O. and Smith, D. R.,** Potato spindle tuber viroid. XIII. Inhibition of replication by actinomycin D, *Virology*, 63, 421, 1975.
85. **Takahashi, T. and Diener, T. O.,** Potato spindle tuber viroid. XIV. Replication in nuclei isolated from infected leaves, *Virology*, 64, 106, 1975.
85a. **Niblett, C. L., Dickson, E., Fernow, K. H., Horst, R. K., and Zaitlin, M.,** Cross protection among four viroids, *Virology*, 91, 198, 1978.
86. **Singh, R. P., Michniewicz, J. J., and Narang, S. A.,** Isolation of two distinct forms of potato spindle tuber metavirus RNA and their finger printing patterns, *Fed. Proc.*, 34, 638, 1975.
87. **Gross, H. J., Domdey, H., and Sänger, H. L.,** Comparative oligonucleotide fingerprints of three plant viroids, *Nucleic Acids Res.*, 4, 2021, 1977.
87a. **Domdey, H., Jank, P., Sänger, H. L., and Gross, H. J.,** Studies on the primary and secondary structure of potato spindle tuber viroid: products of digestion with ribonuclease A and ribonuclease T1, and modification with bisulfite, *Nucleic Acids Res.*, 5, 1221, 1978.
87b. **Gross, H. J., Domdey, H., Lossao, C., Jank, P., Raba, M., Alberty, H., and Sänger, H. L.,** Nucleotide sequence and secondary structure of potato spindle tuber viroid, *Nature (London)*, 273, 203, 1978.
89. **Singh, R. P.,** A local lesion host for potato spindle tuber virus, *Phytopathology*, 61, 1034, 1971.
90. **Mühlbach, H.-P. and Sänger, H. L.,** Multiplication of cucumber pale fruit viroid in inoculated tomato leaf protoplasts, *J. Gen. Virol.*, 35, 377, 1977.
91. **Hadidi, A. and Diener, T. O.,** *De novo* synthesis of potato spindle tuber viroid as measured by incorporation of $^{32}P$, *Virology*, 78, 99, 1977.
92. **Goss, R. W.,** Infection experiments with spindle tuber and unmottled curly dwarf of the potato, *Nebr. Agric. Exp. Sta. Res. Bull.*, 53, 1, 1931.
93. **Garnsey, S. M. and Jones, J. W.,** Mechanical transmission of exocortis virus with contaminated budding tools, *Plant Dis. Rep.*, 51, 410, 1967.
94. **Hollings, M., Stone, O. M., and Bouttell, G. C.,** Carnation Italian ringspot virus, *Ann. Appl. Biol.*, 65, 299, 1970.
95. **Hörst, R. K., Langhans, R. W., and Smith, S. H.,** Effects of chrysanthemum stunt, chlorotic mottle, aspermy and mosaic on flowering and rooting of chrysanthemums, *Phytopathology*, 67, 9, 1977.
96. **Bigornia, A. E.,** Evaluation and trends of researches on the coconut cadang-cadang disease, *Philipp. J. Coconut Stud.*, 5.
97. **Hörst, R. K.,** Detection of a latent infectious agent that protects against infection by chrysanthemum chlorotic mottle viroid, *Phytopathology*, 65, 1000, 1975.
98. **Lawson, R. H.,** Cineraria varieties as starch lesion test plants for chrysanthemum stunt virus, *Phytopathology*, 58, 690, 1968.
99. **Singh, R. P.,** Experimental host range of the potato spindle tuber 'virus', *Am. Potato J.*, 50, 111, 1973.
100. **Lee, C. R. and Singh, R. P.,** Enhancement of diagnostic symptoms of potato spindle tuber virus by manganese, *Phytopathology*, 62, 516, 1972.
101. **Singh, R. P., Lee, C. R., and Clark, M. C.,** Manganese effect on the local lesion symptom of potato spindle tuber 'virus' in *Scopolia sinensis*, *Phytopathology*, 64, 1015, 1974.
102. **Hollings, M. and Stone, O. M.,** Attempts to eliminate chrysanthemum stunt from chrysanthemum by meristem-tip culture after heat-treatment, *Ann. Appl. Biol.*, 65, 311, 1970.
103. **Stace-Smith, R. and Mellor, F. C.,** Eradication of potato spindle tuber virus by thermotherapy and axillary bud culture, *Phytopathology*, 60, 1857, 1970.
104. **Singh, R. P.,** Infectivity of potato spindle tuber viroid in potato plant parts, *Phytopathology*, 67, 15, 1977.

105. **Semancik, J. S., Tsurada, D., Zaner, L., Geelen, J. L. M. C., and Weathers, J. G.,** Exocortis disease: subcellular distribution of pathogenic (viroid) RNA, *Virology*, 69, 669, 1976.
106. **Semancik, J. S. and VanDerWoude, W. J.,** Exocortis viroid: cytopathic effects at the plasma membrane in association with pathogenic RNA, *Virology*, 69, 719, 1976.
107. **Fernow, K. H.,** Tomato as a test plant for detecting mild strains of potato spindle tuber virus, *Phytopathology*, 57, 1347, 1967.
108. **Fernow, K. H., Peterson, L. C., and Plaisted, R. L.,** The tomato test for eliminating spindle tuber from potato planting stock, *Am. Potato J.*, 46, 424, 1969.
109. **Roistacher, C. N., Calavan, E. C., Blue, R. L., Navarro, L., and Gonzales, R.,** A new more sensitive citron indicator for detection of mild isolates of citrus exocortis viroid (CEV), *Plant Dis. Rep.*, 61, 135, 1977.
110. **Dickson, E., Niblett, C. L., Robertson, H. D., Horst, R. K., and Zaitlin, M.,** Mild and severe strains of potato spindle tuber viroid have only minor sequence differences, *Nature*, 277, 60, 1979.
111. **Easton, G. D. and Merriam, D. C.,** Mechanical inoculation of the potato spindle tuber virus in the genus *Solanum*, *Phytopathology*, 53, 349, 1963.
112. **O'Brien, M. J. and Raymer, W. B.,** Symptomless hosts of the potato spindle tuber virus, *Phytopathology*, 54, 1045, 1964.
113. **Singh, R. P., Benson, A. P., and Salama, F. M.,** Sheyenne tomato variety as an indicator for potato spindle tuber virus, *Am. Potato J.*, 41, 304, 1964.
114. **Singh, R. P. and Bagnall, R. H.,** Solanum Rostratum Dunal., a new test plant for the potato spindle tuber virus, *Am. Potato J.*, 45, 335, 1968.
115. **Singh, R. P. and O'Brien, M. J.,** Additional indicator plants for potato spindle tuber virus, *Am. Potato J.*, 47, 367, 1970.

115a. **Niblett, C. L., Dickson, E., Horst, R. K., and Romaine, C. P.,** Symptomatology on additional hosts of four viroids and an efficient purification procedure, *Phytopathology*, in press .

115b. **Singh, A. and Sänger, H. L.,** Chromatographic behavior of viroids of exocortis disease of citrus and of spindle tuber disease of potato, *Phytopathol. Z.*, 87, 143, 1976.

116. **Horst, R. K., Geissinger, C. M., and Staszewicz, M.,** Treatments that improve mechanical transmission of chrysanthemum chlorotic mottle virus, *Acta Hortic.*, 36, 59, 1974.
117. **Singh, R. P., Michniewicz, J. J., and Narang, S. A.,** Piperonyl Butoxide, a potent inhibitor of potato spindle tuber viroid in *Scopolia sinensis*, *Can. J. Biochem.*, 53, 1130, 1975.
118. **Singh, R. P.,** Piperonyl butoxide as a protectant against potato spindle tuber viroid infection, *Phytopathology*, 67, 933, 1977.
119. **Quak, F.,** Meristem culture and virus-free plants, in *Plant, Cell, Tissue, and Organ Culture*, Reinert, J. and Bajaj, Y. P. S., Eds., Springer-Verlag, Berlin, 1977, 598.
120. **Morris, T. J. and Semancik, J. S.,** Nucleotide composition of RNA by polyacrylamide gel electrophoresis, *Anal. Biochem.*, 61, 48, 1974.
121. **Niblett, C. L., Hedgcoth, C., and Diener, T. O.,** Base composition of potato spindle tuber viroid, in Beltsville Symposium on Virology in Agriculture, U.S. Department of Agriculture, Beltsville, 1976, 27.
122. **Randerath, E., Yu, C.-T., and Randerath, K.,** Base analysis of ribopolynucleotides by chemical tritium labelling: a methodological study with model nucleosides and purified tRNA species, *Anal. Biochem.*, 48, 172, 1972.
123. **Diener, T. O.,** Isolation of exonuclease-resistant ribonucleic acid from healthy and potato spindle tuber virus-infected tomato leaves, *Phytopathology*, 60, 1014, 1970.
124. **McClements, W. L.,** Electron Microscopy of RNA: Examination of Viroids and a Method for Mapping Single-Stranded RNA, Ph.D. thesis, University of Wisconsin, Madison, 1975.
125. **McClements, W. L. and Kaesberg, P.,** Size and secondary structure of potato spindle tuber viroid, *Virology*, 76, 477, 1977.
126. **Shatkin, A. J.,** Capping of eucaryotic mRNAs, *Cell*, 9, 645, 1976.
127. **Lee, Y. F., Nomoto, A., Detjen, B. M., and Wimmer, E.,** A protein covalently linked to poliovirus genome RNA, *Proc. Natl. Acad. Sci. U.S.A.*, 74, 59, 1977.
128. **Öberg, B. and Philipson, L.,** Binding of histidine to tobacco mosaic virus RNA, *Biochem. Biophys. Res. Commun.*, 48, 927, 1972.
129. **Hall, T. C., Wepprich, R. K., Davies, J. W., Weathers, L. G., and Semancik, J. S.,** Functional distinctions between the ribonucleic acids from citrus exocortis viroid and plant viruses: cell-free translation and aminoacylation reactions, *Virology*, 61, 486, 1974.
130. **Semancik, J. S.,** Detection of polyadenylic acid sequences in plant pathogenic RNAs, *Virology*, 62, 288, 1974.
131. **Modak, M. J., Marcus, S. L., and Cavalierie, L. F.,** A new sensitive method for detecting polyadenylate in viral and other ribonucleic acids using *Escherichia coli* deoxyribonucleic acid polymerase I, *J. Biol. Chem.*, 249, 7373, 1975.

132. **Hadidi, A., Modak, M. J., and Diener, T. O.,** Preparation and characterization of DNA complementary to potato spindle tuber viroid, in Beltsville Symposium on Virology in Agriculture, U.S. Department of Agriculture, Beltsville, Maryland, 1976, 30.
133. **Diener, T. O. and Smith, D. R.,** Potato spindle tuber viroid. IX. Molecular-weight determination by gel electrophoresis of formylated RNA, *Virology*, 53, 359, 1973.
134. **Sogo, J. M., Koller, Th., and Diener, T. O.,** Potato spindle tuber viroid. X. Visualization and size determination by electron microscopy, *Virology*, 55, 70, 1973.
135. **Diener, T. O., Schneider, I. R., and Smith, D. R.,** Potato spindle tuber viroid. XI. A comparison of the ultraviolet light sensitivities of PSTV, tobacco ringspot virus, and its satellite, *Virology*, 57, 577, 1974.
136. **Owens, R. A., Erbe, E., Hadidi, A., Steere, R. L., and Diener, T. O.,** Separation and infectivity of circular and linear forms of potato spindle tuber viroid, *Proc. Natl. Acad. Sci. U.S.A.*, 74, 3859, 1977.
136a. **Hadidi, A. and Diener, T. O.,** In vivo synthesis of potato spindle tuber viroid: kinetic relationship between the circular and linear forms, *Virology*, 86, 57, 1978.
137. **Henco, K., Riesner, D., and Sänger, H. L.,** Conformation of viroids, *Nucleic Acids Res.*, 4, 177, 1977.
137a. **Klump, H., Riesner, D., and Sänger, H. L.,** Calorimetric studies on viroids, *Nucleic Acids Res.*, 5, 1581, 1978.
138. **Szybalski, W.,** Use of cesium sulfate for equilibrium density gradient centrifugation, in *Methods in Enzymology*, Vol. XII-B, Grossman, L. and Moldave, K., Eds., Academic Press, New York, 1968, 330.
139. **Robertson, H. D., Webster, R. E., and Zinder, N. D.,** A nuclease specific for double-stranded RNA, *Virology*, 12, 718, 1967.
140. **Robertson, H. D. and Dunn, J. J.,** Ribonucleic acid processing activity of *Escherichia coli* ribonuclease III, *J. Biol. Chem.*, 250, 3050, 1975.
141. **Scheffler, I. E., Elson, E. L., and Baldwin, R. L.,** Helix formation by d(TA) oligomers. II. Analysis of the helix-coil transitions of linear and circular oligomers, *J. Mol. Biol.*, 48, 145, 1970.
141a. **Semancik, J. S., Conejero, V., and Gerhart, J.,** Citrus exocortis viroid: survey of protein synthesis in *Xenopus laevis* oocytes following addition of viroid RNA, *Virology*, 80, 218, 1977.
142. **Davies, J. W., Kaesberg, P., and Diener, T. O.,** Potato spindle tuber viroid. XII. An investigation of viroid RNA as a messenger for protein synthesis, *Virology*, 61, 281, 1974.
143. **Kamen, R. I.,** Structure and function of the Q$\beta$ RNA replicase, in *RNA Phages*, Zinder, N. D., Ed., Cold Spring Harbor Laboratory, Cold Spring Harbor, N. Y., 1975, 203.
144. **Sanger, F., Air, G. M., Barrell, B. G., Brown, N. L., Coulson, A. R., Fiddes, J. C., Hutchison, C. A. III, Slocombe, P. M. and Smith, M.,** Nucleotide sequence of bacteriophage $\phi$X174 DNA, *Nature (London)*, 265, 687, 1977.
145. **Duda, C. T., Zaitlin, M., and Siegel, A.,** *In vitro* synthesis of double-stranded RNA by an enzyme system isolated from tobacco leaves, *Biochim. Biophys. Acta*, 319, 62, 1973.
146. **Robertson, H. D.,** Functions of replicating RNA in cells infected by RNA bacteriophages, in *RNA Phages*, Zinder, N. D., Ed., Cold Spring Harbor Laboratory, Cold Spring Harbor, N.Y., 1975, 113.
146a. **Grill, L. K. and Semancik, J. S.,** RNA sequences complementary to citrus exocortis viroid in nucleic acid preparations from infected *Gynura aurantiaca*, *Proc. Natl. Acad. Sci. U.S.A.*, 75, 896, 1978.
147. **Biebricher, C. K. and Orgel, L. E.,** An RNA that multiplies indefinitely with DNA-dependent RNA polymerase: section from a random copolymer, *Proc. Natl. Acad. Sci. U.S.A.*, 70, 934, 1973.
148. **Geelen, J. L. M. C., Weathers, L. G., and Semancik, J. S.,** Properties of RNA polymerases of healthy and citrus exocortis viroid-infected *Gynura aurantiaca* DC, *Virology*, 69, 537, 1976.
149. **Scolnick, E. M., Aaronson, S. A., Todaro, G. J., and Parks, W. P.,** RNA dependent DNA polymerase activity in mammalian cells, *Nature (London)*, 229, 318, 1971.
150. **Semancik, J. S. and Geelen, J. L. M. C.,** Detection of DNA complementary to pathogenic viroid RNA in exocortis disease, *Nature (London)*, 256, 753, 1975.
151. **Robertson, H. D. and Dickson, E.,** RNA processing and the control of gene expression, in *Process. RNA, Brookhaven Symp. Biol.*, 26, 240, 1974.
152. **Dickson, E. and Robertson, H. D.,** Potential regulatory roles for RNA in cellular development, *Cancer Res.*, 36, 3387, 1976.
153. **Reanny, D. C.,** A regulatory role for viral RNA in eukaryotes, *J. Theor. Biol.*, 49, 461, 1975.
154. **Chow, L. T., Gelinas, R. E., Broker, T. R., and Roberts, R. J.,** An amazing sequence arrangement at the 5′ ends of adenovirus 2 messenger RNA, *Cell*, 12, 1, 1977.
155. **Dickson, E.,** A carrier-free method for the extraction of sub-microgram quantities of RNA from polyacrylamide gels, in preparation
156. **Dickson, E.,** unpublished observations.

# CROWN GALL: TRANSFER OF BACTERIAL DNA TO PLANTS VIA THE Ti-PLASMID

**J. Schell***

## TABLE OF CONTENTS



## I. INTRODUCTION: CROWN GALL, A NATURAL INSTANCE OF GENETIC ENGINEERING OF PLANTS BY BACTERIA

Crown gall is a neoplastic disease of plants. The causative agents of these plant tumors, are a group of gram-negative soil bacteria belonging to the genus *Agrobacterium*. Upon wounding and infection of plants, these bacteria can transform plant cells into autonomously growing tumor cells. The disease is very widespread in nature, affecting most dicotyledonous plants both in cultivated and in virgin areas. Monocotyledonous plants rarely, if ever, form crown gall. That bacteria were responsible for the crown gall disease was already firmly established in 1911.[1] However, the elucidation and understanding of the precise role the bacteria played in bringing about this phenomenon took a very long time and is still a matter of very active research and speculation.

The major steps in the development of our understanding were as follows: in a number of publications Braun and co-workers[2] established that the tumorous character of crown gall tissues could be maintained indefinitely in vitro in the absence of the bacterium. Thus, they focused attention to a hypothetical tumor-inducing principle (TIP) produced by the bacteria and responsible for the induction, if not for the maintenance, of the tumorigenic condition of the crown gall plant cells.[3,4] Although most of the major components of *A. tumefaciens* have, over the years, been implicated as TIP,

* The following people (in alphabetical order) contributed to this paper and should be considered to be the authors: De Beuckeleer, M., De Block, M., De Greve, H., Depicker, A., De Vos, G., De Vos, R., De Wilde, M., Dhaese, P., Dobbelaere, M.-R., Engler, G., Genetello, C., Hernalsteens, J. P., Holsters, M., Jacobs, A., Messens, E., Schell, J., Seurinck, J., Silva, A., Van Haute, E., Van Montagu, M., Van Vliet, F., Villarroel, R., and Zaenen, I. The article was prepared by J. Schell.

no rigorous and reproducible experiments were produced until very recently to elucidate the nature of the TIP. One of the problems was that most of the thinking was oriented to look at the problem primarily from the plant's point of view. The observations of Petit and collaborators[5] at Versailles were the basis for a more fundamental implication of the bacteria in the crown gall phenomenon. These authors[5] demonstrated in 1970 that there are at least two different forms of pathogenic agrobacteria. The distinction was based on the metabolism of two arginine derivatives that had previously been detected in crown gall tissues. Perhaps the most striking phenotypic difference between normal (untransformed) and crown gall cells is the presence in the latter of *N*-α-(D-1-carboxyethyl)-L-arginine (octopine) or *N*-α-(1,3-dicarboxypropyl)-L-arginine (nopaline). The type of arginine derivative synthesized in the tumor was found to be specified by the particular strain of the bacteria inducing the tumor and to be independent of the host-plant on which the tumors were induced. Furthermore, these authors demonstrated that *Agrobacterium* strains that induce the synthesis of octopine in crown gall cells can selectively use this product, but not nopaline, as sole energy, carbon, and/or nitrogen source, whereas *Agrobacterium* strains that induce the synthesis of nopaline in crown galls can selectively use it, but not octopine, as energy, carbon and/or nitrogen source.

Thus, a genetic linkage between oncogenicity and "opine" metabolism was established and it was suggested that genetic information could be transferred from bacterium to plant. This correlation has been repeatedly confirmed and extended.[6,7] In fact it is now clear that "opines" should be given an operational definition rather than a chemical one, as will be discussed later. Indeed several new "opines" have recently been discovered, such as histopine[8] and also "agropine",[9] which appears to be a $C_{11}H_{17}NO_7$ compound.

The next critical step was the discovery in our laboratory of the Ti-plasmids and the demonstration of their involvement in oncogenicity.[10-13] This discovery was soon confirmed and extended in other laboratories[14] and has turned out to be the basis for a precise identification of the elusive TIP (see next Section). Work in our laboratory[11] had also demonstrated that earlier claims that bacteriophages might be the TIP were unfounded. Previous to our discovery of the Ti-plasmids, Kerr[15,16] had observed that oncogenicity could be transferred from one strain of *Agrobacterium* to another by inoculating both strains together or in succession onto the same plant. After development of a crown gall and provided the non-oncogenic acceptor was kept on the crown gall for several weeks, transfer of oncogenicity was readily observed. It was, therefore, assumed that oncogenicity was somehow correlated with an infectious entity.

Since the discovery of the Ti-plasmid and its central role in the crown gall phenomenon, several very important observations have allowed the formulation of a general concept describing and explaining this neoplastic transformation of plant cells. The first of these observations was the demonstration that the genes controlling opine catabolism in agrobacteria and the genes determining opine synthesis in transformed plant cells are both located on the Ti-plasmid. In other words, both the capacity to catabolize opines and the capacity to specifically induce their synthesis in transformed plant cells are determined entirely by the type of Ti-plasmid present in agrobacteria.[17-20]

These observations provided genetic evidence in favor of a model involving the Ti-plasmid in a DNA-transfer mechanism from bacterium to plant. That such a Ti-DNA transfer actually occurs was subsequently demonstrated first by Chilton et al.[21] with DNA hybridization experiments. The same group also demonstrated transcription of part of this T-DNA in the crown gall cells.[22] Finally, the demonstration that the T-DNA (i.e., the Ti-plasmid DNA present in transformed plant cells) was actually re-

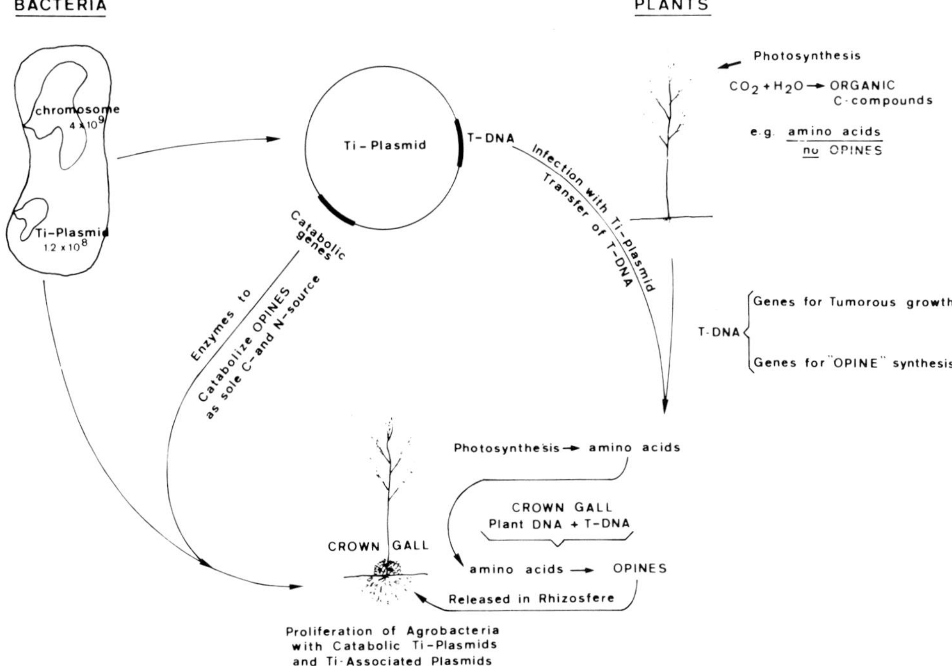


FIGURE 1. The "genetic colonization" concept. For explanations see Section I.

sponsible for the synthesis of the specific opine was provided by work in our laboratory and will be described further in the next section. All these observations can be fitted into a general concept that we have called "genetic colonization".[23]

Agrobacteria, and possibly other bacteria, have evolved an intricate mechanism by which they can transfer a specific genetic information to plants in such a way that the transformed plant cells express a number of new phenotypes, such as uncontrolled proliferation and the synthesis and probable release of "opines". The bacteria, but not the plant, benefit from this transformation because they can selectively utilize the "opines" for their own growth and proliferation. The surface of crown galls therefore could provide an ecological niche in which the bacteria harboring plasmids with genes specifying opine catabolism and utilization — as energy, carbon and/or nitrogen source — have an important selective advantage over other soil bacteria.

We are, therefore, apparently dealing here with a hitherto unsuspected type of parasitism. The parasite manages to introduce some specific genetic information into the genome of its host. Thus, it forces its host to synthesize products (opines) that only the parasite can utilize. The parasite achieves two important goals in this way: it manages to tap some of the photosynthetic capacity of its host and it does this in such a way that it achieves a selective advantage over the other competing organisms. Indeed, only the crown gall-inducing agrobacteria and some of their non-oncogenic relatives carry the plasmid-linked genes with which the opines can be degraded. How the agrobacteria achieve this "genetic colonization" is summarized in Figure 1 and will be described further in the next section.

One advantage of this concept is that it allows some operational definitions of:

1. Crown gall: plant cells that contain and express a DNA segment derived from bacterial Ti-plasmids specifying neoplastic growth and opine synthesis. The fact

that we have isolated some Ti-mutants that induce tumors in which no opines are synthesized (see Section II.D) can be seen as an extension of this definition.
2. "Opines": products specifically synthesized by crown gall plant cells. The synthesis of these products should be directly or indirectly controlled by DNA derived from Ti-plasmids and agrobacteria should be able to selectively use these products as energy, carbon, and/or nitrogen sources. Opines may or may not be able to induce conjugative transfer of the Ti-plasmids (see Section II.F).
3. Ti-plasmids: a group of bacterial conjugative plasmids with two major features: (a) they harbor genes that allow their bacterial hosts to use opines as energy, carbon and/or nitrogen source or more generally to ecologically benefit from opines and (b) they harbor and are able to transfer part of their genes (DNA) to plant cells. At least two types of functions are controlled by this transferred DNA (T-DNA): tumorous growth of the transformed plant cells and opine synthesis. There are also what we would like to call "Ti-associated" plasmids in agrobacteria. These plasmids can specify the catabolism and utilization of opines and allow their bacterial host to benefit from the transformation brought about by Ti-plasmids.

## II. STRUCTURE AND FUNCTION OF THE Ti-PLASMIDS

### A. The TIP of Agrobacterium is Carried by Large Plasmids

Strong evidence that genes essential for crown gall induction are located on relatively large plasmids rapidly followed our discovery of such plasmids in a number of oncogenic *Agrobacterium* strains.[10,11] That these "Ti-plasmids" are responsible for the oncogenic properties of *A. tumefaciens* was established in several ways:

1. Loss of the Ti-plasmid from oncogenic strains results in the loss of oncogenicity.[12,14]
2. Introduction of a Ti-plasmid in a non-oncogenic acceptor strain by conjugation or transformation confers the capacity to induce crown gall tumors[13,14] All results reported since then have confirmed these observations.[7,17,18,24-26]
3. Deletion and transposon insertion mutants of various Ti-plasmids have been isolated. *Agrobacterium* strains carrying some of these Ti-plasmid mutants were shown to be non-oncogenic.[27] These mutants will be described in more detail under Section II.D.

Ti-plasmids have been found in all the oncogenic *A. tumefaciens, A. rhizogenes* and *A. rubi* strains investigated. In fact, it has recently been demonstrated that the developmental differences between the three variants (crown gall, hairy root, and cane gall) are probably also specified by the respective Ti-plasmids carried by the so-called *tumefaciens, rhizogenes,* and *rubi* strains. Indeed the property (e.g., to induce hairy root disease) can be cotransferred with the Ti-plasmid from a *rhizogenes* donor to a non-oncogenic *Agrobacterium* acceptor.[47] There are also indications that the host range of different *A. tumefaciens* strains is determined by the type of Ti-plasmid they carry.[48] The plasmids from different strains have been characterized:

1. By measuring the length of circular molecules after spreading on a water hypophase by the Kleinschmidt technique[10]
2. By comparing fingerprints of plasmid DNAs digested with restriction endonucleases[18,26,28]

3. By DNA-DNA hybridization between different plasmids[29-31,32]
4. By electron-microscopic studies of heteroduplex molecules formed after denaturation and reannealing of different plasmids[20,33]

All these studies have established that both oncogenic and non-oncogenic *Agrobacteria* contain large plasmids with molecular weights ranging between 90 and 182 million. Several strains have been found to contain more than one such large plasmid and in several cases only one of the plasmids was associated with oncogenicity. Plasmids from non-oncogenic strains may either show little homology to the Ti-plasmids even though they code for nopaline utilization[28] or are largely homologous to some Ti-plasmid.[49]

Furthermore, the Ti-plasmids that have been studied fall in three classes: (1) "octopine" Ti-plasmids which code for octopine (and related opines such as lysopine, histopine, and probably agropine) metabolism; (2) "nopaline" Ti-plasmids which code for nopaline and ornaline metabolism;[50] and (3) Ti-plasmids that probably code for an as yet not identified "cryptopine".[51]

All the "octopine" type Ti-plasmids that have been studied thus far show a very high degree of homology (close to 100%) whereas different "nopaline" Ti-plasmids show much more diversification in terms of overall homology, restriction endonuclease fingerprints, and DNA rearrangements. The homology between these three classes of Ti-plasmid is limited to a few regions separated by large segments of nonhomologous DNA.[31,32] Recent work in our laboratory has established that genes involved directly or indirectly in oncogenicity are located in regions of homology (see Figure 4, Section II.D)

## B. Genetic and Physical Demonsration of DNA Transfer

As explained in the introduction, the first genetic evidence in favor of a DNA transfer model came from the demonstration that genes controlling opine synthesis in transformed plant cells were localized on the Ti-plasmid.[17-20]

The obvious way to explain the involvement of the Ti-plasmids both in oncogenicity and in control of opine synthesis was to assume that at least part of the Ti-plasmid was somehow transferred to the transformed plant cells where it was maintained and expressed. In order to test whether this assumption was correct, the following strategy was applied: first a genetic identification of which segment(s) of the Ti-plasmids was responsible for the control of oncogenicity and for the determination of opine synthesis; subsequently it was determined whether or not this DNA was present in the transformed plant cells.

To genetically identify the various functions determined by the Ti-plasmids combination of insertion and deletion mutants of the Ti-plasmids was used (see Section II.D). In particular the IncP-plasmid RP4 and the transposons Tn1 and Tn7[34,35] was used to produce insertions and deletions. Two such insertion mutants are of particular interest here. One of them is a cointegrate between the nopaline Ti-plasmid from strain C58 and the IncP-plasmid RP4 (pTiC58::RP4). This plasmid carries all the genetic markers of both parental molecules and these markers are transferred by conjugation as one unit. The pTiC58::RP4 cointegrate readily dissociates with exact separation of the genetic markers of pTiC58 and of RP4. The cointegrate state is non-oncogenic, whereas after dissociation the oncogenicity of pTiC58 is restored. These observations could be explained by assuming that RP4 reversibly integrated into a Ti-DNA segment that is essential for oncogenicity. By restriction endonuclease fingerprint analysis and by Southern gel blotting hybridizations[46] it was possible to localize the insertion site of RP4 on a physical map of the TiC58 plasmid and demonstrate that this insertion had

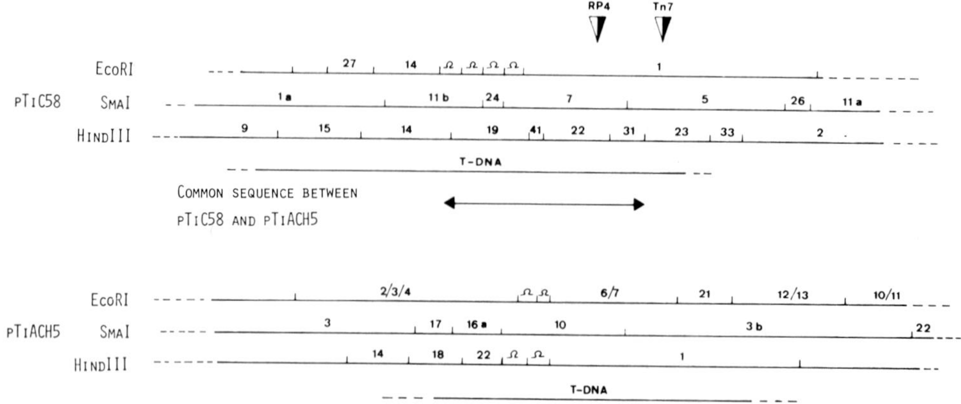


FIGURE 2. Schematic representation of the "common" regions and of the T-DNA of "nopaline" and "octopine" Ti-plasmids.[31] The detailed restriction map of the TiC58 Eco RI fragment 1 and the evidence for the precise localization of the RP4 integration site in the cointegrated TiC58::RP4 plasmid will be described elsewhere.[57] Arguments for the extent and localization of the octopine T-region and for the relevant restriction map were taken from Chilton et al.[21,32] and from our own data.

occurred in a DNA segment that is highly conserved in all Ti-plasmids.[31] These observations led to the postulation that this conserved or "common" DNA segment was indeed involved in oncogenicity.[20] These findings were corroborated and extended by Chilton and collaborators[32] who, furthermore, demonstrated that this "common DNA" overlapped with the segment of the Ti-plasmid from the octopine strain B6-806 that they had previously found to be present in a crown gall tissue induced with this strain.[21,22]

The other insertion mutant in the Ti-plasmid that is of particular interest here was obtained by the isolation of a large number of independent Tn7 insertions in the nopaline Ti-plasmids pTiC58 and pTiT37 (see Section II.D) and by screening strains harboring these Ti::Tn7 plasmids for the capacity to induce crown gall tumors in which no nopaline would be synthesized. Several such mutants were found and one of them was extensively studied. The rationale behind these experiments was as follows: if opine synthesis in transformed plant cells is indeed determined by Ti-plasmid genes and if, as expected, the synthesis of opines is independent from the control of the neoplastic growth in these plant cells, one would expect that Ti-plasmid mutants could be isolated that would still induce tumors but no opine synthesis. Furthermore one would predict that if DNA transfer from bacterium to plant did indeed occur and was responsible for the crown gall phenomenon, transformed plant cells should harbor precisely that segment of the Ti-plasmid that contains the genes necessary to specify opine synthesis.

All these predictions turned out to be correct. Again by restriction endonuclease fingerprint analysis and by heteroduplex analysis with electron microscope we were able to localize the Tn7 insertion that had mutated TiT37 to an ONC$^+$ NOS$^-$ phenotype (i.e., a plasmid able to induce tumors but no nopaline synthesis) on a fragment homologous to Hind III fragment 23 of pTiC58. A physical map of this region of the TiT37 and TiC58 plasmids was constructed and is presented in Figure 2. As can be seen from this map, both the integration of RP4 yielding a reversible ONC$^-$ phenotype (on Hind III fragment 22) and the insertion of Tn7 yielding an ONC$^+$ NOS$^-$ phenotype (on Hind III fragment 23) have occurred on the same segment of the Ti-plasmid, thus identifying this region of the Ti-plasmid as the probable T-DNA (i.e., Ti-DNA segment transferable to plants).

In subsequent experiments we were able to show that this segment of the Ti-plasmid

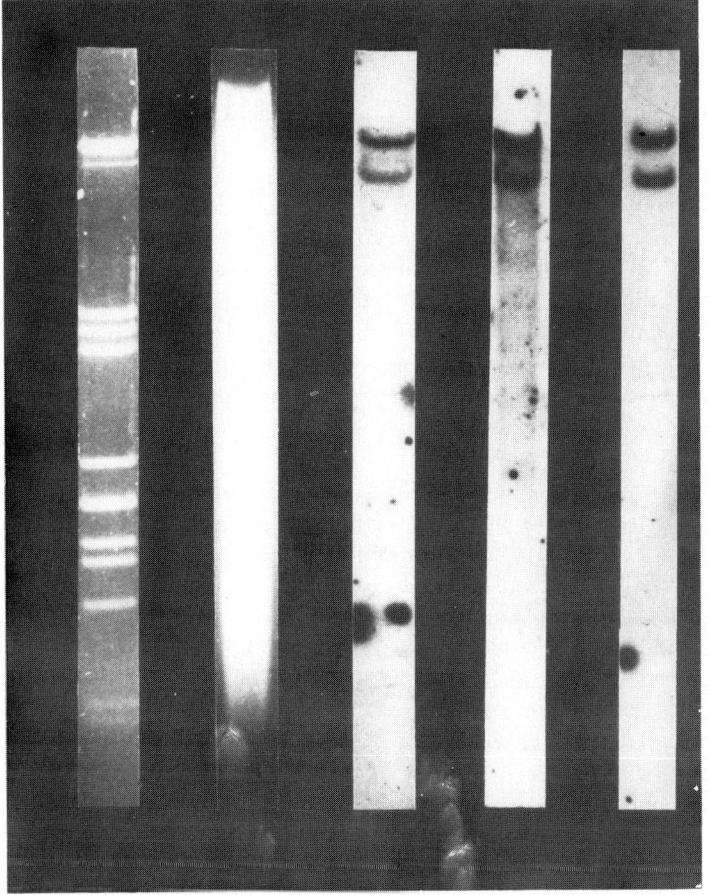


FIGURE 3. Gel blotting hybridization between (Southern[46]) total DNA extracted from a T37-induced tobacco teratoma tissue culture digested with restriction endonuclease Eco RI and radioactive probes consisting of the TiC58 Hind III fragments 22, 31, and 23 cloned in the vector pBR322. The procedure used was similar to that described in Reference 31. The left lane is a cold gel pattern of phage λ DNA digested with Pst I restriction endonuclease, used as markers for molecular weight determination of the fragments. The second lane from the left is the cold gel pattern of the Eco RI digested teratoma DNA. The last three lanes are the actual Southern gels respectively hybridized with the Hind III fragments 22, 31, and 23. The numbering of the Hind III fragments are those from the TiC58 plasmid. The TiT37 plasmid yields a somewhat different fragment pattern, but was shown to be completely homologous with TiC58 in the T-DNA region shown in Figure 2.

is indeed present in T37 transformed tobacco crown gall cells. We obtained a cloned teratoma culture, induced on tobacco (*Nicotiana tabacum* cv Havana) cells by strain T37, from Dr. A. C. Braun of the Rockefeller University.[36] DNA was prepared from this culture, digested with the restriction endonuclease Eco RI, and hybridized by a blotting technique[46] with the Hind III fragments 22, 31, and 23 from the nopaline plasmid C58. As can be seen in Figure 3, all three of these Ti fragments hybridized to the same Eco RI bands of the digested crown gall DNA. These experiments have been reported and briefly described previously.[23,37] Fragments 22 and 31 contain the Ti-

DNA that had previously been found to be highly conserved in Ti-plasmids[31,32] (see above), whereas fragment 23, which is immediately adjacent to this segment and was shown to specify nopaline synthesis in crown gall cells, presents homology only to other "nopaline" Ti-plasmids but not to "octopine" or "cryptopine" Ti-plasmids.[31,32]

These experiments demonstrate that the highly conserved or "common" DNA found in all Ti-plasmids is not only part of the T-DNA in octopine crown galls as shown by Chilton et al.[21,32] but also in nopaline crown galls and firmly establish that DNA transfer from bacteria to plants does occur and involves a specific segment of the Ti-plasmid (the T-DNA) that carries information controlling both oncogenicity and opine synthesis. Both these functions of the T-DNA are independent and appear to be of different evolutionary origin. Indeed, opine synthesis can be inactivated by an insertion mutation in fragment Hind III - 23 without affecting oncogenicity. Furthermore, the part of the T-DNA that controls opine synthesis is nonhomologous in different types of Ti-plasmids, whereas the part of the T-DNA specifically involved in oncogenicity (see Section II.D) appears to be common to all Ti-plasmids.

### C. The T-DNA of Ti-Plasmids

With T-DNA that segment of the Ti-plasmid that is present in transformed plant cells is designated. In the previous section the evidence establishing the existence and nature of the T-DNA has been summarized. However, some important questions are at present still unanswered or under study. (1) Is the T-DNA integrated in the plant DNA or does it form an independently replicating unit? The observations illustrated in Figure 3 indicate that the T-DNA is probably integrated. Indeed, the three Ti Hind III fragments (22, 31, and 23) that were used as probes and which form a contiguous segment of T-DNA hybridized to two fragments of the Eco RI digest of the crown gall DNA with molecular weights of, respectively, ± 6.5 and 9.5 megadaltons. No hybridization was observed with the Hind III fragment 33 which is immediately adjacent to fragment 23 (see Figure 3). Since there are no Eco RI sites either in the Hind III fragment 23 nor in fragment 33, the cut that delineates one end of the crown gall DNA fragments that contain the T-DNA must have occurred outside the T-DNA and hence probably in some plant DNA segment into which the T-DNA is presumably integrated. It is not certain whether the 6.5 and 9.5 Eco RI crown gall fragments that contain the T-DNA represent partial digests or two different sites of integration of the T-DNA in the plant DNA. The most likely explanation appears to be that we are dealing with two different sites of integration of the T-DNA in plant DNA. It is possible that a fraction of the transformed cell population has undergone a small deletion (of 3 megadaltons) of the plant DNA immediately adjacent to the T-DNA. An alternative explanation could be that the two fragments represent two different sizes of an independently replicating, nonintegrated T-DNA. In this case we would have to postulate that the right end (in Hind III-fragment 23) will be hooked up to the left end of the T-DNA. However, this seems unlikely because (a) no band was observed when undigested T37 crown gall DNA was hybridized with the same Hind III-22, 23, and 31 fragments, and (b) no unique fragment of either a Hind III or an Eco RI digest of the crown gall was found that would hybridize both to probes that contained the left end of the T-DNA (Hind III-fragments 9, 15, 14) and to a probe containing the right end of the T-DNA (Hind III-fragment 23). (2) Is the T-DNA integrated in the chromosomal or plastid DNA of the plant cells? No convincing evidence has been reported up to now to answer this question. (3) Is the T-DNA of a given Ti-plasmid of a unique size or can its size vary in different clones of transformed plant cells? Preliminary reports made by Merlo[52] indicate that the size of the T-DNA found in three independently derived tobacco crown gall lines is probably different. All three tumor lines contained a common portion of the Ti-plasmid and one of these tumor lines contained in addition a contiguous fragment of plasmid sequences. Merlo[52] also observed that differential amplifications of portions of the T-DNA appeared to have occurred to various extents in the different cell lines.

### D. The Genetic and Functional Organization of a Ti-Plasmid

Various functions such as oncogenicity, opine synthesis (in transformed plant cells) and catabolism, conjugation (see Section II.F), sensitivity to agrocin 84,[38] host range, and developmental properties of transformed cells (see Section II.A) are determined by the Ti-plasmids. In order to study the functional organization of the various Ti genes involved in these processes we isolated a large number of mutant Ti-plasmids inactivating one, or several, of these functions.

Mutants of the Ti-plasmids were obtained with the use of "transposons". Antibiotic resistance transposons insert in a target DNA at many different sites. They are, thus, efficient mutagens causing mutant phenotypes both by inserting into genes and by producing polar effects.[39] The sites of these insertions can easily be mapped by means of electron microscopic heteroduplex analysis and by analysis of the fingerprints obtained after digestion of the Ti-plasmids with various restriction endonucleases. The transposons Tn7 that confer trimethoprim and streptomycin/spectinomycin resistance[40] and Tn1 that confer ampicillin resistance were used in our studies. Insertions of Tn7 in the nopaline Ti-plasmid, TiC58, were isolated after conjugational transfer of the Ti-plasmid from a donor strain harboring a chromosomal Tn7 insertion to a plasmid-free receptor strain. Selection was made for receptor colonies showing the Tn7 coded resistance to streptomycin and spectinomycin. The only way the Tn7 genes could transfer from donor to receptor under these conditions was by insertion into the Ti-plasmid of the donor prior to conjugational transfer. Tn1 insertions in the TiC58 plasmid were isolated in an analogous way. The source of the Tn1 in this case was a deleted, nontransmissible derivative of the inc P-plasmid RP4 isolated in our laboratory. The Ti::Tn7 and Ti::Tn1 transconjugants that were thus isolated were tested for the expression of the Ti-plasmid coded phenotypes, nopaline catabolism, oncogenicity, nopaline production by tumors, and agrocin 84 sensitivity. For all these phenotypes mutations resulting from the Tn7 and Tn1 insertions were isolated.

The site of insertion of Tn7 or Tn1 in many of these mutant Ti-plasmids was mapped both by electron-microscopic heteroduplex analysis and by restriction endonuclease fingerprint analysis. As a result these Ti functions have been localized on the physical map of TiC58. The sites of the Tn1 and Tn7 insertions were hot spots for the origin of deletions. We thus succeeded in obtaining deletions in the TiC58 plasmid abolishing one or several of the known functions. These deletions have also been mapped on the physical map of TiC58 by electron-microscopic heteroduplex analysis and by restriction endonuclease fingerprint analysis. All these results have been summarized in Figure 4.

The main conclusions from this work can be summarized as follows:

1. The genes controlling nopaline synthesis (NOS) in the transformed plant cells and the genes determining the catabolism of nopaline by agrobacteria (NOC) are functionally different and localized on different but adjacent segments of the Ti-plasmid. It is remarkable that insertion and deletion mutants inactivating nopaline catabolism ($NOC^-$) have been mapped in different sites covering a fairly large region of the Ti-plasmid (±7% of the total map). This would indicate that genes controlling nopaline catabolism are part of a fairly large transcription unit and can easily be inactivated by polar effects.

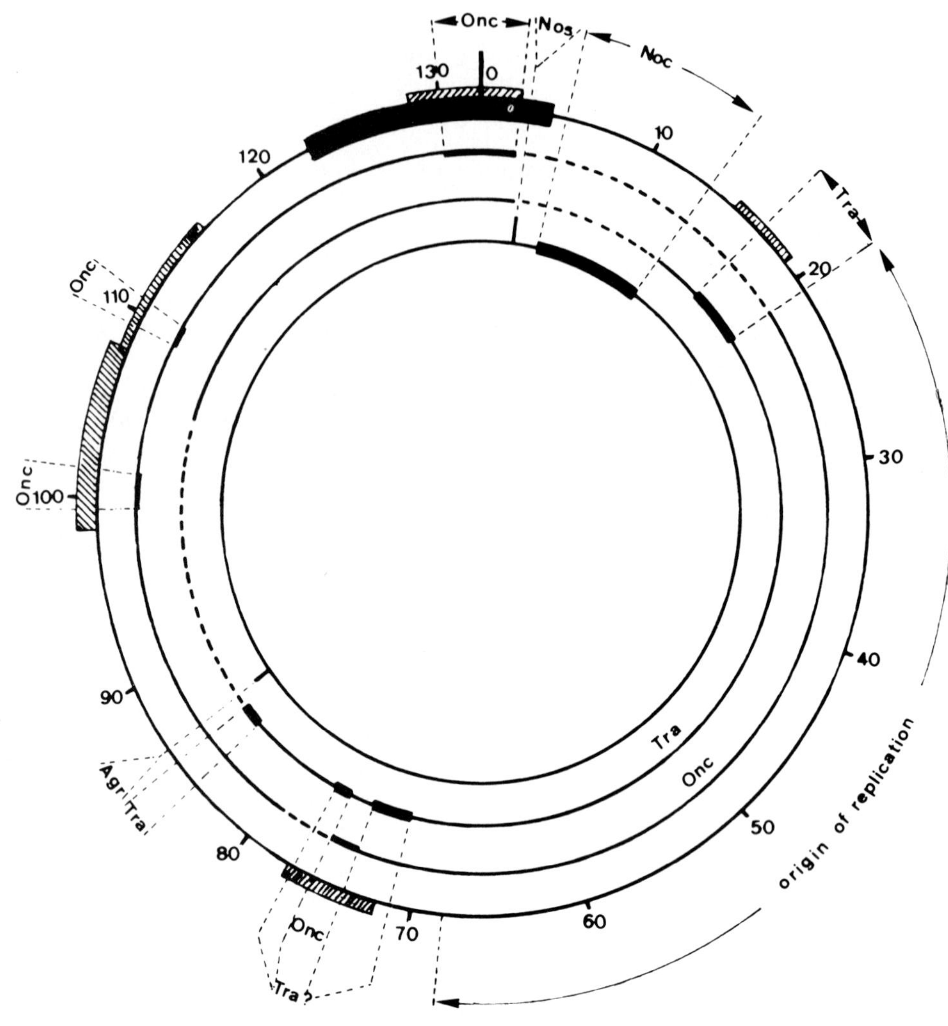


FIGURE 4. Functional organization of the "nopaline" plasmid from strain C58. Schematic representation of the localization of the various functions known to be determined by pTiC58. The regions of homology were determined by gel blotting[46] hybridizations between pTiC58 and pTiAch5 as described elsewhere.[31] Tn7, Tn1, and RP4 insertions, inactivating one or the other of the known functions, were mapped by electron-microscopic analysis of restriction endonuclease digest fingerprints of the various insertion and deletion mutants. For further explanations, see Section II.D. Regions that can be deleted without affecting oncogenicity have been indicated on the circle called ONC. Regions that can be deleted without affecting conjugational transfer have been indicated on the circle called TRA.

2. Genes controlling oncogenicity (ONC) either directly or indirectly have been mapped in at least two different regions of the Ti-plasmid. One of these regions overlaps with the Ti segment previously identified as being part of the T-DNA (see Sections II.B and II.C) since it covers the Hind III fragments 22 and 31. It is, therefore, very likely that this region is directly involved in the establishment and/or maintenance of the tumorous growth pattern of transformed plant cells.

The fact that other regions of the Ti-plasmid carry genes essential for oncogenicity strongly points to the possibility that several sets of genes of the Ti-plasmid are involved in the transfer of Ti-plasmid DNA from bacteria to plant and possibly in the integration of the T-DNA in the plant DNA (see Section II.E).

3. At least two and possibly three different regions of the Ti-plasmid carry genes (TRA) involved in the conjugational transfer of the Ti-plasmid from one *Agrobacterium* to another. Up to now we have not found insertion mutants that simultaneously inactivate conjugation and oncogenicity. This would indicate that the mechanisms involved in the transfer of Ti-plasmids between bacteria (TRA) and between bacterium and plant (ONC) are largely, if not entirely, independent. Interestingly, one of the TRA regions is located in a segment adjacent to the genes involved in nopaline catabolism (NOC). This location of TRA genes fits with our observations[41] that the conjugative properties of Ti-plasmids are normally repressed and can be induced by opines, and that the regulation of the opine catabolism genes and some of the TRA genes involves a common repressor.
4. The genes controlling agrocin 84 sensitivity (AGR) are localized in the vicinity of one of the TRA regions. This might indicate that this region of the Ti-plasmid contains genes determining some cell wall structures of *A. tumefaciens*.

In Figure 4 we have indicated the various regions that show base sequence homology between the nopaline plasmid TiC58 and octopine-type Ti-plasmids. These studies were performed in our laboratory by hybridizing cloned fragments of the TiC58 plasmid to an octopine-type Ti-plasmid using the gel blotting technique of Southern.[31,32] A very striking picture emerges from these studies. Three different types of DNA sequences can be identified in this Ti-plasmid:

1. A highly conserved DNA sequence, the so-called "common DNA", that was found to be present in all Ti-plasmids and that is present in transformed plant cells (in other words, it was found to be part of the T-DNA)[31,32] (see also Section II.B). This segment probably contains the genes directly involved in the establishment and/or maintenance of the tumorous growth pattern of transformed plant cells.
2. Several regions that are at least partially homologous between octopine and nopaline Ti-plasmids. In all of these regions mutants were found that inactivate oncogenicity and/or transfer functions.
3. In between these regions of homology, we found that the TiC58 contains very large DNA segments that are partially or completely homologous to other nopaline-type plasmids but show no homology with octopine-type plasmids. Perhaps the most striking of these regions is the one immediately to the right of the "common DNA" sequence. Both in octopine Ti-plasmids[21] and in nopaline Ti-plasmids[53] this segment was found to be part of the T-DNA. Furthermore, we have demonstrated that this segment contains the genes specifying nopaline synthesis in transformed plant cells (see Section II.B).

One must, therefore, conclude that Ti-plasmids are of mixed origin; the genes directly or indirectly involved in oncogenicity and most of the genes involved in conjugational transfer appear to be of common origin. In view of the incompatibility between octopine and nopaline Ti-plasmids it is likely that genes and DNA sequences involved in replication are of common origin. On the oher hand the genes involved in opine synthesis (in transformed plant cells) and in opine catabolism are evidently nonallelic and of different origin. The T-DNA apparently consists of functionally and evolutionarily different parts. One part, which is common to all Ti-plasmids, is directly involved in oncogenicity. Another part, which is covalently linked to the first one, is involved in opine synthesis. At least two, and probably three, different opine-specifying DNA sequences exist. They must have evolved independently and somehow got linked up with the common DNA sequence that is directly involved in oncogenicity. A strong argument in favor of this model was provided by a deletion mutant of TiC58. In this deletion most, if not all, of the NOC region and also most, if not all, of the noncommon part of the right end of the T-DNA is lost. A C58 strain carrying this deleted plasmid is still able to induce tumors, although these tumor cells do not synthesize nopaline. The insertion of a Tn7 transposon in this noncommon part of the T-DNA (in the Hind III fragment 23) resulted in a strain able to induce tumors without nopaline synthesis. These observations convincingly show that this noncommon part of the T-DNA is involved in opine synthesis and is not essential for oncogenicity. The left end of the T-DNA also consists of DNA that is not homologous in octopine and nopaline Ti-plasmids. The function of this part of the T-DNA is not yet understood.

## E. Speculations and Facts about the Mechanisms Involved in Oncogenicity

In the previous section the evidence indicating that several regions of the Ti-plasmid apparently specify functions essential for oncogenicity was described. One can speculate about these functions if one considers the various steps that have to occur in order to bring about the stable transformation. A first step could be the site of attachment of the oncogenic bacteria to receptor sites on the walls or membranes of susceptible plant cells. The Lippincotts[42,43] have described much of the evidence in favor of such a step. It was also reported[44] that the Ti-plasmid was involved in the determination of structures, on the bacterial cell wall, which are essential to adherence. One of the ONC regions of the Ti-plasmid might play a role here.

Since the bacterial cell wall is involved in this attachment process, it is very likely that some chromosomal genes of *Agrobacterium* will also play a role in this step.

A second step would probably be the transfer of some, or the whole, of the Ti-plasmid from bacterium to plant. To our knowledge no experiments were reported to study this point. We think it likely that this step will be controlled by one of the ONC regions genetically identified on the Ti-plasmid, and that subsequently the T-DNA of the Ti-plasmid somehow gets integrated in the plant DNA. At the present time the mechanism of integration is not known. How many sites are available for integration or whether the integration takes place in plastid or in chromosomal DNA is also not known.

Finally, the T-DNA must be expressed. Direct evidence for transcription of the T-DNA in crown gall cells was provided by Drummond et al.[22] and Kemp and Ledeboer.[54] It is not clear yet how much and by which mechanism this transcription occurs. Does the T-DNA contain the right signals for m-RNA initiation and processing or are these signals provided by the plant DNA segment into which the T-DNA integrated? These questions are, at present, unresolved.

## F. The Ti-Plasmids as Conjugative Catabolic Plasmids

The first indications that the Ti-plasmids are conjugative plasmids, i.e., plasmids which can promote their conjugational transfer from a donor to a receptor bacterium, came from the observation[13] that the *in planta* transfers of oncogenicity observed by Kerr[15,16] were due to Ti-plasmid transfers. However, these transfers could be detected only after several weeks of mixing donor and acceptor bacteria on crown galls, and the mechanism of this transfer was therefore unclear and the subject of many speculations.

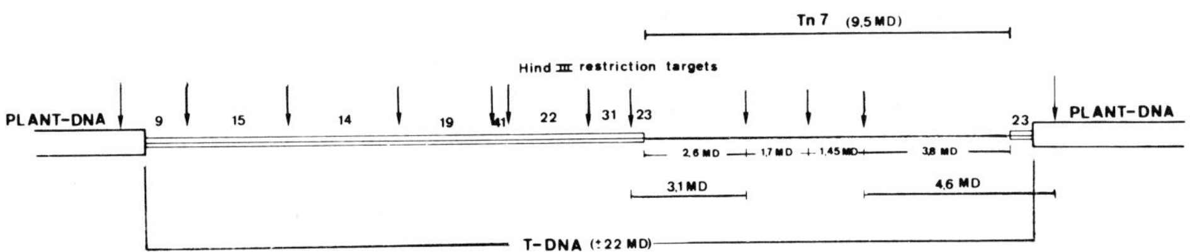


FIGURE 5. Schematic representation of the structure of the TiT37 NOS⁻::Tn7 DNA present in the tobacco crown gall tissue induced with pTiT37 NOS⁻::Tn7. The evidence is based on gel blotting[46] hybridizations similar to those shown in Figure 3. The crown gall DNA was digested with Hind III and Eco RI restriction endonucleases and hybridized with radioactive probes consisting of Col E1::Tn7 DNA and pBR322 + TiC58 T-DNA.

Recently, an explanation was found for the *in planta* transfers when it was discovered that the conjugative properties of Ti-plasmids are normally repressed on artificial laboratory growth media but can be induced by opines.[18,25] In the case of octopine Ti-plasmids, octopine was found to be an efficient inducer of conjugative activity. A further study of the regulation of the expression of the conjugative properties of these plasmids revealed that a common regulator was controlling the expression of both the genes involved in octopine catabolism and the TRA genes involved in conjugation.[41] The Ti-plasmids must, therefore, be seen as catabolic plasmids (i.e., plasmids that carry genetic information to allow their host to perform some uncommon catabolic activity) and give us a first example of conjugative activity in bacteria completely dependent on induction by the sustrate. Recently, the catabolic nature of Ti-plasmids was further illustrated by the finding[55] that Ti-plasmids also control some steps in the degradation of arginine.

## III. GENETIC ENGINEERING WITH Ti-PLASMIDS

With the realization that the Ti-plasmid is in fact a natural vector able to promote transfer, integration, and expression of foreign DNA in plants, it became evident that the question should be asked whether or not the properties of this system could be used to introduce, at will, foreign DNA into various plants.[45] (See the chapters by Bingham and Hull.)

The general strategy used to test this possibility was to incorporate, by either in vivo or in vitro methods, genes and DNA sequences into the transferable DNA segment (the T-DNA) of the Ti-plasmid and subsequently to test whether the added DNA is also transferred to the transformed plant cells. As described in Sections II.B and II.D, we were able to introduce the transposon Tn7 into various segments of Ti-plasmids. In particular, one insertion mutant was of interest because the insertion had produced a Ti-plasmid able to induce tumors but no nopaline synthesis.

As described in Section II.D, we were able to demonstrate that the Ti DNA segment into which this insertion had occurred (the Hind III fragment 23) was part of the T-DNA region. As illustrated in Figure 5, we thus had inserted a group of genes, with a total molecular weight of ± 9.6 megadalton, into the T-DNA of the TiT37 plasmid. Since strains carrying this plasmid were still able to induce tumors, this plasmid was an obvious choice to test the capacity of Ti-plasmids to introduce foreign DNA into plants.

Recent results in our laboratory have confirmed that this is the case. Indeed the DNA from crown galls induced by the TiT37 NOS⁻::Tn7 plasmid on tobacco was shown to contain the Tn7 DNA. This was achieved by digesting the crown gall DNA with the restriction endonuclease Hind III and hybridizing the fragments with a radioactive probe DNA consisting of Col E1::Tn7 DNA using the Southern gel blotting technique.[46] All the expected fragments of the Tn7 DNA were thus found to be present in the transformed plant DNA as illustrated in Figure 5. By hybridizing the same Hind III digest against a radioactive probe containing the Hind III fragment 23 of the T-DNA, we were similarly able to demonstrate that the Tn7 DNA, present in the plant DNA, was still integrated in the Hind III-23 fragment of the T-DNA. These experiments thus provided the first evidence that we had succeeded in creating plant cells containing a specific set of genes that we had chosen to introduce in the plant cells via the Ti-plasmid and demonstrated the usefulness of the Ti-plasmid as vectors for genetic engineering of plants.

One might object that it is of little practical value to engineer plant tumors. However, recent observations, originally made by Braun and Wood,[36] have demonstrated that whole plants can be regenerated from transformed teratoma cultures. In these studies, tobacco teratoma-derived tumor shoots were isolated and grafted to cut stem tips of normal tobacco plants of a morphologically distinct cultivar. This way shoots were obtained that developed quite normally and ultimately flowered and set viable seed. It was found that the leaves of these grafts were normally organized but were still transformed in the sense that when such specialized cells were isolated and planted on a basic culture medium, they grew as crown gall cells and synthesized nopaline. Gordon and collaborators[56] and also our group in Gent studied these tissues and we found that they still contained T-DNA from the TiT37 plasmid, thus confirming that these plants were still transformed. Interestingly, the T-DNA appeared to be lost by meiosis. Indeed, haploid tissue obtained from the anthers of the regenerated plants as well as tissues obtained from plants grown from the seeds of the regenerated plants required hormones for growth in vitro and did not contain any T-DNA.

Although many questions remain, hope is justified that new plants could be created by genetic engineering using the Ti-plasmid, or a vector derived from it, to introduce the chosen genetic information in the plants.

## REFERENCES


1. **Smith, E, F., Brown, N. A., and Townsend, C. O.,** Crown-gall of plants: its cause and remedy, *U.S. Dep. Agric. Bur. Plant Ind. Bull.*, 213, 1, 1911.
2. **Braun, A. C. and White, P. R.,** Bacteriological sterility of tissues derived from secondary crown-gall tumors, *Phytopathology*, 33, 85, 1943.
3. **Braun, A. C.,** Thermal studies on the factors responsible for tumor initiation in crown-gall, *Am. J. Bot.*, 34, 234, 1947.
4. **Braun, A. C. and Mandle, R. J.,** Studies on the inactivation of the tumor-inducing principle in crown-gall, *Growth*, 12, 255, 1958.
5. **Petit, A., Delhaye, S., Tempé, J., and Morel, G.,** Recherches sur les guanidines des tissus de crown gall. Mise en évidence d'une relation biochimique specifique entre les souches d'*Agrobacterium* et les tumeurs qu'elles induisent, *Physiol. Vég.*, 8, 205, 1970.
6. **Lippincott, J. A., Beiderbeck, R., and Lippincott, B. B.,** Utilization of octopine and nopaline by *Agrobacterium*, *J. Bacteriol.*, 116, 378, 1973.
7. **Kerr, A. and Roberts, W. P.,** *Agrobacterium:* correlations between and transfer of pathogenicity, octopine and nopaline metabolism and bacteriocin 84 sensitivity, *Physiol. Plant Pathol.*, 9, 202, 1976.
8. **Kemp, J. D.,** A new amino acid derivative present in crown-gall tumor tissue, *Biochem. Biophys. Res. Commun.*, 74, 862, 1977.
9. **Firmin, J. L.,** personal communication, 1978.
10. **Zaenen, I., Van Larebeke, N., Teuchy, H., Van Montagu, M., and Schell, J.,** Supercoiled circular DNA in crown-gall inducing *Agrobacterium* strains, *J. Mol. Biol.*, 86, 109, 1974.

11. **Schell, J.,** The role of plasmids in crown gall formation by *A. tumefaciens*, in *Genetic Manipulations with Plant Material*, Ledoux, L., Ed., Plenum Press, New York, 1975, 163.
12. **Van Larebeke, N., Engler, G., Holsters, M., Van Den Elsacker, S., Zaenen, I., Schilperoort, R. A., and Schell, J.,** Large plasmid in *Agrobacterium tumefaciens* essential for crown gall-inducing ability, *Nature (London)*, 252, 169, 1974.
13. **Van Larebeke, N., Genetello, C., Schell, J., Schilperoort, R. A., Hermans, A. K., Hernalsteens, J. P., and Van Montagu, M.,** Acquisition of tumor-inducing ability by non-oncogenic agrobacteria as a result of plasmid transfer, *Nature (London)*, 255, 742, 1975.
14. **Watson, B., Currier, T. C., Gordon, M. P., Chilton, M. D., and Nester, E. W.,** Plasmid required for virulence of *Agrobacterium tumefaciens*, *J. Bacteriol.*, 123, 255, 1975.
15. **Kerr, A.,** Transfer of virulence between isolates of *Agrobacterium*, *Nature (London)*, ??3, 1175, 1969.
16. **Kerr, A.,** Acquisition of virulence by non-pathogenic isolates of *Agrobacterium radiobacter*, *Physiol. Plant Pathol.*, 1, 241, 1971.
17. **Bomhoff, G., Klapwijk, P. M., Kester, H. C. M., Schilperoort, R. A., Hernalsteens, J. P., and Schell, J.,** Octopine and nopaline synthesis and breakdown genetically controlled by a plasmid of *Agrobacterium tumefaciens*, *Mol. Gen. Genet.*, 145, 177, 1976.
18. **Genetello, C., Van Larebeke, N., Holsters, M., Depicker, A., Van Montagu, M., and Schell, J.,** Ti plasmids of *Agrobacterium* as conjugative plasmids, *Nature (London)*, 265, 561, 1977.
19. **Montoya, A. L., Chilton, M. D., Gordon, M. P., Sciaky, D., and Nester, E. W.,** Octopine and nopaline metabolism in *Agrobacterium tumefaciens* and crown-gall tumor cells: role of plasmid genes, *J. Bacteriol.*, 129, 101, 1977.
20. **Schell, J. and Van Montagu, M.,** Transfer, maintenance and expression of bacterial Ti-plasmid DNA in plant cells transformed with *A. tumefaciens*, *Brookhaven Symp. Biol.*, 29, 36, 1977.
21. **Chilton, M. D., Drummond, H. J., Merlo, D. J., Sciaky, D., Montoya, A. L., Gordon, M. P., and Nester, E. W.,** Stable incorporation of plasmid DNA into higher plant cells: the molecular basis of crown gall tumorigenesis, *Cell*, 11, 263, 1977.
22. **Drummond, M. H., Gordon, M. P., Nester, E. W., and Chilton, M. D.,** Foreign DNA of bacterial plasmid origin is transcribed in crown gall tumours, *Nature (London)*, 269, 535, 1977.
23. **Schell, J., Van Montagu, M., De Beuckeleer, M., De Block, M., Depicker, A., De Wilde, M., Engler, G., Genetello, C., Hernalsteens, J. P., Holsters, M., Seurinck, J., Silva, A., Van Vliet, F., and Villarroel, R.,** Interactions and DNA transfer between *Agrobacterium tumefaciens*, the Ti-plasmid and the plant host, *Proc. R. Soc. London, Ser. B*, 204, 251, 1979.
24. **Chilton, M. D., Farrand, S. K., Levin, R., and Nester, E. W.,** RP4 promotion of transfer of a large *Agrobacterium* plasmid which confers virulence, *Genetics*, 83, 609, 1976.
25. **Kerr, A., Manigault, P., and Tempé, J.,** Transfer of virulence *in vivo* and *in vitro* in *Agrobacterium*, *Nature (London)*, 265, 560, 1977.
26. **Van Larebeke, N., Genetello, C., Hernalsteens, J. P., Depicker, A., Zaenen, I., Messens, E., Van Montagu, M., and Schell, J.,** Transfer of Ti plasmids between *Agrobacterium* strains by mobilisation with conjugative plasmid RP4, *Mol. Gen. Genet.*, 152, 119, 1977.
27. **Hernalsteens, J. P., Engler, G., Van Larebeke, N., Van Montagu, M., and Schell, J.,** Studies on large DNA plasmids of *Agrobacterium tumefaciens*, *Arch. Int. Physiol. Biochim.*, 83, 368, 1975
28. **Sciaky, D., Montoya, A. L., and Chilton, M. D.,** Fingerprints of *Agrobacterium* Ti plasmids, *Plasmid*, 1, 238, 1978.
29. **Currier, T. C. and Nester, E. W.,** Evidence for diverse types of large plasmids in tumour-inducing strains of *Agrobacterium*, *J. Bacteriol.*, 126, 157, 1976.
30. **Merlo, D. and Nester, E. W.,** Plasmids in avirulent strains of *Agrobacterium*, *J. Bacteriol.*, 129, 76, 1977.
31. **Depicker, A., Van Montagu, M., and Schell, J.,** Homologous DNA sequences in different Ti-plasmids are essential for oncogenicity, *Nature (London)*, 275, 150, 1978.
32. **Chilton, M. D., Drummond, M. H., Merlo, D. J., and Sciaky, D.,** Highly conserved DNA of Ti plasmids overlaps T-DNA maintained in plant tumours, *Nature (London)*, 275, 147, 1978.
33. **Engler, G., Villarroel, R., Van Montagu, M., and Schell, J.,** in preparation.
34. **Holsters, M., Silva, A., Genetello, C., Engler, G., Van Vliet, F., De Block, M., Villarroel, R., Van Montagu, M., and Schell, J.,** Spontaneous formation of cointegrates of the oncogenic Ti-plasmid and the wide-host-range P-plasmid RP4, *Plasmid*, 1, 456, 1978.
35. **Hernalsteens, J. P., De Greve, H., Van Montagu, M., and Schell, J.,** Mutagenesis by insertion of the drug resistance transposon Tn7 applied to the Ti plasmid of *Agrobacterium tumefaciens*, *Plasmid*, 1, 218, 1978.
36. **Braun, A. C. and Wood, H. N.,** Suppression of the neoplastic state with the acquisition of specialized functions in cells, tissues, and organs of crown-gall teratomas of tobacco, *Proc. Natl. Acad. Sci. U.S.A.*, 73, 496, 1976.

37. **Belgian Crown-gall Research Group,** Transfer of genes into plants via the Ti-plasmid of *A. tumefaciens*, in *Gene Function*, Vol. 51, Rosenthal, S., et al., Eds., Proc. 12th FEBS Meeting, Pergamon Press, Oxford, 1978, 521.
38. **Engler, G., Holsters, M., Van Montagu, M., Schell, J., Hernalsteens, J. P., and Schilperoort, R. A.,** Agrocin 84 sensitivity: a plasmid determined property in *Agrobacterium tumefaciens*, *Mol. Gen. Genet.*, 138, 345, 1975.
39. **Kleckner, N.,** Translocatable elements in procaryotes, *Cell*, 11, 11, 1977.
40. **Barth, P. T., Datta, N., Hedges, R. W., and Grinter, N. J.,** Transposition of a deoxyribonucleic acid sequence encoding trimethoprim and streptomycin resistances from R483 to other replicons, *J. Bacteriol.*, 125, 800, 1976.
41. **Petit, A., Tempé, J., Kerr, A., Holsters, M., Van Montagu, M., and Schell, J.,** Substrate induction of conjugative activity of *Agrobacterium tumefaciens* Ti plasmids, *Nature (London)*, 271, 570, 1978.
42. **Lippincott, J. A. and Lippincott, B. B.,** Morphogenetic determinants as exemplified by the crown-gall disease, *Physiol. Plant Pathol.*, 4, 356, 1976.
43. **Lippincott, J. A. and Lippincott, B. B.,** Cell walls of crown-gall tumors and embryonic plant tissues lack *Agrobacterium* adherence sites, *Science*, 199, 1075, 1978.
44. **Whatley, M. H., Margot, J. B., Schell, J., Lippincott, B. B., and Lippincott, J. A.,** Plasmid or chromosomal determination of *Agrobacterium* adherence specificity, *J. Gen. Microbiol.*, 107, 395, 1978.
45. **Schell, J. and Van Montagu, M.,** The Ti plasmid of *Agrobacterium tumefaciens*, a natural vector for the introduction of *Nif* genes in plants?, in *Genetic Engineering for Nitrogen Fixation*, Hollaender, A., Ed., Plenum Press, New York, 1977, 159.
46. **Southern, E. M.,** Detection of specific sequences among DNA fragments separated by gel electrophoresis, *J. Mol. Biol.*, 98, 503, 1975.
47. **Hernalsteens, J. P. and Schilperoort, R.,** unpublished.
48. **Loper, J. E. and Kado, C.,** personal communication.
49. **Depicker, A., Van Montagu, M., and Schell, J.,** unpublished data.
50. **Kemp, J.,** personal communication.
51. **Tempe, J.,** personal communication.
52. **Merlo, D.,** personal communication.
53. **Depicker, A., Van Montagu, M., and Schell, J.,** unpublished.
54. **Kemp, J. and Ledeboer, A.,** personal communication.
55. **Petit, A., Tempe, J., Kerr, A., and Ellis, J.,** personal communication.
56. **Gordon, M.,** personal communication.
57. **De Wilde, M. et al.,** in preparation.

# PROSPECTS FOR NOVEL GENETIC MODIFICATION IN PLANTS USING SEXUAL AND SOMATIC CELL METHODS

**E. T. Bingham**

## TABLE OF CONTENTS



## I. INTRODUCTION

There is much interest currently in novel approaches to genetic modifications of plants owing to the potential applications in crop improvement and food production. Several factors support optimism that accelerated progress in genetic modification will be realized. Foremost is power of traditional breeding methods for directed genetic modification. These methods already have a distinguished record of accomplishment.

Essentially all important food, forage, and fiber crops have been bred for high levels of not only yield and quality, but resistance to many diseases and insect pests. Additionally, in many crops unique genetic modifications have been made to permit mechanical harvest (tomato and many other vegetable and fruit crops), improve natural seed drying (corn), nutritional quality (high lysine corn, high protein oats, and increased levels of vitamin A in carrots), photo-insensitive flowering for wide adaptation (wheat, rice, and soybeans), adaptation to very specialized soil and climatic conditions (alfalfa and many other field crops), and increased animal digestibility (forage grasses). Modifications are sometimes even made for esthetic reasons (burpless cucumbers, transparent sweet corn silks which are inconspicuous in the processed corn, and many types of seedless characteristics).

The result is that as many as 10 to 15 qualitative (single gene) and 5 to 10 quantitative (several gene) traits may be bred into modern cultivars. This is usually accomplished by starting with improved parents, which may already possess all but two or three of the required qualitatiive and all but one of two of the quantitative traits. Thus, much crop improvement can be made using traditional breeding and these methods are becoming more powerful each year through a deeper understanding of the genetics, cytogenetics, physiology, and taxonomy of crop and related species.

To insure that genetic resources are available to maintain future breeding programs and hopefully prevent genetic vulnerability to diseases and insects, most developed countries have had national programs to find and maintain genetic diversity of major crop species and their wild relatives. Such programs are now being undertaken at an international level and involve both developed and underdevelped nations because of the recent awareness that plant genetic resources are being lost due to destruction of wild plant habitat by expanding human populations and elimination of many diverse native cultivars due to extensive cultivation of "green revolution" wheat and rice cultivars.

Added to the solid base of traditional breeding methods are the prospects of exciting new cellular approaches to plant modification suggested by recent advances in the culture of plant tissues, cells, and protoplasts. Advancements include regeneration of whole plants from cells in culture of several important crop species, reported success of variant selection at the cellular level in a few species, and somatic cell hybridization and recovery of hybrid plants in *Nicotiana* and in *Petunia*. Variant selection at the cellular level should be efficient for traits associated with the biochemistry of the cells. Research is already underway to identify cell variants resistant to toxic compounds, such as herbicides, and to identify overproducers of amino acids. Further, prospects appear good to combine plant cell culture and genetic engineering through recombinant DNA techniques.

Somatic hybridization followed by regeneration of the hybrid cells into whole plants has thus far only been accomplished with species which will hybridize sexually, but holds the potential for producing hybrids beyond the limits of sexual compatibility. This would represent unprecedented germplasm transfer, which might in turn be exploited in plant breeding.

Some whole plant and cellular approaches to genetic modification, such as the induction of mutations, appear to be competitive alternatives for traits in which cellular and whole plant phenotypes are similar. In the vast majority of characters associated with crop improvement, however, idealized selection schemes appear to apply to either the whole plant or to the cell, but seldom to both. For example, in corn and in most crop species, morpholgical traits associated with plant yield such as leaf angle, ear number, kernels per row, and number of kernel rows are not biochemically identified and probably will continue to require whole plant selection techniques. Some biochemical processes which take place in the storage organ also may not lend themselves to cellular screening, but this will have to be determined by research.

On the other hand, procedures such as identification of overproducers of amino acids, somatic hybridization, transformation, and genetic engineering of DNA appear best suited for selection at the cellular level. Thus, genetic modification using cells offers the potential of increased efficiency and the opportunity to use microbial selection procedures for biochemically identified characters.

In the long run, after the potentials of the cellular approach have been tested, whole plant and cellular methods will surely be complementary. As long as plants are to be grown in soil and in outdoor ecosystems, new cultivars or existing ones with only one gene altered by either sexual or somatic methods will require screening and testing in the conventional manner. The most efficient methods for desired modifications will be determined as they always have been, by processes akin to natural selection.

Broad conceptual changes could demand accelerated effort with the most appropriate method. For example, in the recent past most breeding has been done under optimized growth conditions for the plant. This included irrigation, optimum fertilizer, and chemical control of some weeds, insects, and diseases. In other words, we have modified the environment to suit the plant. Seemingly inevitable constraints on uses of energy and chemicals suggest breeding should also be done to suit the plant to the environment.

This chapter represents the attempt of a plant breeder who uses both sexual and somatic methods to integrate the recent advances in several areas into the total package of crop improvement. Prospects for novel modification using both whole plant and cellular systems will be discussed and alternatives compared.

## II. MANIPULATION OF SINGLE MUTANT GENES

### A. Introduction

The concept of a single gene will be discussed for simplicity, although expression of some desired traits may require simultaneous manipulation of both structural and regulatory genes. Where an economically useful character is concerned, the plant breeding procedures required to incorporate it into a cultivar will be the same whether it was obtained using cellular methods or whole plants.

In self-pollinating species, it is most efficient to isolate the desired gene or produce it by mutation in an existing cultivar which possesses all the other necessary yield, quality, and resistance traits. If the gene must be isolated in a different line and then transferred to the cultivar, 6 to 8 generations of backcrossing (about 3 to 4 years) will be required. In cross-pollinating species such as corn or in selfers such as sorghum and wheat where hybrids may be desired, several sexual generations over 3 to 4 years will always be required to transfer a given gene into two or more parents of the hybrid cultivar.

In cross-pollinated species such as alfalfa and most forage grasses where the cultivar is produced by intercrossing from four to several hundred parents, transfer of the gene to all parents may still be done in 3 to 4 years, but will obviously require more work. Thus, in most cases, mutant isolation or gene transfer are in themselves not means to an end, but one step in the total breeding procedure, which is then followed by two to several years of regional testing before the new cultivar is released.

### B. Cellular Approach

Advantages of the cellular approach for the selective identification of certain classes of spontaneous and induced mutations have now been extensively reviewed.[1-5] The ideal situation is thought to be where callus can be dispersed and grown in liquid medium to produce a suspension culture of rapidly dividing, chromosomally stable, single cells or small aggregates, which when plated at a density to minimize cross feeding will divide and produce a callus which will regenerate into a whole plant.

The suitability of such cells to microbial mutagenesis and selective screening techniques is evident. They can be exposed to agents of stress, nutritional analogues, and toxic substances. The cells which survive potentially contain the beneficial traits, and hopefully can be selectively isolated as whole plants. Moreover, the cellular system insures that the desired mutant is carried in all cells of the regenerated plant. Heritability of spontaneous or induced mutants in seeds or whole plants requires their origin in a sporogenous cell layer.

The argument for using cells in selection regimes is compelling, but their use is limited to species for which regeneration techniques are known. In species which have not been regenerated or where the efficiency is low, the use of excised embryo or even seedling may prove to be a workable alternative.[6] In cases where the desired trait may exist as part of the natural variation in the species and where it must be identified at the cellular level (as may be the case for some biochemical traits), cellular testing and conventional breeding strategy could be combined. Callus and cell cultures could be established from seedlings which would be maintained for the duration of the cellular test and then kept or discarded on the basis of the cellular phenotype. Thus, we may not be limited to idealized systems for certain types of cellular research.

Higher plants with the haploid or gametic chromosome number occur rarely when a cell in a gametophyte develops directly into a plant. The term haploid is synonymous with monoploid when dealing with diploid species. Monoploid plant cell cultures would appear most suitable for microbial methods of variant selection, since all mutations, recessive or dominant, can be detected. Historically, the majority of induced mutations have been recessive, suggesting that monoploids should be used whenever possible in variant selection. Monoploid cell cultures are being used in *Petunia*[7] and *Nicotiana sylvestris*,[8,9] and monoploids are available in several other species, including *Datura*,[10] *Nicotiana* spp,[10] barley, rice, maize, and several other crop species.[11,12]

Many crop species are polyploids, e. g., tobacco, cotton, wheat, oats, potato, alfalfa, and many grasses, in which case the haploid plants actually have two or three monoploid sets of chromosomes and possess some degree of genetic redundancy. There are several reports of successful variant selection, however, using cells from such haploids[13-16] and cells from diploids.[17] Several of these variants have been apparent overproducers of amino acids[14,17] and the mutations involved may have been dominants. In yeast, regulation of overproduction of amino acids is a dominant trait.[18] Such dominant plant genes would presumably be expressed in diploid or polyploid cells and may be isolated using appropriate selective procedures.

The powerful selection regimes used in microorganisms to obtain regulatory mutants may serve as a goal in higher plant work. In bacteria, mutants which make several times the normal amount of certain amino acids may be selected using amino acid analogues.[19,20] Such regulatory mutants for overproduction of nutritionally important compounds are selected with relative ease in bacteria.[21] The regulatory controls altered include feedback inhibition and repression. Very little is known about the regulatory mechanisms in higher plants, but some information concerning the effects of regulatory mutants is accumulating concerning simple eukaryotes such as fungi.[18,22,23]

Considerable research effort has gone into the isolation of amino acid overproducers in plants with relatively good success.[6,9,14,17] The general scheme is to select lines which have relaxed feedback control of the biosynthesis of specific amino acids. Cells[9,14,17] or excised embryos[6] are cultured on media containing analogues of specific amino acids or inhibitory levels of the end product amino acid. Normal cells are inhibited, whereas variants which overproduce the amino acid are capable of growth.

Research on regulation of amino acid synthesis is pertinent to our goals in alfalfa breeding. In alfalfa, the product consumed is herbage and an elevated level of an unbound amino acid should be utilizable in herbage whereas it may not be incorporated in a seed crop. Alfalfa leaf protein rivals animal protein in terms of the amino acid requirements for human and other monogastic animal nutrition, but it is slightly low in methionine.[24] Hence, we have initiated a variant selection program in alfalfa to identify overproducers of methionine. Two years were required, however, to breed a model line which would regenerate after being in suspension culture. Cultivated alfalfa is an autotetraploid and the genetic redundancy could hinder expression of mutations, even so-called dominant mutations. For example, in some traits one dominant allele with three recessives at the same locus is not always sufficient for full phenotypic expression. Therefore, diploid plants were desired and were obtained by maternal haploidy.[25] We then needed a line which would regenerate after cells had been in suspension culture, and this was developed by conventional breeding and selection.[26]

We believe the concept of breeding for optimum stocks for cellular experiments will necessarily precede the full application of cellular methods in other crop species.

Mutagenesis of cultured cells can be accomplished with both ionizing radiation and chemical mutagens. Reactions of cell suspensions of sugar cane to gamma-irradiation have been reported.[27] Chemical mutagens seem preferable and indeed are the main group used in published and ongoing research because they are much easier to handle in the liquid media in which cells are grown, and do not require special equipment.

At this time there is a small, but growing, list of induced mutants from cell and tissue culture. These include certain amino acid and vitamin auxotrophs in cell lines of tobacco and fern, a soybean cell line resistant to 5-bromodeoxyuridine, a NaCl-tolerant line of mutant tobacco, and an auxin-autotrophic cell line of the maple tree.[2] In most cases, the whole plants were not regenerated or fully analyzed. Hence, it is not known whether resistance was due to a desired point mutation, alteration in chromosome number, or physiological adaptation.

In a recent review, Nabors[1] presents a strong argument for the usefulness of spontaneous mutations in variant selection in plant cell cultures. Under the basic assumption that the likelihood of a particular mutation is on the order of $1 \times 10^{-5}$ per gene copy per generation and that 100 m$\ell$ of a rapidly growing suspension culture of tobacco contains about $1 \times 10^{7}$ cells, there appears to be potential for efficient screening for desired spontaneous variants. A 100 $\ell$ suspension culture usually contains more than $10^{10}$ cells, and the possibility should exist for recovering even recessive mutant phenotypes in diploid cells. Monoploid cells would, of course, be the most efficient for the identification of recessive mutations. The notion of using spontaneous mutations is attractive to the plant breeder because most induced mutations are deleterious, and mutagenesis would increase the frequency not only of the desired mutant, but also the load of deleterious ones. Research is clearly needed to compare the efficiency of spontaneous and induced mutation cell culture screening for given traits.

Selection for herbicide resistance would seem to be a model use for tissue and cell culture, especially when mode of herbicidal action is at cellular level. Early in our tissue culture research,[96] an alfalfa clone was identified which was tolerant as callus to normally toxic levels of 2,4-dichlorophenoxy acetic acid (2,4-D). It was identified, but not selected in culture, although the principle should be the same. This line and two controls were cloned by shoot cuttings and whole plants treated with normal herbicide levels of 2,4-D in a greenhouse experiment. The line which was resistant or tolerant as callus was not more tolerant as a whole plant than the controls. Our conclusion was that this might be expected in some cases since the one site of herbicidal action is in meristematic tissue. Zenk[9] cited work from his laboratory, where a haploid (monoploid) cell line of *Nicotiana sylvestris* was isolated which was resistant to toxic levels of 2,4-D. He indicated that the mechanism of resistance in their case was due to increased capacity of the resistant strain to metabolize 2,4-D. Such a mechanism could be effective in the whole plant if herbicide was broken down before reaching the meristematic regions. Zenk[9] also reported finding a soybean suspension culture line resistant to the herbicide atrazine. Other apparently spontaneous mutant lines isolated in culture include three in tobacco[15,16,28,29] and one in *Capsicum*.[29]

## C. Whole Plant Approach

Mutation induction and screening using whole plants has a proven potential for genetic modification. Investigation of induced mutations for plant improvement began in the 1930s soon after the demonstration of the mutating effect of X-rays in *Drosophila* and in *Hordeum*.[30] One of the first cultivated crops in which a useful mutant was induced (1934) was tobacco, which seems ironic since tobacco has been the model species in recent cellular research. In 1942, a mutant was produced in barley by X-irradiation which was resistant to mildew[31] and later mutation breeding based on X-radiation became widespread. Since that time, many different physical and chemical agents have been used for mutant induction.[32] Barley, which is a diploid ($2n = 14$), has been a key crop species in much of the basic research on mutagenic efficiency and specificity. Efficiency of mutagenesis is lowered in polyploids, but useful mutants have still been produced in polyploids such as wheat[33] and peanut.[34]

Progress continues to be made in increasing the efficiency of mutant induction and screening whole plants and some significant advances can be seen in the following examples. Nilan[32] and his colleagues increased the effectiveness of ethyleneimine and sodium azide, and increased the efficiency of methylnitrosourea by adjusting the pH of the treatment solution applied to barley seeds. Responses to pH were found to be associated with cell membrane permeability. Treatment at a pH higher than seven increased concentrations of mutagen in cells and thus increased mutation. Use of seeds in mutation research permits greater control of cell cycles than can be achieved in whole plant tissues and further increases in mutation have been obtained by treatments at different stages of the cell cycle. Generally, the maximum rates of mutation frequency with the least lethal or physiological effects occur when treatments are applied at the onset of DNA synthesis.

Increased efficiency of mutation induction using physical mutagens has not been as pronounced as with chemicals. Effectiveness of X-irradiation and gamma rays, however, has been increased by manipulation of oxygen, water, and temperature.[35-37] Concepts and methods of increasing efficiency of mutation induction developed using seeds and whole plants should be applicable to the emerging cellular research on mutant induction.

In mutagenesis, it would be desirable if modifications could be selectively induced at specific loci. While chemical mutagens are often highly specific at the molecular level, resulting in nucleotide deletions, additions, substitutions, or modifications, the problem remains to cause these changes only in desired cistrons. Since susceptibility of some bacterial genes to mutation is genetically regulated, the hope remains that analogous specificities can arise, or be engineered, in plants. Thus far, this hope has not been realized.[37]

It is generally accepted that mutation rates are low and that most mutations are deleterious recessives. Some new concepts in mutation are emerging, however, based on studies in molecular genetics and on the rates of evolution of molecules. One new concept is that there is relatively high frequency of selectively neutral or nearly neutral mutations in nature.[38] Another new concept is that the neutral or near neutral genetic changes constitute a tremendous store of genetic variability, which permits sequential mutational changes culminating in new functions or forms.[38] Thus, a rare, sudden change in phenotype may not be due to a single large mutational event, but due to a near neutral modification in such a sequence as to result in major phenotypic change.

A survey of the proceedings of the meetings of the International Atomic Energy Agency over the past several years indicates that seeds are most frequently used for mutation induction, followed by gametophytes, especially pollen. Mutant induction in pollen has the advantages of eliminating chimeras in the fertilization products of the treated generation and the elimination of gross chromosomal abnormalities through pollen inviability. Recently, an elegant system of producing and identifying recessive male sterile mutants on specific chromosomes of wheat was devised by Driscoll.[33] It involved treating euploid pollen with X-rays and applying it to plants deficient for chromosome 5B. The $F_1$ progeny were monosomic for the irradiated 5B chromosome and induced recessive male sterile mutants on this chromosome were detected directly. Gametophytes, however, are not as suitable for chemical mutagenesis as seeds and usually have to be treated on the plant, which immediately presents space and timing problems not encountered with seed. Seeds of most species can be treated in large numbers at any time of the year.

An interesting and potentially efficient approach was discussed by Gaul et al.[39] This involves treating barley seeds with ethyl methane sulfonate (EMSO) and then redrying so that they may be stored, shipped, and machine planted after treatment. This approach is attractive in that it permits selection based on large field populations, and is free of the current practical limitations of cellular techniques, such as regeneration.

Mutations contributing to improvement of the protein quality of seeds are of great current interest. Brock and Langridge[38] surveyed the protein synthesizing pathways of higher plants and concluded that it should be possible by mutation to modify certain aspects, thereby altering the amount or type of a specific seed protein. Pathways most likely to be modified included organic nitrogen formation, transamination, and more free amino acid accumulation. Recent achievements in mutation induction in this area include six high lysine barley lines.[40]

In 1974, a total of 98 registered crop cultivars had reportedly been produced by mutation breeding since 1950[30] and the total now probably exceeds 100 cultivars. In the same time period, however, between 5000 and 10,000 cultivars have been produced by conventional breeding using natural variation. Obviously, most plant breeding is done using natural variation existing in the respective crop species. The magnitudes of the numbers of cultivars developed by the respective methods should be kept in mind by those making predictions about the application of cellular methods for induction and selection of genetically modified plants.

## III. TRANSFER OF ALIEN GENES

### A. Cellular Approach

The term "alien" has been used in plant breeding to describe genetic material transferred sexually from a donor species to a recipient species where pairing and genetic recombination between donor and recipient chromosomes ordinarily does not take place. Recently, in the area of genetic engineering, alien has implied insertion of eukaryotic germplasm into a prokaryote, or vice versa. In this section, it will be used in reference to transfer of genetic information from one plant (or organism) to another where there are wide species or generic differences between the donor and recipient.

Transformation of genetic information is a powerful tool in bacterial genetic systems. There is currently interest in trying to harness this phenomenon for use in modifying higher plant genetic systems. Indeed, there are now several reports of transformation-like events in higher plants, but reproducibility reportedly has not been good,[4] and application remains as a potential. For the transformation to be heritable in sexually reproduced plants, the exogenous DNA must function normally in mitotic and meiotic divisions.

Hess[41] reported using DNA from a red-flowered *Petunia* variety to transform a white-flowered form into a red-flowered form. The red-flowered trait was heritable, suggesting that it was somehow incorporated or attached to a chromosome. Several other transformation-like events have been reported for *Petunia*,[42] as has the DNA-mediated genetic correction (transformation) of a thiamineless condition in *Arabidopsis*.[43] Uptake of *Escherica coli* DNA by protoplasts has been reported in soybean, carrot,[44] and tomato.[45] Finally, the uptake and expression of bacteriophage DNA cell culture has been demonstraed in barley[46] and in tomato and Arabidopsis,[47] and in Sycamore.[48] These suggestions that DNA from transducing bacteriophages can be taken up, transcribed, and translated in plant cells, coupled with recent advances in genetic engineering using restriction enzyme digestion, creates a potential for unprecedented genetic modification of certain traits.

This possibility of using recombinant DNA procedures in modifying plant genetic systems is receiving much current attention. Initial components of the recombinant process are bacterial plasmid DNA, eukaryotic DNA isolated by density gradient centrifugation, and by restriction endonucleases. These components are combined and incubated (about 37°C) to digest and break up the respective DNA molecules, and cooled (to about 5°C) for 24 hr to allow association of cohesive ends of the fragmented DNAs. The hybridization reaction is then completed by the covalent joining of two or more DNA molecules. The hybrid molecules may be propagated and thus eukaryotic genes may be cloned if the hybrid segment is joined to a vector DNA molecule (plasmid) capable of independent propagation in a bacterium such as *E. coli.* Certain bacteriophage DNAs and certain plasmid DNAs have been used as vectors,[49] and biochemical techniques have been developed that are generally applicable for joining covalently any two DNA molecules.[50] There are now several reports of functional genetic expression of eukaryotic DNA in bacterial systems.[51-54] Additionally, Carlson[46] observed that bacteriophage genes can be transcribed and translated in plant cells. These methods, used in proper sequence, promise to permit the transference of specific genes from one plant to another. The biochemical expression of these genes may be selected in the bacterial vector if the promoter and operator regions are preserved. If genetic engineering involving recombinant DNA can be harnessed for plant improvement, it would seem superior to other novel cellular approaches owing to the precision with which single genes can be selected for modification. All cellular approaches, however, will probably be limited to relatively simple biochemical steps for the foreseeable future.

A frequently discussed use of recombinant DNA research is the development of new ways to increase biological nitrogen fixation. This concept was considered among others at a recent symposium titled: Genetic Engineering for Nitrogen Fixation.[55] The reported consensus of the participants was that recombinant DNA techniques and other forms of genetic engineering are unlikely to result in greater biological nitrogen fixation and crop productivity in the near future. This conclusion seems justified for goals such as transfer of the symbiotic nitrogen fixing capacity from legumes to cereals. This type of goal seems premature because the genetic system involved is not yet identified and is probably very complex, involving the legume's anatomy, physiology, and bacterial infection specificity. On the other hand, there are genetically realistic goals using existing symbiotic systems which may well be accomplished. For example, it should be feasible to use bacterial genetics and plant breeding to breed a more efficient strain of *Rhizobium* bacteria and a plant with more nodules, and combine the two using existing seed inoculation methods.

Research on all the aspects of nitrogen fixation and utilization is increasing, and whether the goals seem realistic or not, much basic knowledge will be obtained on the bacterial and plant systems involved. Such basic knowledge may be expected to contribute directly or indirectly to plant improvement.

### B. Whole Plant Approach

Gene transfer using sexual processes of whole plants is frequently used in species where wide hybrids (interspecific and/or intergeneric) can be obtained and where there is enough chromosome homology between the respective genomes to permit incorporation of the desired gene by sexual recombination. Embryo culture[56] has classically been used to obtain many wide hybrids, which otherwise would not be recovered due to failure of endosperm development, which in turn results in embryo abortion.[57] In cases where seed development takes place after wide hybridization, but where hybrid seedlings are chlorophyll deficient, grafting[58] has been used to grow and obtain crosses on wide hybrids. Where pairing and recombination does not take place between donor and recipient species, irradiation has been used to produce chromosome breaks and incorporate small chromosome segments including the desired alien gene from the donor species.[59]

Embryo culture and irradiation which may be required in achieving alien gene incorporation have been available for years and are probably at least as efficient as potential cellular methods of gene manipulation, yet they have been used far less in crop improvement than might be anticipated. Probably fewer than a dozen cultivars of the thousands that have been bred have embryo culture or irradiation-induced gene transfer in their background. This is because the desired trait can so often be found in the natural variation available in a given species. Nevertheless, the methods are available for use in special cases, often the transfer of disease resistance that is not present in the cultivated species.

Increasing basic knowledge about the reproductive biology of wheat has permitted use of new and potentially very efficient methods of surmounting the obstacle of incorporation of alien variation. Moreover, the concepts delimited in wheat should be applicable to some other important economic species. The key factor was the simultaneous discovery of genetic control of chromosome pairing specificity in wheat by Riley and Chapman[60] and by Sears and Okamoto.[61] The discovery was dependent on the availability of sophisticated cytogenetic stocks in wheat which were deficient in turn for each of the 21 pairs of chromosomes in wheat. The gene regulating pairing specificity was discovered by virtue of its absence in stocks deficient for chromosome 5 of the B genome. In the normal case when 5B is present, only strict homologues may pair, thus preventing intergenomic as well as alien chromosome pairing and recombination. When chromosome 5B is removed, however, or the pairing gene is suppressed by a dominant allele which was subsequently found in a related species, pairing specificity is lowered and some intergenomic and alien pairing and combination takes place.

The advantages of a genetic recombination approach to alien gene incorporation are that the incorporation event is not random, but in a specific area of the chromosome, and potentially a segment as small as a single gene may be incorporated. Genetic control of chromosome pairing specificity has been used in wheat[62] to incorporate the gene for yellow rust resistance from *Aegilops comosa,* a wild relative of cultivated wheat. A cultivar named "Compair" carrying the alien resistance was eventually bred and released to the public.[62]

## IV. MANIPULATION OF GENOMES

### A. Cellular Approach

Somatic hybridization of cell protoplasts in vitro is of particular significance in higher plants because it is presently possible in some species to recover the hybrid product as a whole plant. The prospects of using this method for novel hybridizations exceeding the normal sexual compatibilities are evident, and will be discussed first. Perhaps less evident is the potential of using this method as an alternative to sexual union in intraspecific hybridizations to maximize heterozygosity and hybrid vigor. This will be discussed at the end of the section.

The initial step of producing plant protoplasts, that of enzymatically removing the cell wall with a combination of cellulases and pectinases, has been accomplished in a wide range of species,[63] and whole plants have been regenerated from protoplasts in at least six crop species.[64] Fusion is also not a major problem, having been induced in almost all mixtures of protoplasts attempted by using polyethylene glycol, which causes protoplasts to aggregate.[64] Exotic interkingdom fusions have been reported between human cells (HeLa) and tobacco protoplasts.[65] A problem in the full use of somatic fusion to obtain plant hybrids at will is the failure of most hybridized protoplasts to divide and form a callus colony from which the hybrid plant might be regenerated. Selective regeneration of the hybrid product is ideal and has been accomplished thus far in cultivated tobacco[66] and its relatives,[67,68] in *Petunia*,[69] and in the liverwort *Sphaerocarpus*.[70] There have been far more reviews dealing with the potentials of somatic hybrids than such actual hybrids, hence we will only review a few of the details which apply directly to crop improvement. Progress is being made, but general application of the method is probably years away.

There has been much discussion about the need for selective media on which hybrids will callus and regenerate while donor cells will not. It is significant that Power et al.[69] who recently hybridized *Petunia hybrida* and *P. parodii* using leaf protoplasts, succeeded without a single perfect selective screen by using a combination of naturally occurring differences in the sensitivity of the cultured protoplasts to media and drugs. From a plant breeding standpoint, if indeed the hybrid were valuable enough, a conventional breeder would not hesitate to proceed even if no selective scheme were available, and isolate his hybrid from among donor and hybrid plants regenerated. If both donors are capable of regeneration and are biologically compatible enough to form a synkaryon, hybrids may predominate due to hybrid vigor.

In research to confirm and extend the original somatic hybrid between *Nicotiana glauca* and *N. langsdorffii*, Smith et al.[68] used the original materials and methods in a larger experiment. A total of 174 calli were obtained on the hormoneless medium which selects against the parental types. A total of 23 verified hybrid plants were obtained from 19 calli which regenerated. These hybrids, however, did not possess the euploid amphidiploid chromosome number (42), but instead ranged from 56 to 64 chromosomes. In *Petunia*, somatic hybrids included amphidiploids and aneuploids with fewer than the amphidiploid chromosome number.[69] Data are too limited at this point to assess the extent aneuploidy may be a problem in the application of somatic hybridization to directed plant improvement.

How much of an impact will exotic somatic hybrids have on crop improvement when they can be made at will? An examination of some wide crosses using the sexual system should assist in making a rational judgment. Triticale is a truly man-made plant species derived by crossing wheat (*Triticum aestivum* or *T. turgidum*) and rye (*Secale cereale*) and doubling the chromosome number. The hybrids date back to 1875 and the first amphiploid hybrid was reported in 1891.[70] Triticale plants are very impressive vegetatively, but until very recently their heads or ears have not been completely filled with kernels, owing to partial sterility, and even recently developed lines with fully fertile heads may have kernels with shrunken endosperm, which results in reduced grain weight.

Part of the problem may be attributed to the fact that wheat is naturally self-pollinated and the genomes are adjusted to the associated homozygosity, while rye is naturally an outcrossing species whose chromosomes are most meiotically stable in the presence of heterozygosity.[71] Since Triticale is predominantly self-pollinating, the rye chromosomes are somewhat unstable. There has been much effort expended recently towards the development of high yielding Triticales by hybridizing different wheat-rye combinations and different amphiploid levels; some Triticales evidently are more productive than wheat in certain environmental niches.[70,72,73] However, presently available Triticales evidently cannot compete successfully with wheat in the U.S.

The classical amphidiploid Raphanobrassica which is an intergeneric hybrid between *Raphanus sativus* and *Brassica oleracea* appears to be on its way becoming a productive cultivated fodder crop, some 50 years after it was first produced.[74] Recently, McNaughton[75] has bred several new Raphanobrassicas using both diploid and autotetraploid parents of the fodder types which represent a large gene pool of material from which to select.

In both Triticale and Raphanobrassica, literally hundreds of hybrid combinations have had to be made and tested to find ones with proper agronomic characteristics. Somatic hybridization, if fully workable, should be an efficient method of producing hundreds of hybrid combinations, but could not assist in the greatest time-consuming aspect: testing the agronomic potential of hybrids. Also, many years have elapsed between the original hybridization and a commercial product. Thus, mere efficient exotic hybridization, whether sexual or somatic, is not likely to have an immediate impact on the food supply for the current population of the world.

Selective and sequential elimination of the chromosomes of one parent is known to occur after certain intergeneric hybridizations involving barley (*Hordeum vulgare*) and a wild relative *H. bulbosum*.[76,77] Elimination takes place in early embryogenesis and results in haploid embryos which have to be salvaged by embryo culture. Recently, the same phenomenon has been reported after barley × rye hybridizations.[78] One has to wonder whether this may not be a problem in plant somatic hybridizations. It would be advantageous if the same materials could be used in somatic hybridizations as soon as possible to test the concept, but this may be years away in the species involved.

Somatic hybridization of genomes transfers all the genetic variation from donor cells (parents) to the hybrid cell (offspring), something which is not done in normal sexual reproduction. Therefore, a practical use of somatic hybridization may be to produce intraspecific hybrids of a vegetatively reproduced polyploid crop species, such as the cultivated potato. Evidence is accumulating which suggests that maximum heterosis in autopolyploids requires maximum heterozygosity.[79] For example, in an autotetraploid, the order of desirability of different genotypic structures at a locus is:

$$a_ia_ja_ka_l > a_ia_ia_ja_k > a_ia_ia_ja_j > a_ia_ia_ia_i$$

Thus, the tetra-allelic state is the most heterotic. The problem is that this condition cannot be achieved simultaneously at all loci by normal sexual reproduction, because gametes carry two alleles which will often be the same allele, e.g., $a_ia_i$. It is easy to envision, however, how somatic hybridization of somatic cells of two different diploid hybrids, e.g., $a_ia_j$ and $a_ka_l$, could fix the heterozygosity in a somatic hybrid $a_ia_ja_ka_l$. This fact coupled with vegetative reproduction would permit the unlimited reproduction of the maximal heterotic genotype.

In genetic terms, normal gametes transmit only the additive portion of genetic variance from parent to offspring; nonadditive intra- and inter-locus genetic variance is not normally transmitted. Hybridization of somatic cells or certain restitution gametes with the somatic chromosome number (2n gametes) can uniquely transmit both the additive and nonadditive genetic components. Further discussion of this phenomenon follows in the whole plant section.

### B. Whole Plant Approach

Novel approaches to genetic modification in plant breeding and new interpretations of polyploid evolution are emerging from recent awareness of the importance of gametes with the unreduced chromosome number (2n gametes). Such gametes occur naturally at a low frequency in essentially all species surveyed, probably owing to developmental accidents. They can occur at a comparatively high frequency in plants carrying mutant genes for meiotic events. Certain 2n gametes, namely those formed by restitution of the somatic chromosome number by failure of the first division of meiosis, rival somatic hybridization in transmitting genetic variation from parent to offspring. They transmit all the somatic genetic variance from the centromeres to the first crossovers on the chromosomes and all the additive and half the nonadditive beyond the first crossover.[80] Thus, 2n gametes, as well as somatic hybridization, offer a potential for maximizing heterosis. The potential of 2n gametes in potato breeding has been recognized[81,82] and already is being exploited through the use of first division restitution gametes. Moreover, concepts developed in potato are directly applicable to many other polyploid crops and similar work is known to be underway in forage legumes and grasses.

In evolutionary theory, which is often directly applicable to germplasm transfer in plant breeding, 2n gametes are finally being recognized as the principal mechanism in polyploid evolution. Harlan and de Wet[83] recently summarized the occurrence and probable role of 2n gametes in plant polyploid series and concluded that almost all polyploids arise by way of 2n gametes, other mechanisms such as spontaneous somatic doubling being negligible. Combining the knowledge of 2n gametes and effective pollination control has resulted in efficient, large scale transfer of exotic germplasm from wild diploids to cultivated tetraploid crops.[84,85]

## V. CHROMOSOME MANIPULATIONS

### A. Cellular Approach

The most frequent spontaneous genetic modification in cell cultures is polyploidy, followed in frequency by aneuploidy.[86,87] Such modifications are generally disadvantageous, leading to genetic redundancy and genetic imbalance, which is not desired in variant selection. Aneuploid cells, however, may lead to interesting mutant types, such as a deficiency aneuploid tobacco which flowered under a short day condition.[88]

Euploid and aneuploid oat plants regenerated from embryo-derived-callus[89] provide a model for the use of tissue culture in producing aneuploids. Although most of the callus-derived plants were euploid, several were aneuploid and the presence of univalents at meiosis suggests some were deficiency aneuploid. Oats, a hexaploid species, is tolerant of such chromosomal imbalance and cytogeneticists may find this method an efficient source of aneuploids.

In sugar cane, an asexually reproduced crop of high polyploidy, tissue and cell culture has already been used to produce or select variants, and to isolate plants with homogeneous cells from plants which were chromosomal mosaics composed of cells of different chromosome numbers. Heinz, Mee, and Nickell[90] observed cell cultures with aneuploid numbers involving the loss or gain of individual chromosomes. Where variants can be reproduced vegetatively, such variants can potentially be used to increase the dosage of genes and perhaps gene products for desired traits, such as sugar production or pest resistance. In sugarcane, plants regenerated from cells in culture have included plants that differed in chromosome number, growth, and yield potential,[27] and resistant variants from a cultivar which was sensitive to sugarcane mosaic virus.[91]

Polyploidization in culture can be minimized, but is difficult to eliminate, especially in species where 2,4-D is used as an auxin source. In several species, 2,4-D, which enhances mitotic activity, increases the incidence of polyploidy.[92,93] In species such as alfalfa, where induction of callus formation on a medium containing 2,4-D is required for optimal regeneration, a background level of polyploidization must currently be tolerated. It can be used to advantage, however, in producing a polyploid series in one genetic background for physiological and breeding research. Such ploidy series have been produced in alfalfa due to spontaneous doubling in culture[94] and in *Datura* due to spontaneous doubling and probable fusion of haploid microspores.[95]

### B. Whole Plant Approach

The concept of chromosome manipulations in plants using available methods is being used quite broadly among crop plants and no review will be attempted here. For example, trisomics and interchange stocks are used routinely in maize for location of genes and cytogenetic research. In wheat, complete sets of monosomic and nullisomic-tetrasomic compensation series are available and allow genetic analysis of genes on individual chromosomes. Furthermore, similar stocks are being developed in numerous crop species and the potential for new approaches to genetic manipulation is great. Triploids, haploids, and meiotic mutants remain the successful workhorses in most chromosome manipulation research and are being used in an ever increasing number of species.

## VI. SUMMARY AND CONCLUSIONS

Novel genetic modification in plants appears assured owing to increasing sophistication and efficiency of both whole plant and cellular methods. Manipulation of single mutant genes, alien genes, whole genomes, and chromosomes has been examined. For some time to come, most genetic modification will probably derive from whole plant methods since they are broadly applicable to essentially all higher plant species. Cellular methods are just now moving from the model experiment phase in a very few species to similar applications in important food and forage crop species. Cellular methods have to be optimized for each species, hence broad application is some time away. Additionally, most genetic modifications associated with improvement of plant quality, quantity, adaptation, and pest and pathogen resistance involve individual and integrated morphological and physiological systems which probably necessitate selection using whole plants. Cellular techniques at fruition should dominate selection for those biochemical characters whose cellular phenotypes reflect the desired condition in the whole plant.

Somatic hybridization of plant cells is currently possible using related or totally unrelated species. In fact, plant and animal somatic cells have been hybridized. To date, however, regeneration and recovery of higher plant hybrids has been possible only in the Solanaceae, using species which also could be sexually hybridized. Whether somatic hybrids beyond the limits of sexual hybridization will be useful or teratological forms has not yet been established.

Somatic hybridization has the unique capability of transmitting all genetic variation from donor plants to somatic hybrids — something not possible using normal sexual reproduction. Therefore, it may eventually be very useful in maximizing hybrid vigor in vegetatively reproduced species such as potato, where diploid donor lines are available and the cultivated form is tetraploid. Hence, intraspecific as well as interspecific and intergeneric somatic hybridization may prove useful in novel genetic modification.

Interestingly, the model system of transmitting diploid genetic variation to a cultivated tetraploid hybrid is also possible using a certain class of sexual gametes with the unreduced chromosome number. Moreover, this system is already in use in potato improvement and its potential is being examined in other polyploid crop species.

Chromosome doubling and instability of chromosome numbers, common abnormalities of cells cultured in vitro, have been put to novel use in obtaining polyploids and aneuploids, respectively. Aneuploids have manifold uses in plant research, and often are an important step in genetic modification. Their production in culture is potentially useful since it will bypass the sexual process, which sometimes eliminates aneuploid gametes. Thus, novel genetic modification is expected on many fronts using the most efficient methods; hopefully, cellular and whole plant methods can complement each other in an integrated program of plant improvement.

## REFERENCES

1. **Nabors, M. W.,** Using spontaneously occurring and induced mutations to obtain agriculturally useful plants, *BioScience,* 26, 761, 1976.
2. **Bottino, P. J.,** The potential of genetic manipulation in plant cell cultures for plant breeding, *Radiat. Bot.,* 15, 1, 1975.
3. **Carlson, P. S. and Palacco, J. C.,** Plant cell cultures: genetic aspects of crop improvement, *Science,* 188, 622, 1975.
4. **Chaleff, R. S. and Carlson, P. S.,** Somatic cell genetics of higher plants, *Annu. Rev. Genet.,* 8, 267, 1974.
5. **Smith, H. H.,** Model systems for somatic cell plant genetics, *BioScience,* 24, 269, 1974.
6. **Green, C. E. and Phillips, R. L.,** Potential selection systems for mutants with increased lycine, threonine, and methionine in cereal crops, *Crop Sci.,* 14, 827, 1974.
7. **Binding, H., Binding, K., and Staub, J.,** Selektion in Gewebekulturen mit haploiden zellen, *Naturwissenschaften,* 57, 138, 1970.
8. **Audus, L. J.,** *Plant Growth Substances,* Leonard Hill, London, 1972.
9. **Zenk, M. H.,** Haploids in physiological and biochemical research, in *Haploids in Higher Plants,* Kasha, K. J., Ed., University of Guelph, Ontario, 1974, 339.
10. **Sunderland, N.,** Anther culture as a means of haploid induction, in *Haploids in Higher Plants,* Kasha, K. J., Ed., University of Guelph, Ontario, 1974, 339.
11. **Kimber, G. and Riley, R.,**Haploid angiosperms, *Bot. Rev.,* 29, 480, 1963.
12. **Kasha, K. J., Ed.,** *Haploids in Higher Plants,* University of Guelph, Ontario, 1974.
13. **Carlson, P. S.,** Induction and isolation of auxotrophic mutants in somatic cell cultures of *Nicotiana tabacuum, Science,* 168, 487, 1970.
14. **Carlson, P. S.,** Methionine sulfoximine-resistant mutants of tobacco, *Science,* 180, 1366, 1973.
15. **Heimer, Y. M. and Filner, R.,** Regulation of the nitrate assimilaton pathway of cultured tobacco cells. II. Properties of a variant cell line, *Biochim. Biophys. Acta,* 215, 152, 1970.
16. **Nabors, M. W., Daniels, A., Nadolyn, L., and Brown, C.,** Sodium chloride tolerant lines of tobacco cells, *Plant Sci. Lett.,* 4, 155, 1975.
17. **Widholm, J. M.,** Cultured carrot cell mutants: 5 methyltryptophan-resistance trait carried from cell to plant and back, *Plant Sci. Lett.,* 3, 323, 1974.
18. **Rasse-Messenguy, F. and Fink, G. R.,** Feedback-resistant mutants of histidine biosynthesis in yeast, in *Genes, Enzymes and Populations,* Srb, A. M., Ed., Plenum Press, New York, 1973, 85.
19. **Umbarger, H. E.,** Regulation of amino acid metabolism, *Annu. Rev. Biochem.,* 38, 323, 1969.
20. **Arditti, R. R., Eron, L., Zubay, G., Tocchini-Valentini, G., Connaway, S., and Beckwith, J.,** *In vitro* transcription of the *lac* operon genes, *Cold Spring Harbor Symp. Quant. Biol.,* 35, 437, 1970.
21. **Umbarger, H. E.,** Metabolite analogues as genetic and biochemical probes, *Adv. Genetic.,* 16, 119, 1971.
22. **Gross, S. R.,** Genetic regulation mechanisms in fungi, *Annu. Rev. Genet.,* 3, 395, 1969.
23. **Metzenberg, R. L.,** Genetic regulatory systems in Neurospora, *Annu. Rev. Genet.,* 6, 111, 1972.
24. **Akeson, W. R. and Stahlman, M. A.,** Leaf protein concentrates: a comparison of protein production per acre of forage with that from seed and annual crops, *Econ. Bot.,* 20, 244, 1966.
25. **Bingham, E. T.,** Isolation of haploids of tetraploid alfalfa, *Crop. Sci.,* 11, 433, 1971.
26. **McCoy, T. J. and Bingham, E. T.,** Regeneration of diploid alfalfa plants from cells grown in suspension culture, *Plant Sci. Lett.,* 10, 59, 1977.

27. **Nickell, L. G. and Heinz, D. J.**, Potential of cell and tissue culture techniques as aids in economic plant improvement, in *Genes, Enzymes and Populations*, Srb, A. M., Ed., Plenum Press, New York, 1973, 109.
28. **Lesure, A. M.**, Selection of markers of resistance to base-analogues in somatic cell cultures of *Nicotiana tabacum, Plant Sci. Lett.*, 1, 375, 1973.
29. **Dix, P. J. and Street, H. E.**, Sodium chloride-resistant cultured cell lines from *Nicotiana sylvestris* and *Capsicum annuum, Plant Sci. Lett.*, 5, 231, 1975.
30. **Sigurbjornsson, B. and Micke, A.**, Philosophy and accomplishments of mutation breeding, in *Polyploidy and Induced Mutations in Plant Breeding*, International Atomic Energy Agency, Vienna, 1974, 303.
31. **Stadler, L. J.**, Mutations in barley induced by X-rays and radium, *Science*, 68, 186, 1928.
32. **Nilan, R. A.**, Increasing the effectiveness, efficiency, and specificity of mutation induction in flowering plants, in *Genes, Enzymes and Populations*, Srb. A. M., Ed., Plenum Press, New York, 1974, 139.
33. **Driscoll, C. J.**, Induction and screening of chromosomal male sterile mutants for use in the production of hybrid wheat, in *Polyploidy and Induced Mutations in Plant Breeding*, International Atomic Energy Agency, Vienna, 1974, 139.
34. **Gregory, W. C.**, Mutation breeding, in *Plant Breeding*, Frey, K. J., Ed., Iowa State University Press, Ames, Iowa, 1966, 189.
35. **Conger, B. V., Constantin, M. J., and Carabia, J. V.**, Seed radiosensitivity: wide range in oxygen-enhancement ratio after gamma-irradiation of eight species, *Int. J. Radiat. Biol.*, 22, 225, 1972.
36. **Klein, R. M. and Klein, D. T.**, Post-irradiation modulation of ionizing radiation damage to plants, *Bot. Rev.*, 37, 397, 1971.
37. **Smith, H. H.**, Comparative genetic effects of different physical mutagens in higher plants, in *Induced Mutations and Plant Improvement*, International Atomic Energy Agency, Vienna, 1972, 75.
38. **Brock, R. D. and Langridge, J.**, Prospects for genetic improvement of seed protein in plants, in *Induced Mutations and Plant Improvement*, International Atomic Energy Agency, Vienna, 1972, 3.
39. **Gaul, H., Frimmel, G., Gichner, T., and Ulonska, E.**, Efficiency of mutagenesis, in *Induced Mutations and Plant Improvement*, International Atomic Energy Agency, Vienna, 1972, 121.
40. **Ingversen, J., Andersen, A. J., Doll, H., and Köie, B.**, Selection and properties of high lysine barley, in *Nuclear Techniques for Seed Protein Improvement*, International Atomic Energy Agency, Vienna, 1973, 193.
41. **Hess, D.**, Versuche zur Transformation un lioheren Pflanzen: genetische Charakterisierung einger mutmasslich transformierter Pflanzen, *Z. Pflanzenphysiol.*, 63, 31, 1970.
42. **Hoffman, F. and Hess, D.**, Die aufnahme radioaktiv Markierter DNS in isolierte Protoplasten von *Petunia hybrida, Z. Pflanzenphysiol.*, 69, 81, 1973.
43. **Ledoux, L., Huart, R., and Jacobs, M.**, DNA-mediated genetic correction of thiamineless *Arabidopsis thaliana, Nature (London)*, 249, 17, 1974.
44. **Ohyama, K., Gamborg, O. L., and Miller, R. A.**, Uptake of exogenous DNA by plant protoplasts, *Can. J. Bot.*, 50, 2077, 1972.
45. **Gahan, P. B., Anker, P., and Strour, M.**, An autoradiographic study of bacterial DNA in *Lycopersicon esculentum, Ann. Bot. London*, 37, 681, 1973.
46. **Carlson, P. S.**, The use of protoplasts for genetic research, *Proc. Natl. Acad. Sci. U.S.A.*, 70, 598, 1973.
47. **Doy, C. H., Gresshoff, P. M., and Rolfe, B. G.**, Biological and molecular evidence for the transgenosis of genes from bacteria to plant cells, *Proc. Natl. Acad. Sci. U.S.A.*, 70, 723, 1973.
48. **Johnson, C. B., Grierson, D., and Smith, H.**, Expression of lambda-plac5 DNA in cultured cells of higher plants, *Nature (London)*, 244, 105, 1973.
49. **Cameron, J. R., Panasenko, S. M., Lehman, I. R., and Davis, R. W.**, *In vitro* construction of bacteriophage lambda carrying segments of the *Escherichia coli* chromosome: selection of hybrids containing the gene for DNA ligase, *Proc. Natl. Acad. Sci. U.S.A.*, 72, 3416, 1975.
50. **Jackson, D. A., Symons, R. H., and Berg, P.**, Biochemical method for inserting new genetic information into DNA of simian virus 40: circular SV 40 DNA molecules containing lambda phage genes and the galactose operon of *Escherichia coli, Proc. Natl. Acad. Sci. U.S.A.*, 69, 29044, 1972.
51. **Morrow, J. F., Cohen, S. N., Chang, A. C. Y., Boyer, H. W., Goodman, H. M., and Helling, R. B.**, Replication and transcription of eukaryotic DNA in *Escherichia coli, Proc. Natl. Acad. Sci. U.S.A.*, 71(5), 1743, 1974.
52. **Thomas, M., Cameron, J. R., and Davis, R. W.**, Viable molecular hybrids of bacteriophage lambda and eukaryotic DNA, *Proc. Natl. Acad. Sci. U.S.A.*, 71(11), 4579, 1974.
53. **Polisky, B., Bishop, R. J., and Gelfand, D. H.**, A plasmid cloning vehicle allowing regulated expression of eukaryotic DNA in bacteria, *Proc. Natl. Acad. Sci. U.S.A.*, 73(11), 3900, 1976.

54. **Struhl, K., Cameron, J. R., and Davis, R. W.**, Functional genetic expression of eukaryotic DNA in *Escherichia coli, Proc. Natl. Acad. Sci. U.S.A.*, 73(5), 1471, 1976.
55. **Marx, J. L.**, Nitrogen fixation: prospects for genetic manipulation, *Science*, 196, 638, 1977.
56. **Webster, G. T.**, Interspecific hybridization of *Melilotus alba* × *M. officinallis* using embryo culture, *Agron. J.*, 47, 138, 1955.
57. **Brink, R. A. and Cooper, D. C.**, The endosperm in seed development, *Bot. Rev.*, 13, 423, 1947.
58. **Smith, W. K.**, Propagation of chlorophyll-deficient sweet clover hybrids as grafts, *J. Hered.*, 34, 135, 1943.
59. **Sears, E. R.**, The transfer of leaf rust resistance from *Aegilops umbellata* to wheat, *Brookhaven Symp. Biol.*, 9, 1, 1956.
60. **Riley, R. and Chapman, V.**, Genetic control of cytologically diploid behavior of hexaploid wheat, *Nature (London)*, 182, 713, 1958.
61. **Sears, E. R. and Okamoto, M.**, Intergenomic relationships in hexaploid wheat, *Proc. Intern. Congr. Genet. 10th, Montreal*, Vol. 2, University of Toronto Press, Toronto, 1959, 258.
62. **Riley, R., Chapman, V., and Johnson, R.**, The incorporation of alien disease resistance in wheat by genetic interference with the regulation of meiotic chromosome synapsis, *Genet. Res.*, 12, 199, 1968.
63. **Gamborg, O. L., Constabel, F., Fowke, L., Kao, K. N., Ohyama, K., Kartha, K., and Pelcher, L.**, Protoplast and cell culture methods in somatic hybridization in higher plants, *Can. J. Genet. Cytol.*, 16, 737, 1974.
64. **Gamborg, O. L., Constabel, F., Kao, K. N., and Ohyama, K.**, Plant proplasts in genetic modifications and production of intergeneric hybrids, in *Modification of the Information Content of Plant Cells*, Markham, R., Davies, D. R., Hopwood, D. A., and Horne, R. W., Eds., North Holland, Amsterdam, 1975, 181.
65. **Jones, C. W., Mastrangelo, I. A., Smith, H. H., Liu, H. Z., and Meck, R. A.**, Interkingdom fusion between human (HeLa) cells and tobacco hybrid (GGLL) protoplasts, *Science*, 193, 401, 1976.
66. **Melchers, G. and Labib, G.**, Somatic hybridization of plants by fusion of protoplasts, I. Selection of light resistant hybrids of "haploid" light sensitive varieties of tobacco, *Mol. Gen. Genet.*, 135, 277, 1974.
67. **Carlson, P. S., Smith, H. H., and Dearing, R. D.**, Parasexual interspecific plant hybridization, *Proc. Natl. Acad. Sci. U.S.A.*, 69, 2292, 1972.
68. **Smith, H. H., Kao, K. N., and Combatti, N. C.**, Interspecific hybridization by protoplast fusion in *Nicotiana, J. Hered.*, 67, 123, 1976.
69. **Power, J. B., Frearson, E. M., George, D., Evans, P. K., Berry, S. F., and Cocking, E. C.**, Somatic hybridization of *Petunia hybrida* and *Petunia parodii, Nature (London)*, 263, 500, 1976.
70. **Briggle, L. W.**, Triticale — a review, *Crop Sci.*, 9, 197, 1969.
71. **Riley, R. and Law, C. N.**, Genetic variation in chromosome pairing, *Adv. Genet.*, 13, 57, 1965.
72. **Gustafson, J. P. and Qualset, C. O.**, Genetics and breeding of 42-chromosome triticale. I. Evidence for substitutional polyploidy in secondary triticale populations, *Crop Sci.*, 14, 248, 1974.
73. **Gustafson, J. P. and Qualset, C. O.**, Genetics and breeding of 42-chromosome triticale. II. Relations between chromosomal variability and reproductive characters, *Crop Sci.*, 15, 810, 1975.
74. **Karpechenko, G. D.**, Hybrids of *Raphanus sativus* L. × *Brassica oleracea* L., *J. Genet.*, 14, 375, 1924.
75. **McNaughton, I. H.**, Synthesis and sterility of *Raphanobrassica, Euphytica*, 22, 70, 1973.
76. **Lauge, W.**, Crosses between *Hordeum vulgare* L. and *H. bulbosum* L., II. Elimination of chromosomes in hybrid tissues, *Euphytica*, 20, 181, 1971.
77. **Subrahmanyam, N. C., and Kasha, K. J.**, Selective chromosomal elimination during haploid formation in barley following interspecific hybridization, *Chromosoma*, 42, 111, 1973.
78. **Fedak, G.**, Haploids from barley × rye crosses, *Can. J. Genet. Cytol.*, 19, 15, 1977.
79. **Dunbier, M. W. and Bingham, E. T.**, Maximum heterozygosity in alfalfa: results using haploid-derived autotetraploids, *Crop Sci.*, 15, 527, 1975.
80. **Mendiburu, A. O. and Peloquin, S. J.**, The significance of 2n gametes in potato breeding, *Theor. Appl. Genet.*, 49, 53, 1977.
81. **Mok, D. W. S. and Peloquin, S.J.**, Breeding value of 2n pollen (diplandroids) in tetraploid × diploid crosses, *Theor. Appl. Genet.*, 46, 307, 1975.
82. **Mendiburu, A. O., Peloquin, S. J., and Mok, D. W. S.**, Potato breeding with haploids and 2n gametes, in *Haploids in Higher Plants*, Kasha, K. J., Ed., University of Guelph, Ontario, 1974, 249.
83. **Harlan, J. R. and de Wet, J. M. J.**, On ö Winge and a prayer, the origins of polyploidy, *Bot. Rev.*, 41, 361, 1975.
84. **Bingham, E. T.**, Transfer of diploid *Medicago* spp. germplasm to tetraploid *M. sativa* L. in 4×-2× crosses, *Crop Sci.*, 8, 760, 1968.
85. **Hanneman, R. E., Jr. and Peloquin, S. J.**,Ploidy levels of progeny from diploid-tetraploid crosses in potato, *Am. Potato J.*, 45, 255, 1968.

86. **Sarcristán, M. D. and Melchers, G.,** The caryological analysis of plants regenerated from tumorous and other callus cultures of tobacco, *Mol. Gen. Genet.,* 105, 315, 1969.
87. **Bayliss, M. W.,** The effect of growth *in vitro* on the chromosome complement of *Dancus carota* (L) suspension cultures, *Chromosoma,* 51, 401, 1975.
88. **Takebe, I., Labib, G., and Melchers, G.,** Regeneration of whole plants from isolated mesophyll protoplasts of tobacco, *Naturwissensch,* 58, 318, 1971.
89. **Cummings, D. P., Green, C. E., and Stuthman, D. D.,** Callus induction and plant regeneration in oats, *Crop Sci.,* 16, 465, 1976.
90. **Heinz, D. J., Mee, G. W. P., and Nickell, L. G .,** Chromosome numbers of some *Saccharum* species hybrids and their cell suspension cultures, *Am. J. Bot.,* 56, 450, 1969.
91. **Coleman, R. E.,** New plants produced from callus tissue culture, *Sugarcane Research Report,* U.S. Department of Agriculture, ARS, 38, 1970.
92. **Nitsch, C.,** Pollen culture — a new technique for mass production of haploid and homozygous plants, in *Haploids in Higher Plants,* Kasha, K. J., Ed., University of Guelph, Ontario, 1974, 123.
93. **Binding, H.,** Mutation in haploid cell culture, in *Haploids in Higher Plants,* Kasha, K. J., Ed., University of Guelph, Ontario, 1974, 323.
94. **Bingham, E. T. and Saunders, J. W.,** Chromosome manipulations in alfalfa: scaling the cultivated tetraploid to seven ploidy levels, *Crop Sci.,* 14, 474, 1974.
95. **Sunderland, N., Collins, G. B., and Dunwell, J. M.,** Nuclear fusion in pollen embryogenesis of *Datura innoxia* Mill, *Planta,* 117, 227, 1974.
96. **Saunders, J. W. and Bingham, E. T.,** unpublished.

# *Index*

# INDEX

## A



## B



## C

## D

## E



## F



## G



## H



## I

## K



## L



## M



## N



## O



## P

## R

## S



## T

## V



## W



## X



## Z